Protein Modelling

Gábor Náray-Szabó

Editor

Protein Modelling

 Springer

Editor
Gábor Náray-Szabó
Chemistry Department
Eötvös Loránd University
Budapest
Hungary

ISBN 978-3-319-09975-0   ISBN 978-3-319-09976-7   (eBook)
DOI 10.1007/978-3-319-09976-7

Library of Congress Control Number: 2014954238

Springer Cham Heidelberg New York Dordrecht London

Printed on acid-free paper

Springer is part of Springer Science+Business Media (www.springer.com)

Preface

Almost two decades ago we jointly edited a book (*Computational Approaches to Biochemical Reactivity*) with Arieh Warshel. The volume summarized results in the rapidly developing computational aspects of biochemical reactivity, which were new at that time. Since then interested specialists faced a really spectacular development in the field, which was honoured by awarding the 2013 Nobel Prize in Chemistry to three scientists, among them Arieh. Hundreds of first-class publications and dozens of successes in computer-aided drug design provide evidence for the high-level application of models and methods to practical problems related to the structure and function of proteins. Maybe the most important lesson we could learn from the history of *Protein Modelling* is that there is no single and omnipotent method available for the treatment of various problems, rather different levels of approximations should be applied. More than in case of quantum mechanical modelling of small and medium-sized molecules, the concept of the chemical bond provides a sound basis for computational methods, which can be extended and refined by the use of quantum mechanics. Therefore, *Protein Modelling* is now well established and used to facilitate the thorough understanding and rational design of biochemical processes.

Budapest, Hungary, July 2014   Gábor Náray-Szabó

Contents

Chapter 1
Introduction

Gábor Náray-Szabó

1.1 Introduction

After formulating the basic equation of quantum mechanics, the British physicist Paul Dirac predicted that "The fundamental laws necessary for the mathematical treatment of a large part of physics and the whole chemistry are thus completely known, and the difficulty lies only in the fact that application of these laws leads to equations that are too complex to be solved" [1]. The mathematical difficulty proved to be formidable, since for molecules the numerical work, necessary for the brute-force solution of the highly complex, non-linear coupled set of differential equations increases in a breath-taking pace with the number of its electrons. It is therefore not surprising that for a long time quantum chemistry could provide reliable computational results only for the simplest molecules with a few electrons. Rapid development of computer hardware and software allowed the treatment of larger and larger molecules, thus, beginning in the sixties of the last century theoretical organic chemistry could rely more and more on computational results.

The achievements of quantum chemistry were acknowledged by the Nobel Prize donated in 1998 to John A. Pople and Walter Kohn for the development of computational methods in quantum chemistry. Quantum mechanical computations for medium-size molecular systems became an everyday practice since then serving as a basis for the interpretation of empirical results, in some cases even replacing experiments. Examples for the successful application are given in Tables 1.1 and 1.2.

As it is seen, calculated excitation energies for the psoralen molecule consisting of twenty atoms and ninety six electrons are close to measured ones, in some

G. Náray-Szabó (✉)
Laboratory of Structural Chemistry and Biology, Institute of Chemistry, Eötvös Loránd University, Pázmány Péter St. 1A, Budapest 1117, Hungary
e-mail: naraysza@chem.elte.hu

© Springer International Publishing Switzerland 2014
G. Náray-Szabó (ed.), *Protein Modelling*, DOI 10.1007/978-3-319-09976-7_1

Table 1.1 Calculated and experimental (in parentheses) excitation energies (eV) for psoralen [2]

State	E_{VA} (Abs$_{max}$)	T_e (T_0)	E_{VE} (E_{max})
$^1A'$	3.98 (3.7 – 4.3)	3.59 (3.54)	3.45 (3.03)
$1^1A''$	5.01 (–)	3.1 (–)	2.78 (–)
$1^3A'$	3.27 (–)	2.76 (2.7)	2.29 (2.7)

E_{VA} vertical absorption, T_e adiabatic electronic band origin, E_{VE} vertical emission, Abs$_{max}$ experimental absorption, T_0 experimental band origin, and E_{max} emission maximum

Table 1.2 Calculated (density functional theory with the B3LYP functional) and experimental (exp) bond distances (Å) in some metal carbonyl complexes [3]

	Ni(CO)$_4$ Ni–C	Ni(CO)$_4$ C–O	Cr(CO)$_6$ Cr–C	Cr(CO)$_6$ C–O
B3LYP	1.845	1.137	1.918	1.141
Exp	1.838	1.141	1.926	1.141

cases even complement experiment by providing reliable substitutes for data not available by now. A further example for the successful use of quantum mechanics is given for carbonyl complexes of nickel and chromium, for which theory predicts bond lengths lying very close to the experimental values (Table 1.2).

The spectacular success of small-molecule quantum chemistry was not enough to sceptic experimentalists, who demanded appropriate modelling of real systems and events, like chemical reactions in solution or on solid-state surfaces, enzyme reactions or protein folding. Since models of molecular entities involved in these processes consist of several thousands of non-hydrogen atoms, exact quantum chemical treatment, similar to that in case of small molecules mentioned above, is out of question at present. Therefore theoretical chemists, interested in the solution of this formidable problem, dismissed Dirac and went back to classical chemical models considering a molecule as an ensemble of atoms and bonds. More or less independently from quantum chemistry effective force fields have been developed, first for small molecules [4–6], later for proteins [7, 8]. Application of the force fields as a set of parameters in a simple energy expression, involving atoms and bonds, lead to the development of molecular mechanics, another tool for the description of molecular events depending on energy changes of the system. This approach initiated a breakthrough in the modelling of proteins, since problems related to conformation, folding, ligand binding and the like could be handled at the atomic level. The basic issue of enzyme reactions could be treated by the ingenious combination of quantum mechanical and molecular mechanical models, first for the 1,6-diphenyl-1,3,5-hexatriene molecule [9] and somewhat later for a full enzyme, lysozyme [10]. Now this approach is referred to as the application of multiscale models for complex chemical systems, for which the 2013 Nobel Prize in chemistry has been donated to Martin Karplus, Michael Levitt and Arieh Warshel, apparently the most successful group of scientists in this field. Presently, full and appropriate computer modelling of proteins in the biophase became possible. Similarly as quantum mechanics in small-molecule organic chemistry, multiscale protein modelling became an everyday practice in structural biology.

This book can be considered as some kind of a continuation of a former one published almost 20 years ago [11]. At that time models were less precise, computational results were less reliable, but it already could be seen that the approach is continuously developing, larger and larger systems can be treated, higher and higher accuracy can be attained. In the first chapter by Imre Csizmadia and co-workers the potential of quantum chemical methods applied to small molecular models describing protein conformation and folding will be discussed. Then, in a survey written by Carme Rovira we can learn, how subtle motions at the active site, often coupled to electronic rearrangements, can be followed by quantum chemical calculations. Combined quantum mechanical/molecular mechanical (QM/MM) methods are getting more and more popular, therefore several recent reviews are available [12–18]. In this book a special approach will be discussed by Ferenczy and Náray-Szabó. Molecular mechanical force fields become more and more sophisticated, polarisation can also be treated as discussed by Khoruzhii et al. Ullmann gives a survey on protein electrostatics, a simple and illustrative alternative for the discussion of energetic aspects. Further simplifications in protein force fields are possible, if certain atoms of the protein are grouped in order to be considered as single centres, as discussed by Giorgetti and Carloni, who wrote a chapter on coarse grained models. A very interesting approach can be elaborated if we abandon the demand for atomic resolution. Harris and co-workers report on mesoscale methods that access larger systems, from about ten to some hundreds of a nanometer. Tusnády uses another approach, bioinformatics, for the structure prediction of transmembrane proteins. Modelling structures of these proteins is crucial, as in many cases it is almost the only available computational technique to get structural information about them. Proteins are by far not rigid, therefore dynamic aspects of their function is very important, in some cases crucial (see the chapter by Perczel et al.). Molecular docking, which has become an increasingly important tool for the study of protein-ligand complexes playing an important role in structural biochemistry and drug discovery is discussed by Ramos. Finally, Tarcsay and Keserű provides an example for the extensive application of the multiscale computational methods. They treat an important point in drug design, ADMET (absorption, distribution, metabolism, excretion and toxicity) prediction based on protein structures.

References

1. Dirac P (1929) Quantum mechanics of many-electron systems. Proc R Soc Lond A123:714–733
2. Serrano-Andres L, Serrano-Perez JJ (2012) In: Leszczynski J (ed) Handbook of computational chemistry. Springer, Dordrecht, p 515
3. Jacobsen H, Cavallo L (2012) In: Leszczynski J (ed) Handbook of computational chemistry. Springer, Dordrecht, p 114
4. Westheimer FH, Mayer JE (1946) J Chem Phys 14:733–737

5. Hill TL (1946) J Chem Phys 14:465
6. Drostovsky J, Hughes ED, Ingold CK (1965) J Chem Soc 173–194
7. Némethy ED, Scheraga HA (1965) Biopolymers 4:155–184
8. Lifson S, Warshel A (1967) J Chem Phys 49:5116–5128
9. Warshel A, Karplus M (1972) J Am Chem Soc 94:5612–5625
10. Warshel A, Levitt M (1976) J Mol Biol 103:229–247
11. Náray-Szabó G, Warshel A (eds) (1997) Computational approaches to biochemical reactivity. Kluwer, Dordrecht
12. Friessner RA, Guallar V (2005) Rev Phys Chem 56:389–427
13. Gao JL, Ma SH, Major DT, Nam K, Pu JZ, Truhlar DG (2006) Chem Rev 106:3188–3209
14. Lin H, Truhlar DG (2007) Theoret Chim Acc 117:185–199
15. Senn HM, Thiel W (2009) Angew Chem Int Ed 48:1198–1229
16. van der Kamp MW, Mulholland A (2013) Biochemistry 52:2708–2728
17. Vreven T, Byun KS, Komáromi I, Dapprich S, Montgomery JA, Morokuma K, Frisch MJ (2009) J Chem Theor Comput 2:815–826
18. Zhang R, Lev B, Cuervo JE, Norskov SY, Salahub DR (2010) Adv Quant Chem 59:353–401

Chapter 2
Quantum Chemical Calculations on Small Protein Models

Imre Jákli, András Perczel, Béla Viskolcz and Imre G. Csizmadia

2.1 Ab Initio Quantum Chemistry of Peptides

During the past 50 years (1963–2013) many thousands ab initio computations were published on small peptides. Many molecules contain the acid amide or peptide bond (–CO–NH–) but the smallest molecule is formamide ($HCO–NH_2$). It might be expected that the first ab initio computations were to be carried out on formamide. However, in 1963, when digital computers, such as IBM 709, were not transistorized, therefore, only the iso-electronic formyl fluoride (HCOF) was possible to be subjected to ab initio Molecular Orbital (MO) computations. The results of this first computation were reported in 1963 in the Quarterly Progress

I. Jákli (✉) · A. Perczel
MTA-ELTE Protein Modelling Research Group Pázmány, Péter sétány,
1/A H-1117 Budapest, Hungary
e-mail: jimre@chem.elte.hu

A. Perczel
Laboratory of Structural Chemistry and Biology, Institute of Chemistry, ELTE,
Pázmány Péter sétány, 1/A H-1117 Budapest, Hungary
e-mail: perczel@chem.elte.hu

B. Viskolcz · I.G. Csizmadia
Department of Chemical Informatics, University of Szeged, Boldogasszony sgt. 6,
H-6725 Szeged, Hungary

I.G. Csizmadia
Department of Chemistry, University of Toronto, M5S 3H6, Toronto, ON, Canada
e-mail: icsizmad@jgypk.u-szeged.hu

© Springer International Publishing Switzerland 2014   5
G. Náray-Szabó (ed.), *Protein Modelling*, DOI 10.1007/978-3-319-09976-7_2

Report of MIT [1]. The full research was published [2] in 1966, after the IBM Research Centre generously offered some time on their, transistorized, IBM 7090 computer to finish the largest basis set computation of that time. The original MIT report was kindly reproduced by an ACS journal in the Supplementary Material of the paper published, on the 50th anniversary, in the Journal of Physical Chemistry B [3]. Eventually, the original aim, the formamide molecule ($HCO–NH_2$) was computed afterward when the University Toronto acquired an IBM 7094 transistorized computer and a paper was published [4] in 1968.

Soon after the gate has opened, an unbelievable volume of computation appeared in the literature which created the feeling of a *Molecular Revolution*. From that point onward force-field softwares are continuously redesigned on the basis of ab initio computational data. Without the aspiration for completeness, a brief historic summary is given, in this chapter, about the historic development of Quantum Chemical Calculations on small peptides which may be treated as a prototype model of the protein folding problem. However, more emphasis will be given to the current trends that will effectively influence future directions.

2.2 Relative Stability of Peptide Bonds Formed

The thermodynamic stability of an acid amide or peptide bond may be measured by the reaction heat ($\Delta_r H$) or free energy change ($\Delta_r G$ of its format) with respect to the sum of the appropriate measure(s) of acid and amine residues involved.

$$R_1-\overset{O}{\underset{OH}{\big\|}}\ +\ H_2N-R_2\ \longrightarrow\ R_1-\overset{O}{\underset{\underset{H}{N}-R_2}{\big\|}}\ +\ H_2O \tag{2.1}$$

Figure 2.1 shows that the $\Delta_r H$ formation of amide from α-amino acid ($n = 1$) is the most exothermic process and thus, the most special of the whole series.

Fig. 2.1 Variation of $\Delta_r H$ for the formation of acid amide linkage, according to Eq. (2.1), with the chain length (n), for fatty acids: $H\text{-}(CH_2)_n\text{--}COOH$ and for N-acetyl amino acids: $AcNH–(CH_2)_n–COOH$. The length of the chain is also indicated in a symbolic fashion by a sequence of letters of the Greek alphabet

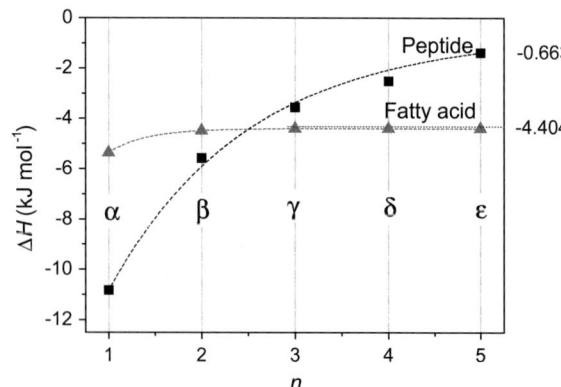

This figure also indicates that the convergence of the two curves at n = 5 is less than 4 kJ mol^{-1}. The convergence limits in Figs. 2.1 and 2.2 are shown on their right hand side. The convergence implies that beyond a certain length the chain makes no noticeable influence on the energetics.

Variation of $\Delta_r H$ for cyclic peptide formation is quite different as shown in Fig. 2.2.

Interestingly enough, both γ- and δ-cyclic peptides show extensive stabilization with respect to their linear form as a 5 or a 6 member ring is formed, respectively.

The peptide bond is a 4π electron containing functional group, as an allyl molecular system would do, the nitrogen lone pair forms a partial double bond with the carbonyl carbon:

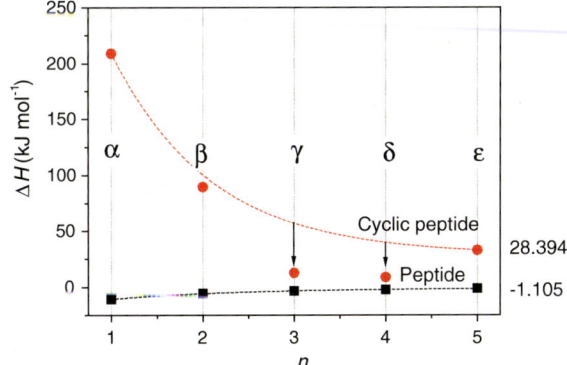

The fact that the acid amide (–CONH$_2$) is close to planarity is frequently rationalized by the above pair of resonance structures. The extent of this conjugation may be characterized numerically by a *conjugativity* value which is frequently referred to as *amidicity*. As shown below, the removal of the conjugation, by the hydrogenation of the C=O double bond, may be used to define the *amidicity* (AM) through the calculation of the enthalpy change (ΔH_{H2}). A relative value of ΔH_{H2} may be used as a percentage value of amidicity defined in terms of the following compounds shown in Fig. 2.3. Calculations of the amidicity scale [5] is given by the following equations:

$$\Delta H_{H2}[I] = H_B - H_A \qquad (2.2)$$

$$[\text{Amidity \%}] = m\,\Delta H_{H2}[I] + [\text{Amidity \%}]_0 \qquad (2.3)$$

Figure 2.4 shows the variation of amidicity both for acyclic (open chain) and cyclic peptides.

Fig. 2.2 Variation of $\Delta_r H$ for the formation of acyclic and cyclic peptides as function of the chain length (n), for amino acids: H$_2$N–(CH$_2$)$_n$–COOH

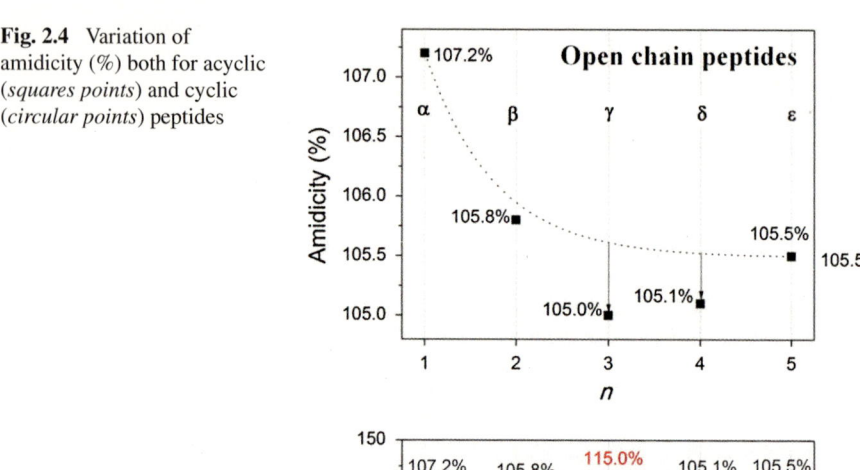

Fig. 2.3 The definition of amidicity

Fig. 2.4 Variation of amidicity (%) both for acyclic (*squares points*) and cyclic (*circular points*) peptides

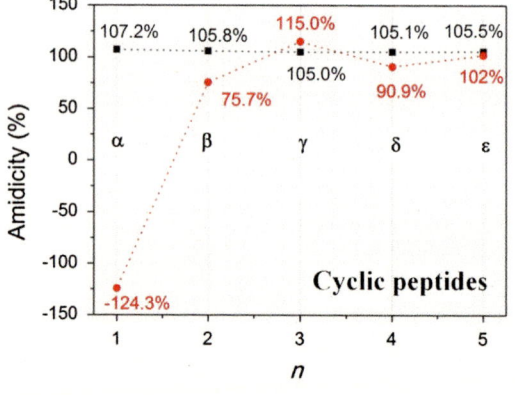

Both the 5- and 6-membered ring, formed by γ- or δ-amino acids respectively show extra stability even though the difference is within 0.5 %. In view of this, it is understandable that glutamic acid can preferably form cyclic amide. In addition, asparagine residue via deamination process can form succinimide (Fig. 2.5).

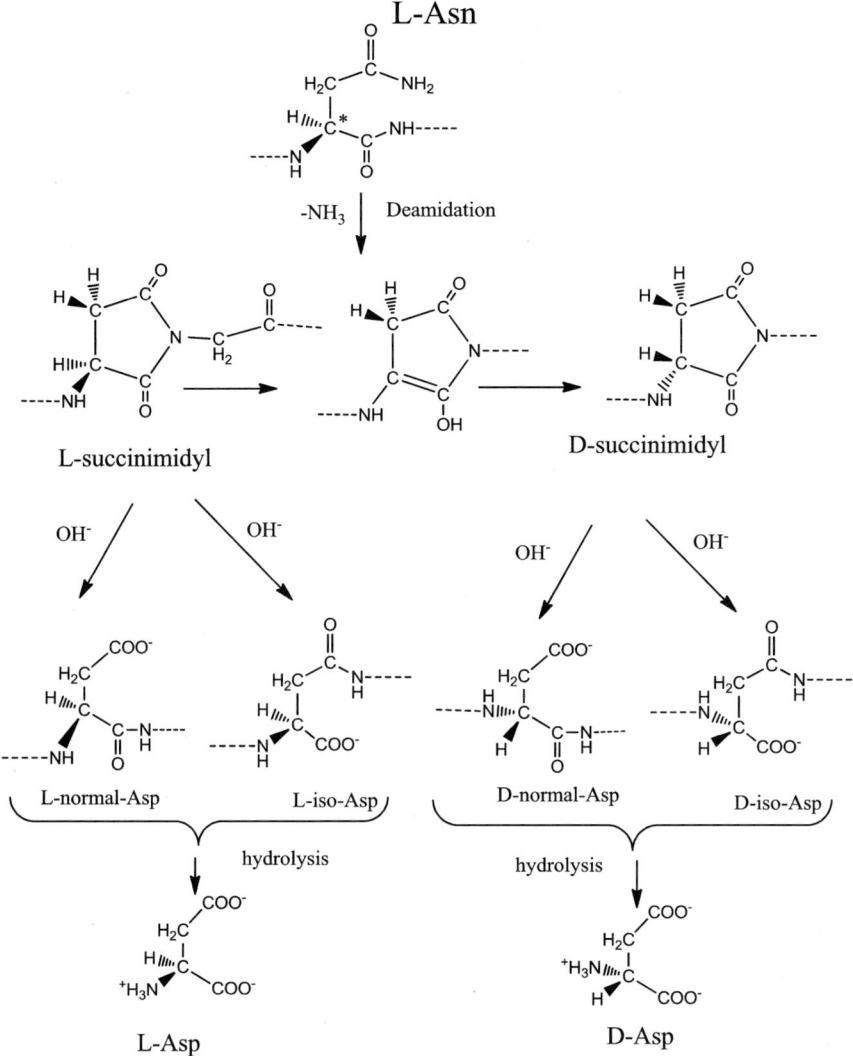

Fig. 2.5 Closed electronic shell mechanism of D-Asp formation from Asn through deamidation and enolization

The asparagine deamidation (Asn $\rightarrow$ Asp) is one of the most important protein degradation pathways and residues of Asp serve as "molecular timers" that can have effects on protein turnover and aging [6–10]. Recently, Trout et al. [11] and subsequently Catak et al. [12] studied, theoretically, the deamidation process of the Asn. The racemization was subsequently explained by enolization of the H–C–C=O moiety [13], as shown in some details in Fig. 2.5.

2.3 Topology of Peptide Conformations

2.3.1 The Concept of the Ramachandran Map

The question of protein folding is a century old problem. About a half a century ago, Ramachandran [14] in India tried to establish what conformations of a single residue diamide were disallowed for steric reason. In the absence of digital computers at that time only ball and stick molecular models were available. The conformational change is measure by two dihedral angles (φ, ψ) associated with the rotation about two bonds connected to the α-carbon as shown below:

I

The potential energy turns out to be a 2D potential energy surface (PES) [15] that is a mathematical function of two independent variables.

$$E = f(\varphi, \psi) \tag{2.4}$$

The shape of the 2D PES can be investigated from its 1D-crossections (Fig. 2.6). The idealized topology of minimum energy points is illustrated by Fig. 2.7.

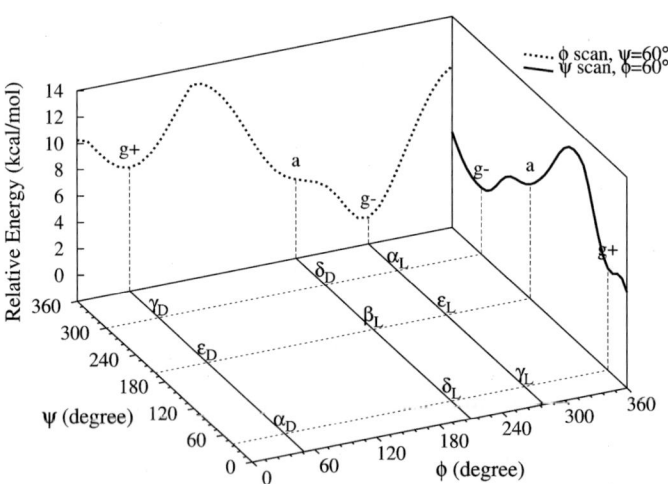

Fig. 2.6 Schematic illustration of how a chiral (PES) may be related to two component chiral potential energy curves

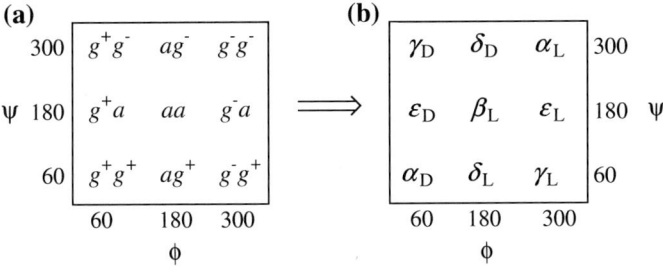

Fig. 2.7 Conformational assignments (**a**) and names (**b**) of peptide conformers on the conformational (PES) of a peptide (P–CONH–CHR–CONH–Q) [16]

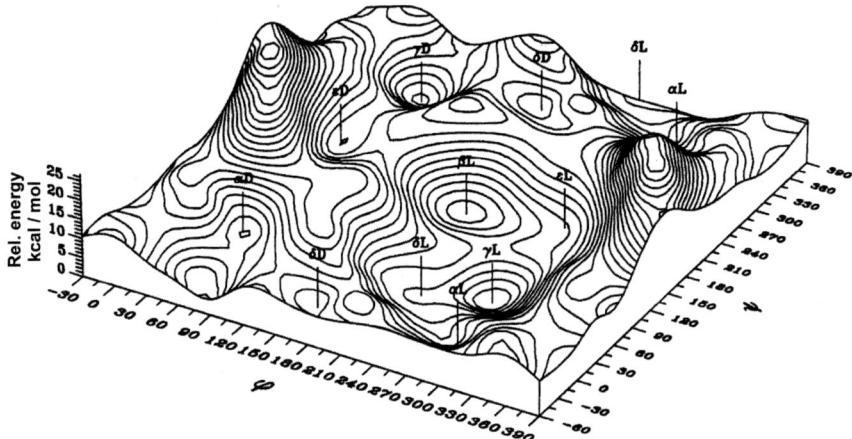

Fig. 2.8 Pseudo-three-dimensional Ramachandran potential energy surface of HCONH–CHCH$_3$–CONH$_2$, presented in the $0° \leq \phi \leq 360°$ and $0° \leq \psi \leq 360°$ range of two independent variables (2D)

2.3.2 Conformational Potential Energy Surfaces (PES)

It may important to amplify that every point on the above surface (Fig. 2.8) is a particular conformation, but as a distinction only the minima are called conformers. The ab initio conformation potential energy surface as any similar surfaces calculated at different levels of theory or by using different methods (e.g. MM, QM/MM, coarse grain), looks like a landscape (Fig. 2.8).

Just like in cartography the landscape may also be shown as a contour diagram. The above surface (Fig. 2.8) represents one of the four equivalent quadrants in Fig. 2.9.

These conformers (Table 2.1 and Fig. 2.10) could be regarded as typical building blocks of folded proteins as the analysis of PDB data [18] have shown it. Most

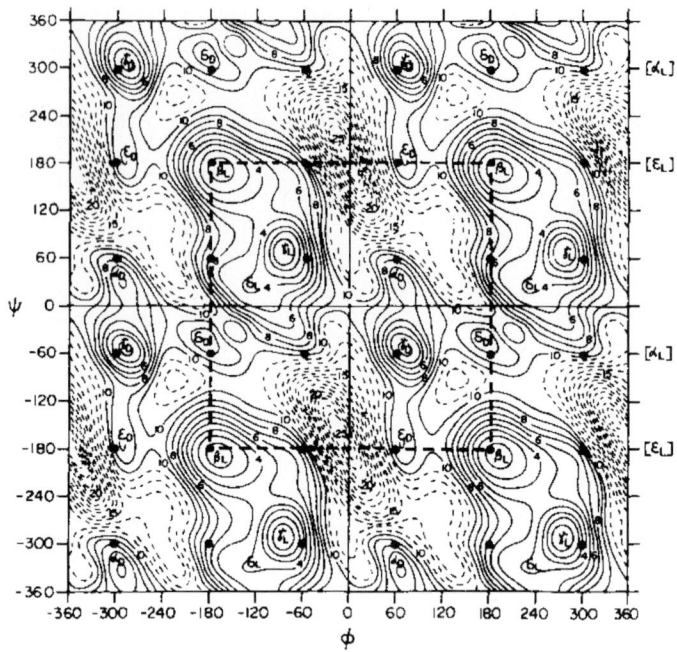

Fig. 2.9 Contour diagram of the 2D Ramachandran potential energy surface of HCONH–CHCH$_3$–CONH$_2$, presented in the $-360° \leq \phi \leq 360°$ and $-360° \leq \psi \leq 360°$ range of independent variables. The central square (*broken lines*) is the IUPAC conventional cut, while the four quadrants are the traditional cuts

Table 2.1 Optimized ϕ, ψ torsional angle pairs for alanine diamide (HCONH–(L)–CHMe–CONH$_2$)

Conformer	Optimized values		Idealized values [16]		Conformational
	ϕ	ψ	ϕ	ψ	Classification
α_L	−66.6	−17.5	−60	−60	g^-g^-
α_D	+61.8	+31.9	+60	+60	g^+g^+
β_L	−167.6	+169.9	−180	+180	aa
γ_L	−84.5	+68.7	−60	+60	g^-g^+
γ_D	+74.3	−59.5	+60	−60	g^+g^-
δ_L	−126.2	+26.5	−180	+60	ag^+
δ_D	−179.6	−43.7	+180	−60	ag^-
ε_L	−74.7	+167.8	−60	+180	g^-a
ε_D	+64.7	−178.6	+60	−180	g^+a

The idealized torsional angle pairs, together with their conformational classification, are also shown for the sake of comparison

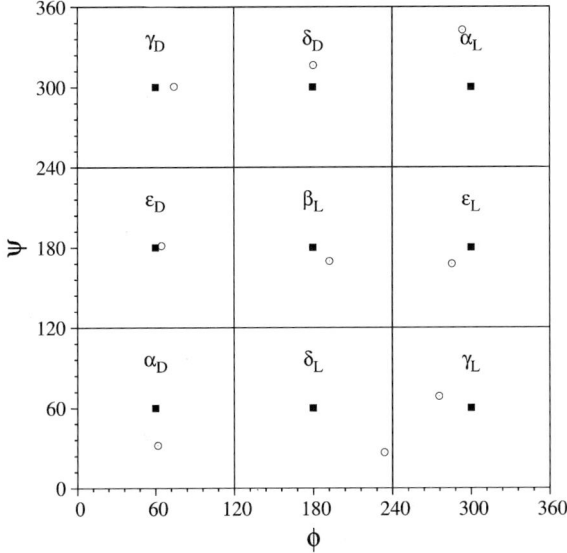

Fig. 2.10 a Schematic illustration of the PES of an amino acid diamide. The idealized positions of the main conformers are marked by *shaded squares*, while the computationally determined positions (actual location depending on methods, type of approach, level of theory etc.) are shown as *open circles*. Shifts but even elimination of some points was observed. The names of the conformers are given as subscripted Greek letters [16, 17]

frequently β_L and its neighboring conformers (e.g. ε_L, γ_L) is a component of the extended β- pleated sheets while α_L is a typical building unit of both 3_{10}- and/or α-helix.

2.3.3 Mathematical Representation of Conformational PES

Peptide folding, just like protein folding, has a conformational aspect. The mathematical representation of the conformational potential energy surface is a tool that may be used to decipher problems related to peptide folding. Consequently, fitting mathematical functions for computed grid points will lead to a mathematical representation of such a conformational problem.

One dimensional trigonometric fit to simple internal rotation energies has been fitted by Pople et al. [19, 20]. For potential energy surfaces (2D) and hypersurfaces (3D or nD) functions of two, three or n independent variable are necessary. These were achieved for relatively simple surfaces [21–23]. More recently the problem has been reinvestigated to see how does the mathematical complexity of Fourier expansion function groups with the complexity of the appearance of the PEC or PES [24]. The three different torsional modes of C–C, N–N and O–O exhibit three different scenarios. The N–N rotation is the most complex because the central anti conformer at 180° becomes a higher energy minimum with respect to the two gauche minima (at around 90° and 270°). In contrast to N–N, rotation about the O–O bond in H_2O_2 the anti conformer becomes a TS and two equivalent conformers were assigned at about 120° and 240°, respectively. Thus, this PEC exhibited intermediate complexity. Ethane has 3 identical minima hydrogen peroxide has 2 identical minima. In contrast to these hydrazine has 3 minima of which 2 are identical lower minima and the 3rd minimum is a higher minimum. All of these apparent complexities are summarized in Table 2.2, in which it is indicated that C–C can be fitted by a single ($m = 1$) cosine term, while peroxide requires already two cosine terms ($m = 2$) and hydrazine must have at least three terms ($m = 3$) in the Fourier expansion (2.5).

$$E(\varphi) = a_0 + \sum_{m=1}^{n} \left(a_j \cos \frac{m 2\pi k \varphi}{360} \right)$$

(2.5)

The three conformational PECs are shown in Fig. 2.11.

Table 2.2 Accuracy (R^2) of fitted functions with increasing number of terms (m) for a C–C, N–N and O–O bonds

Molecule	R^2		
	$m = 1$	$m = 2$	$m = 3$
Ethane	0.9936	1.0000	1.000
Peroxide	0.9427	0.9994	1.000
Hydrazine	0.7739	0.9802	0.9977

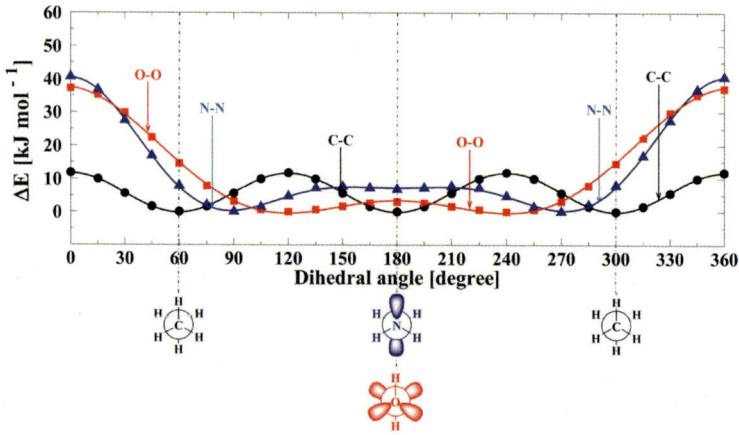

Fig. 2.11 Potential energy curves for rotation about the C–C, N–N and O–O bonds [24]

Thus, with increasing complexity (increasing topological differences of critical points) of the PEC the complexity of the fitted explicit mathematical function must also increase. For a conformation PES, pentane is a classical example since it may be treated as 1,3-dimethyl propane and therefore it could act as a 2D example (Fig. 2.12).

The surface is quite symmetric so a cosine and sine compilation of trigonometric functions (6) with $m, n = 3$ yielded on $R^2 = 0.9347$. With the increase m, n the R^2 value improved slightly. The R^2 value improved with $m, n > 3$ as shown in Table 2.3.

$$E(\varphi, \psi) = a_0 + \sum_{m=1}^{\infty} \sum_{n=1}^{\infty} (a_1 \cos mk_\varphi \varphi \cos nk_\psi \psi$$

$$+ a_2 \cos mk_\varphi \varphi \sin nk_\psi \psi + a_3 \sin mk_\varphi \varphi \cos nk_\psi \psi$$

$$+ a_4 \sin mk_\varphi \varphi \sin nk_\psi \psi) \tag{2.6}$$

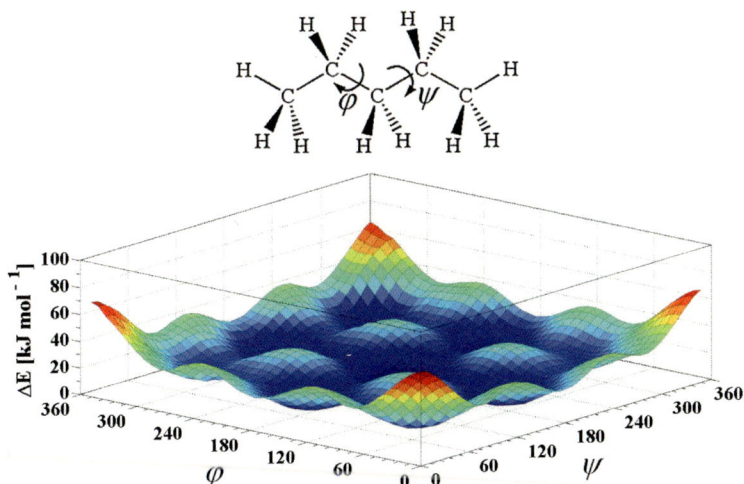

Fig. 2.12 Conformational PES of pentane treated as 1,3 dimethyl-propane

Table 2.3 Slight variation of accuracy (R^2) of fitted functions with increasing number of terms (m, n) for pentane (1,3 dimethyl propane)

Molecule	R^2			
	m, n = 3	m, n = 4	m, n = 5	m, n = 6
Pentane	0.9347	0.9367	0.9371	0.9383

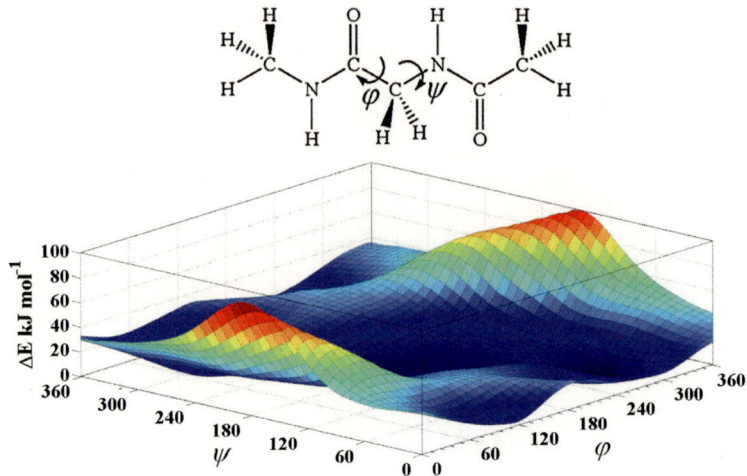

Fig. 2.13 Ramachandram type conformational PES of N-acetyl glycine-N-methylamide

A similar type of fitting for the conformational PES of glycine diamide is already a more complex task: surface shown in Fig. 2.13.

$$
\begin{aligned}
E(\varphi,\psi) = a_0 + \sum_{m=1}^{\infty}\sum_{n=1}^{\infty} A_{m,n} e^{\left(-\left(\frac{(b_m\varphi-\varphi_0)^2}{2\sigma_\varphi^2}+\frac{(b_n\psi-\psi_0)^2}{2\sigma_\psi^2}\right)\right)} \\
+ a_1 \cos mk_\varphi\varphi \cos nk_\psi\psi + a_2 \cos mk_\varphi\varphi \sin nk_\psi\psi \\
+ a_3 \sin mk_\varphi\varphi \cos nk_\psi\psi + a_4 \sin mk_\varphi\varphi \sin nk_\psi\psi \\
+ \sum_{j=1}^{\infty}\sum_{l=1}^{\infty} d_1 \cos(jk_\varphi\varphi - lk_\psi\psi)d_2 \cos(jk_\varphi\varphi - lk_\psi\psi) \\
+ \sum_{j=1}^{\infty}\sum_{l=1}^{\infty} d_3 \cos(jk_\varphi\varphi - lk_\psi\psi)d_4 \sin(jk_\varphi\varphi - lk_\psi\psi) \\
+ \sum_{j=1}^{\infty}\sum_{l=1}^{\infty} d_5 \sin(jk_\varphi\varphi - lk_\psi\psi)d_6 \cos(jk_\varphi\varphi - lk_\psi\psi) \\
+ \sum_{j=1}^{\infty}\sum_{l=1}^{\infty} d_7 \sin(jk_\varphi\varphi - lk_\psi\psi)d_8 \sin(jk_\varphi\varphi - lk_\psi\psi)
\end{aligned}
\tag{2.7}
$$

Even though the fitted analytical Eq. (2.7) is far more complicated than that used for pentane (Eq. 2.6) the R^2 value is about the same: $R^2_{Gly} = 0.9287$ as was in the former case. However, the main problem of such an approach is not that the analytical function gets more and more complicated as the polypeptide growth in size and complexity, but that the PES varies a lot as the level of theory is changed [25, 26]. Thus, as both shape and topology of the appropriate PEHS changes for the very same chemical entity by varying QM method and/or the level of theory applied, the concept of using these PEHS as markers of protein building blocks seems to miss the point. Although very informative, these PEHS can hardly be used as generalized descriptors of folding properties of polypeptides.

2.3.4 Toroidal Representation of Conformational PES

The two independent periodic variables (φ, ψ), which characterize the Ramachandran map, denote circular motions as illustrated by Fig. 2.14. Therefore, to make any arbiter cut off introduces artificial edges and thus, separates otherwise neighboring conformer types from each other.

To obtain the Ramachandran surface in a toroidal coordinate system we may first roll up the 2D PES along one variable, (e.g., ψ) and obtain a cylinder (Fig. 2.15, top). Subsequently the same can be done along the second variable ϕ,

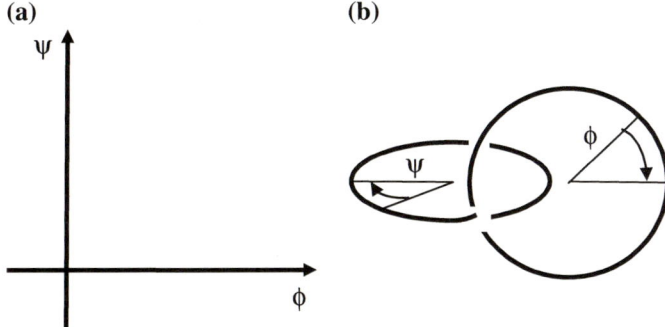

Fig. 2.14 Traditional Cartesian-coordinate system (**a**) and a topologically equivalent circular or toroidal coordinate system (**b**)

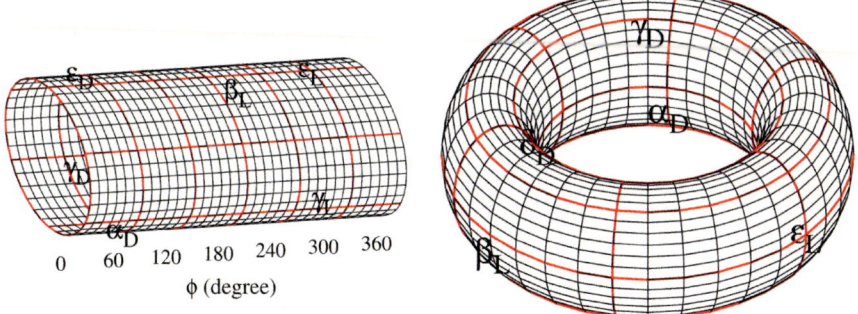

Fig. 2.15 Folding of the Ramachandran-surface coordinates first along ϕ (*top*) and second along ψ (*bottom*), with minima (α_L, β_L, etc.) highlighted

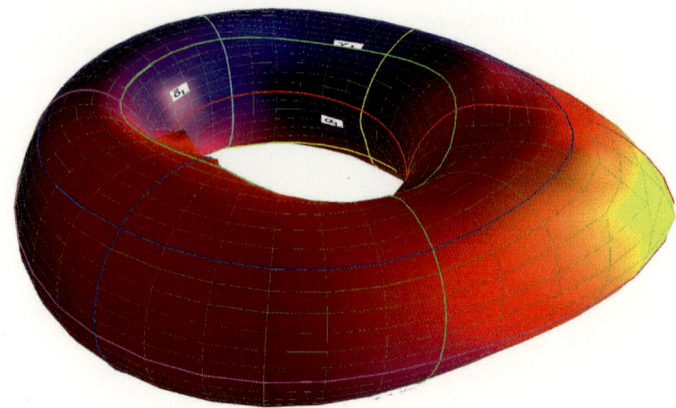

Fig. 2.16 Toroidal representation of the Ramachandran PES for alanine residue in embedded in the AcNH–(L)–CHMe–CONHMe diamide model

to obtain a doughnut-shaped object, such as an O-ring called a 2D-torus (Fig. 2.15, bottom). Location of the nine basic conformers [17], which is visible from the given perspective (Fig. 2.8) are indicated by their subscripted Greek-letter codes occur now on the surface of the torus.

If we wish to illustrate the energy depths or heights, shown in Fig. 2.8, on these topographically equivalent geometrical objects, then we may use an appropriate color code. Also, it is easy to squeeze these geometrical objects at their energy minima and blow them up at points related to energy maxima [27] (Fig. 2.16).

The situation is analogous to view the Earth as a 3D-globe rather than on a 2D map. It can be seen that points that are located on two sides of a 2D-PES map may in fact adjacent on the 3D-toroid. In other words the periodicity of dihedral rotation are explicitly observable.

To summarize both the traditional as well as any more recent representation of the backbone PES of any amino acid diamide, called as Ramachandran surface, are adequate toll for depicting the different conformational building blocks, lego elements, of peptides and proteins. Moreover, conformational interconversions (paths) via TSs as well as avoided regions (mountains) are easy to visualize, using any sensible cuts and representations. On the other hand, these PESs are very sensitive and changes qualitatively to the method, level of theory, concept etc. used to determine them. Thus, just because these PESs are very method dependent, they can hardly be used as numerical and/or analytical representation of the actual peptide model. In other words for the very same peptide model, multiple PESs can be calculated. In conclusion, these PESs are often useful tool to qualitatively interpret conformational properties but are inadequate to stand on their own for a polypeptide model.

2.4 Backbone Conformations of Small Peptides

2.4.1 Single Amino Acid Diamides

There are 21 DNA coded natural amino acids occurring in proteins. A total of 20 of these amino acids have DNA codons and the 21st of them, Selenocystein (Sec), has only RNA codon. For the description of backbone conformations all of these amino acids need the appropriate 2D-Ramachandran potential energy surfaces (PES). Two of these 21 amino acids have exceptional structures: (i) Glycine (Gly) which has no side chain and thus achiral and (ii) Proline (Pro) in which the side chain is connected back to the amino nitrogen atom fixing the ϕ torsional angle at around 70°. For the remaining 19 proteogenic resides side chains are of varying length and therefore have different degree of conformational freedom. The 1, 2, 3 and 4 dihedral angles add 1, 2, 3 and 4 extra dimensions to the basic 2D Ramachanran PES of the residue. In this way we may end up as $2 + 1 = 3D$, $2 + 2 = 4D$, $2 + 3 = 5D$ and $2 + 4 = 6D$ PEHS per residue. Side chains are classified as being, apolar, polar or explicitly charged as function of the pH. Characteristics of the various amino acids are summarized in Table 2.4. Hetero atom(s) of selected side chains are of importance as they can "self" interact and form intramolecular side chain-backbone interaction(s). The nature of these interactions can be observed even in smaller peptides. However, in proteins there are many more type of interactions (e.g. side/chain-side/chain). These inter-residue interactions require larger peptides to be used as model systems as shown below.

The classification of amino acid residues is also possible via their side-chain topologies. In 19 out of the 20 proteinogenic amino acids the first substituting atom are always carbon. In 16 of the 19 cases the C atom is a methylene (CH_2) group, with one substituent on it, while in 3 cases it is a methine (CH) group, equipped with two substituents. These structural features are shown below.

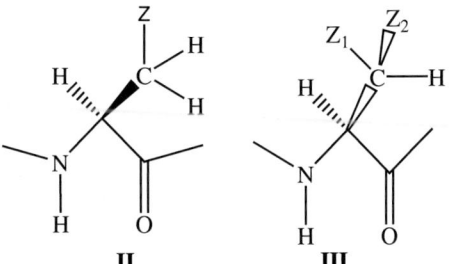

II   **III**

Table 2.4 list of 21 amino acid residues showing their side chains and their dihedral angles of rotation as well as the dimensionality (D) of the full Ramachandran PES.

Table 2.4 List of the 21 proteinogenic α-amino acids, of variable side-chains lengths

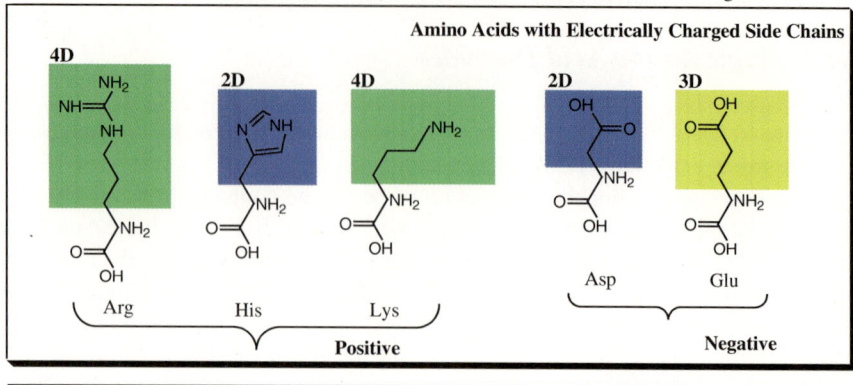

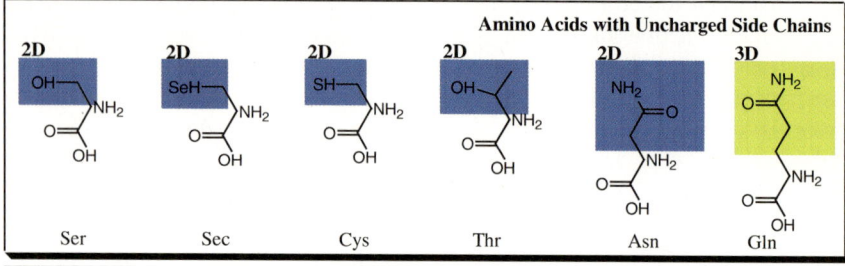

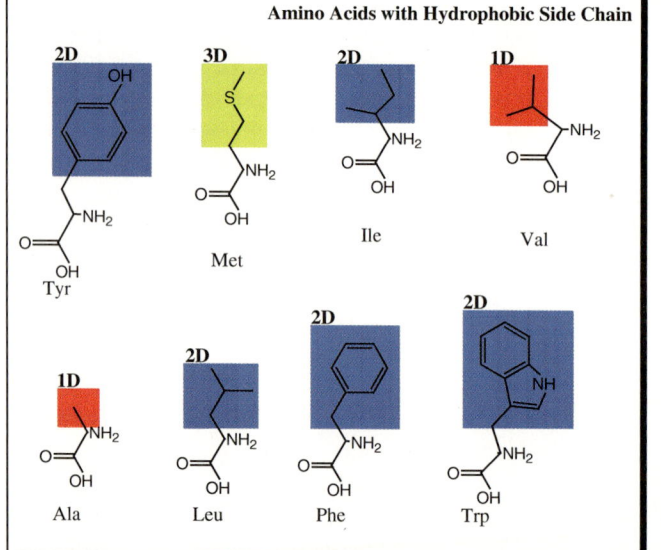

These amino acids can be coupled to each other by forming polypeptides via condensation reaction (Eq. 2.1).

Initially, glycine (Gly) and alanine (Ala) diamide backbone conformers were studied in details [17, 28, 29]. Basis set study for the assessment of the reliability of the used methods of computation were carried out [30]. For selected number of amino acids certain conformers have been identified by matrix isolation infrared spectroscopy [25, 26]. Valine diamide was also studied quite early [31]. Since the α-helical (g⁻g⁻) conformer was not a minimum on the initial PES, therefore various hydrogen bonded structures were generated via the direct solvation method with the inclusion of a single H_2O molecule [32]. This set up was satisfactory to obtain the α-helix minimum on the PES. Serine (Ser) became very popular [33–38] during the years. Cystein (Cys), the sulfur analogue of Ser was investigated considerably later. Actually Cys was computed on its own [39] as well in connection with disulfide bridge formation providing cysteine [40, 41]. The selenium analogue, selenocystein (Sec) was also in focus on its own merit [42] as well for its antioxidant role [43]. Two more sulfur containing amino acids were also investigated, one of them was methionine (Met) [44, 45] and the other one was it demethylated form: homocystein [46]. In addition in their side-chain oxygen containing aminio acid diamides were also studied by various QM methods, for example those included threonine (Thr) [47, 48] and hydroxy-proline [49]. In terms of other polar side chains, asparagine (Asn) [50, 51] and glutamine (Gln) [52] as well as their corresponding acids, Aspartic acid (Asp) [53–56] and glutamic acid (Glu) [52] were also studied.

Conformational properties of aromatic side chain containing amino acids, like phenylalanine (Phe) [57–60] and tyrosine (Tyr) [61] and special apolar amino acid residue like proline (Pro) [62] were also elaborated. As a special case of apolar side chain dehyro-alanine [63] in which a C=C double bond exist between the α and β carbon atoms has also been investigated. Also special attention was given to the question of *trans* → *cis* isomerization of the peptide bond [64, 65].

To discuss all the conclusions of the above individual publication goes beyond the limit of this chapter. The readers are asked to explore them in individual publications. However at least one common conclusion can be drawn from all these conformational stability studies completed on these different side-chain equipped backbone units of peptides and proteins, namely that neither the side chain chemical composition, nor their conformational properties will fundamentally distort the backbone topology of a –CONH–(L)–Xxx–NH– molecular building unit. Thus, the fact that all these proteinogenic residues are α- and L-amino acid residues provides for them a common backbone conformational characteristic.

2.4.2 Dipeptide Diamides

In the early 1990th amino acid dimers concentrated on the simplest dialanine diamide models [66–68], such as -Ala-Ala- and were used to study the first folded systems (e.g. β-turns) by ab initio methods [69]. QM approaches combined with the conformational preferences of the constituent building blocks, known β-turn structures were approved and new forms of turns were revealed. Although found less frequently in proteins, these new β-turns (e.g. $\varepsilon_D\delta_L$, $\varepsilon_D\alpha_L$, $\delta_L\beta_L$) do exists and were successfully assigned in polypeptides and proteins. Thus, multidimensional conformational analysis driven ab initio calculations concluded, that $1 \rightarrow 4$-type intramolecular H-bond of β-turns is more related to the preferred α_L-type sub-structure of the first residue within a β-turn, than to the true nature of the turn-like folded backbone structure. These early QM calculations resulted in already total at least 18 different turn-like foldamers for the For-L-Ala-L-Ala-NH$_2$ model system. Additional β-turn structure analysis revealed [70] the conformational preferences of more complex turn models, such as Pro-Thr [71] and Pro-Pro [72]. These studies revealed how fixing the first torsional angle of the backbone of a dipeptide initiates β-turn formation.

2.4.3 Oligo- and Polypeptide Diamides

Oligopeptides containing 1–5 amino acids were also investigated during the past couple of decades. These studies included either solely alanine residues [73, 74] or Ala in combination with other proteinogenic residues [75]. Antifungal Phe-Arg-Trp [76] was perhaps considered and its conformational preference was established. In addition, the question of beta sheet [77] and α-helix [78–81] stability has also been addressed. Point mutations at the central position of the Pro-Pro-Pro [72] sequence was in focus as the hinge region of immunoglobulin has such sequential properties (e.g. Pro-Xxx-Pro) (Fig. 2.17).

These tripeptide units are already nano-structures in terms of size and they may assume a number of conformations as illustrated by the next figures (Figs. 2.18 and 2.19).

If in the figure below the substituent R is either H or Ph the tripeptide could adopt a relatively long and rather extended backbone structure. However if the central amino acid is Thr, then the internal hydrogen bonds could fold the molecule resulting in a more compact backbone arrangement (Fig. 2.20).

Special attention was given to gluthation [82–85] (Fig. 2.21), as it can form both α- and γ-peptide bond.

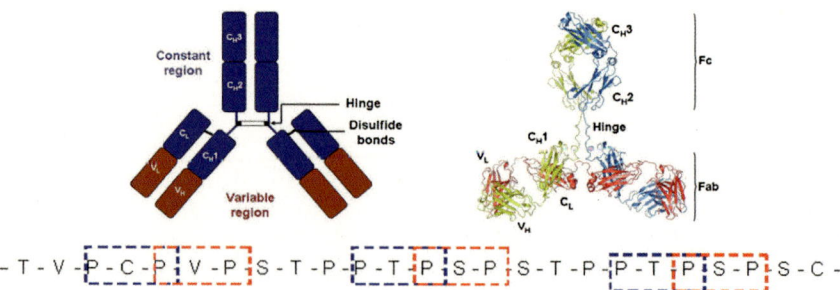

Fig. 2.17 Hinge region location and PXP peptide sequence of immunoglobulin

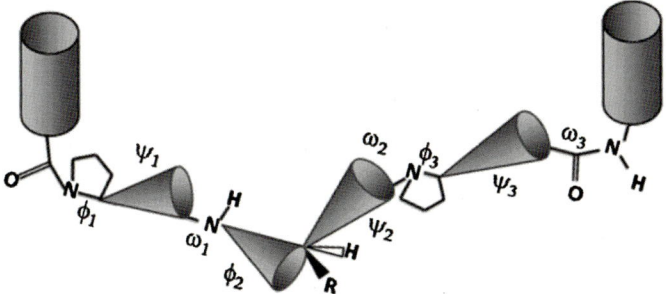

Fig. 2.18 A schematic representation of conformational variations of a PXP tripeptid

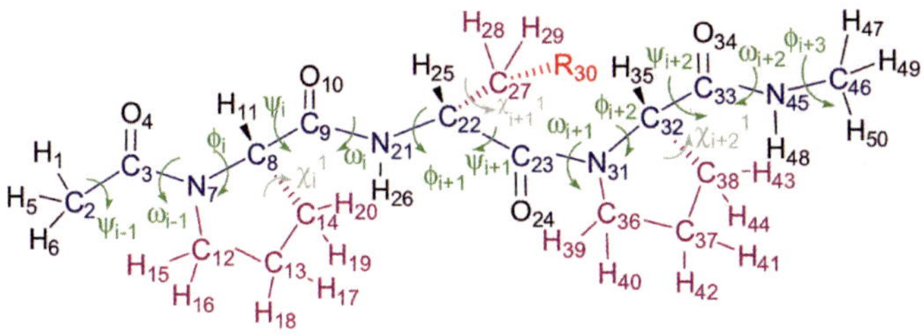

Fig. 2.19 Structural representations of a PXP tripeptide

Fig. 2.20 Side chain/
backbone hydrogen-bonding
networks available in the
different conformers of
HCO-Pro-Thr-Pro-NH$_2$

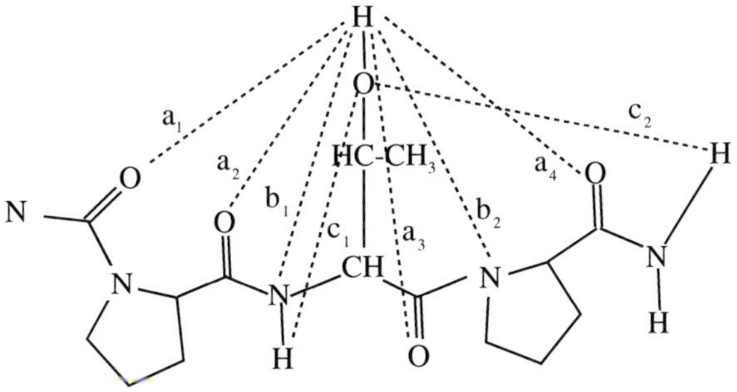

Fig. 2.21 A γ-peptide bond
in Glutathione

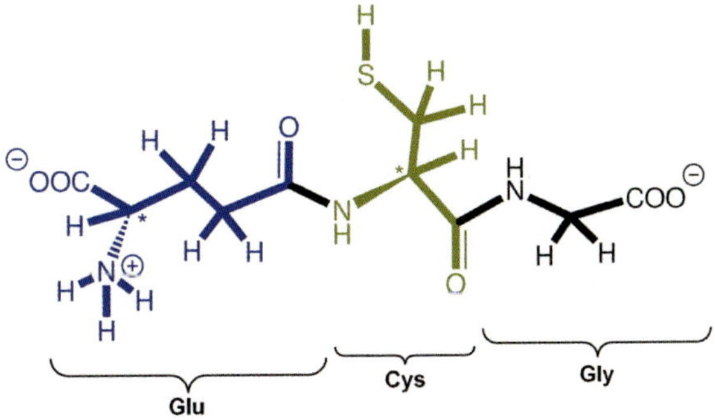

The γ-glutamyl-cystein present in glutathione (γ-Glu-Cys-Gly) is an ancient sequence bit, which has a special role in virtually all living organism as an important antioxidant [86]. It prevents damage to important cellular particles caused by reactive oxygen such as free radicals. In some cases this dipeptide (γ-Glu-Cys) occurs alone in halobacteria [87, 88]. In plants certain mutants occur where the C-terminal glycine is replaced by β-alanine [89], serine [90] or glutamate [91]. All of these suggest that the γ-Gly-Cys moiety was first formed in the ancient period of molecular evolution during which the more stable γ-peptide bond was formed under thermodynamic control.

Collagen build up from -Pro-Gly-HyPro- or POG "triplets" in a repetitive manner has a well characterized triple helical secondary structure, where the chains are stabilized by interchain H-bonds, using the HN of Glys. The relative stability of fully optimized collagen triple helices was determined with respect to a three stranded β-pleated sheet structure by using DFT calculations. In addition the secondary structure preference of Pro, Gly, Ala and HyPro residues was established [92]. De novo calculated collagen structures show a great resemblance to those determined by X-ray crystallography. Interestingly enough the calculated triple helix formation affinities correlate well with the experimentally determined stabilities retrieved from melting point data. The very abundant collagen is not only special by presenting a triple helical structure, but also it is specifically and intensively hydrated [93]. Bella and Berman reported experimentally the structure of the first hydration layer and found that H_2Os form bridges of different length and type around the POG repeats of collagen. Stability and helicity of these hydration layers were computationally determined via 8–12 explicit placed water molecules. Although the stability order of these waters varies from binding places, but they do it in line with the X-ray data. In conclusion, these water binding places on the surface of the triple helix can provide explanation on how an almost liquid-like hydration environment exists between the closely packed tropocollagens [94]. Using the ab initio data it was speculated that these water molecules could serve as reservoirs or buffers providing space for "hole conduction" of water molecules and thus, contribute to the elasticity of collagen known macroscopically for quite some time [95].

Amyloid-like aggregates made of extended like backbone folds of simple polypepetides were also studied by ab initio methods [96]. Accumulated evidences on conformational diseases (e.g. Alzheimer's disease) show the presence of amyloid aggregates, found independent of the primary sequence of the polypeptide chain [97]. Thus the driving force of the conversion from the original to amyloid type foldamer of a primary sequence unit is most probably driven by favorable backbone backbone interactions [98, 99]. In this way most polypeptides and proteins gravitate to an unexpected and highly irreversible thermodynamic minima. They assemble in form of supramolecularnano-systems, within the component macromolecules adopt a common form called as "dead-end" structure(s). Using MDC driven QM calculations on large enough, but still computable polypeptides, it was found that β-pleated sheet structure(s) dominate the "dead-end" molecular foldamer. Several very different building block forms were probed and found both in vacuum and in aqueous environments that their di-, oligo- and polymers make amyloid like fibers. Even in a crystalline state (periodical, tight peptide attachment), the β-pleated sheet assembly remains the most stable superstructure. This theoretical study provides a quantum-level explanation for why proteins can take the amyloid state when local structural preferences jeopardize the functional native and often global folds. The ribbon form of such a nano-assembly is show in Fig. 2.22.

While the β-pleated sheet structure (Fig. 2.22) made of β-layer is typical of an oligopeptide made up of α-amino acid residues, for oligo and polymers of β-amino acids, a self-rapping and thus a spontaneous (energetically favored) nanotubes formation was observed [100]. Octapeptide structure segments of the longer oligopeptide, Penetratin, have been studied computationally as well as by NMR spectroscopy. The computed and the NMR observed structural results agreed well [101].

Certain oligopeptides that have two cysteine residues may be cyclized via disulfide bond formation. In nature, oxytocin is an important example for such cyclic structures (Fig. 2.23a). The relative stability of disulfide bridges that may lead to cyclization or to dimerization (Fig. 2.23b) has been also studied recently [41]. Oligopeptides such as the mammalian neurohypophysial hormone, oxytocin

Fig. 2.22 A schematic "side" view of an amyloid aggregate formed from dozens of polypeptide chains

Fig. 2.23 **a** The structure of oxytocin. **b** Inter-chain disulfide linkage

and vasopressin (Fig. 2.23b) contain two cysteines and by forming an intramolecular cystine they result in a peptidic macrocycle; composed of six residues of 20 atoms in total within the ring. This poses some conformational limitations to them found ideal for binding to their receptors. The relative stability of disulfide bridges that may lead to cyclization or dimerization (Fig. 2.23b) has been also studied recently and found that [41].

2.4.4 Information Accumulation During Polypeptide Folding

The information content change associated with peptide folding has been studied by computing the associated entropy change (ΔS) [80, 102–105] (Figs. 2.24 and 2.25).

The variations of ΔS with extent of polymerization is shown in Fig. 2.26 and the variations of thermodynamic functions in Fig. 2.27. Both figures indicate that the entropy change of poly glycine is faster with increasing degree of polymerization that of poly alanine.

All these entropy focused studies suggest that in folded conformers there is net information accumulation with respect to the unfolded structure.

Fig. 2.24 Variation of relative information content (I/I_0) with ΔS

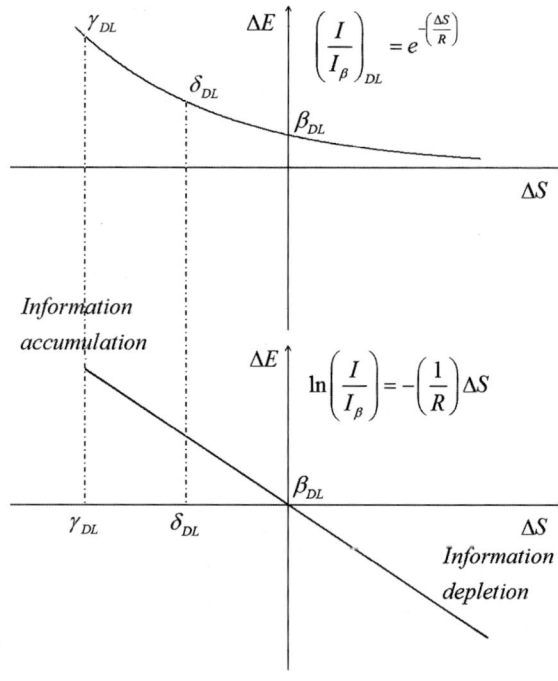

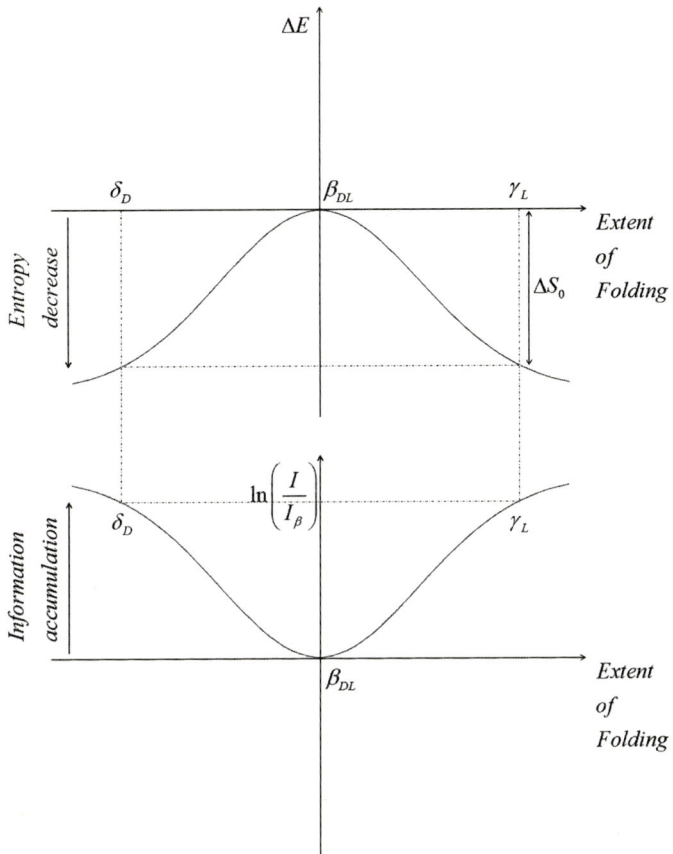

Fig. 2.25 A schematic illustration how ΔS and $\ln(I/I_0)$ varies with peptide folding

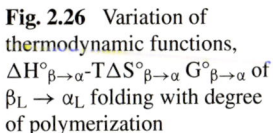

Fig. 2.26 Variation of thermodynamic functions, $\Delta H^\circ_{\beta\to\alpha}$-$T\Delta S^\circ_{\beta\to\alpha}$ $G^\circ_{\beta\to\alpha}$ of $\beta_L \to \alpha_L$ folding with degree of polymerization

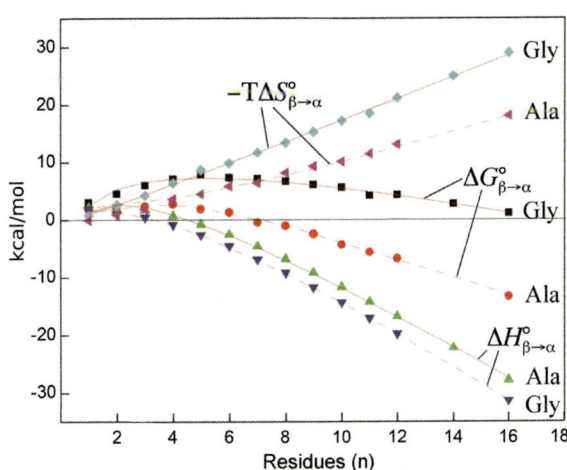

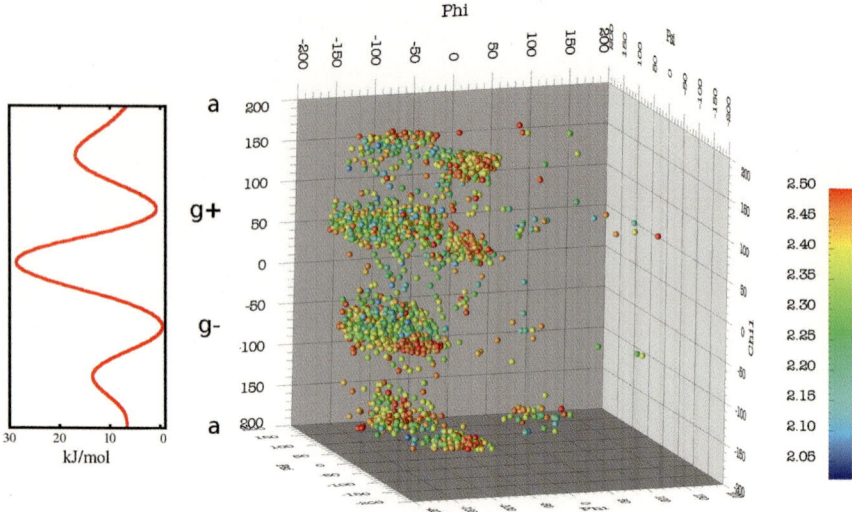

Fig. 2.27 PDB statistics of the distance between the oxygen and the hydrogen in the C–H⋯O sidechain-backbone hydrogen bond in valine residues of proteins. On the *left hand side* the approximate potential energy curve is shown indicating the heavier population of the gauche conformers

2.5 Side Chain Conformations of Small Peptides

Any conformational interconversion of For-L-Ser-NH$_2$ takes place on its 4D hypersurface: $E(\varphi, \psi, \chi_1, \chi_2)$. Both visualizing and making any graphical representation of a 4D-hyperspace are complex. Thus, appropriate 2D-crosssections of the parent 4D hyperspace were calculated and pasted "together" to decipher the topology associated with -Ser- conformers and conformational changes. The analysis of these ab initio conformation energy maps made the tracing of some relaxation paths possible, revealing characteristic side-chain-induced backbone conformational shifts [81]. Interestingly, the partly relaxed conformational hypersurfaces revealed alternative relaxation paths.

2.5.1 A Case Study for Apolar Side Chain (Val)

Beside Alanine the simplest apolar amino acid is Valine which residue requires a 3D PEHS as it has single side-chain dihedral angle beside the backbone φ and ψ variables. Figure 2.27 shows how the energetically more stable g$^+$and g$^-$ side chain rotamers are frequent in proteins, with respect to the less stable and thus less frequent a side chain form.

For-L-Val-NH$_2$ could have in total 27 legitimate minima on its 3D Ramachandran map, $E = E(\varphi, \psi, \chi_1)$. By QM calculations as many as 20 conformers were optimized [31]. A new method was developed for energy partitioning in order to quantify the magnitude of the side chain/backbone interaction probed for the present L-Val model system. Such a side chain/backbone interaction was established by calculating the iPr group for the various backbone conformers of For-L-Val-NH$_2$ relative to that of hydrogen in the corresponding backbone folds of For-Gly-NH$_2$. The comprehensive analysis showed that even an apolar side chain is able to interact with the peptide backbone so "strongly" that it could annihilate otherwise legitimate backbone minima.

Both L-Val and L-Phe residues, prototypes of hydrophobic aliphatic and hydrophobic aromatic amino acid residues, were studied at several basis sets by using different methods (e.g. B3LYP/6-311 ++G**), resulting in a larger dataset compiling results of different levels of theory [106]. Both conformational and energetic properties of these "libraries" were analyzed as a function of the method applied. In addition comparisons of calculated populations of peptide foldamers of these hydrophobic residues were matched with their natural abundance derived from proteins. Analysis concluded that at least for the hydrophobic core of proteins, the conformations of Val (Ile, Leu) and Phe (Tyr, Trp) are controlled by the local energetic preferences of the respective amino acid residues.

2.5.2 A Case Study for Polar Side Chain (Asn and Asp)

Both in the case of Asn as well as Asp residues there are hydrogen bonds between side chains and backbone (Fig. 2.28). Since the H atom of the –COOH group is more protic than that of the –CONH$_2$ moiety, the hydrogen bond is shorter in the case of Asp than in the case Asn. For assessing all H-bonded structures a full 4D potential energy hypersurface would need to be analysed.

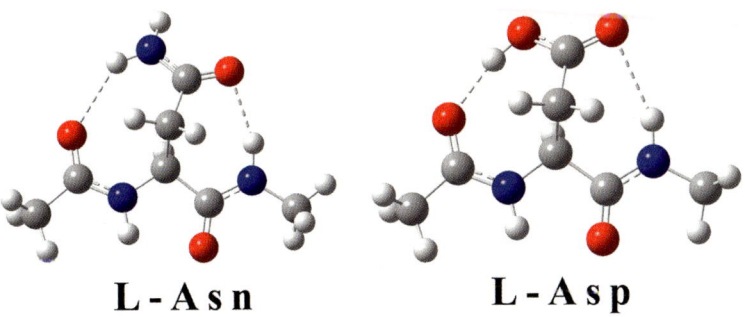

L-Asn   **L-Asp**

Fig. 2.28 Sidechain-backbone N–H···O hydrogen bond in Asn and OH···O hydrogen bond in Asp residues

2.5.3 A Case Study for Protonated Side Chain (His)

Proton affinity and pKa values of N-formyl-L-His-NH$_2$ are found to vary as a function of its backbone and/or side-chain orientation [107]. Examples were presented and confirmed by ab initio calculations, where proteins were crystallized under various pH conditions, resulting in the same histidine residue to adopt different conformers. Furthermore, a hypothesis is given for a protonation-induced conformational modification of the histidine residue in the catalytic triad of chymotrypsin during catalysis, which lowers the pKa value of the catalytic histidine by 1.2 units. Both the experimental and theoretical results support that proton affinity as well as that pKa values of histidine residues are strongly conformationally dependent.

2.6 Peptide Radicals

Accumulated evidence indicates that oxidative stress plays a significant role in a number of diseases such as Parkinson's disease (PD), Alzheimer's disease (AD), Diabetes II and atherosclerosis, just to list a few of the more than 50 examples [108–110]. The mechanisms leading to cellular oxidative stress has been shown to be the result of the excessive production of reactive oxygen species (ROS), that includes non radicals (i.e. H$_2$O$_2$) and free radicals (i.e. OH, O$_2^-$, and NO) as well. Consequently, ROS can interact with different bioactive molecules, to initiate a cascade of events that can lead to cell death [111]. In this way, protein oxidation is a result of hydrogen abstraction by hydroxyl radicals [111]. Under normal conditions, ROS are generated are eliminated by the cell's antioxidant capacity [112]. When the antioxidant capacity is no longer reduce capable to the reduce ROS excess they can accumulate in the cell. This can cause the oxidation of the amino acid residues within the protein backbone, which can result in (a) protein fragmentation (b) a change from the L-configuration to the D-configuration (c) protein aggregation and (d) protein misfolding [113]. Altered protein structure have been observed in a neurodegenerative disease such as Alzheimer's disease, which is generally found in the elderly [114].

It was once thought that all living organisms are composed of only L-amino acids [115]. The discovery of D-aspartic acids (D-Asp) in various human tissues of the elderly people indicates that oxidative stress related to ageing is a main factor in the production of D amino acids. Such a configurational change can result in the accumulation of the D-amino acid and decrease the "original" cnantionerically pure protein concentration [116–119]. In addition, the accumulation of D-amino acids in the brain is affiliated with Alzheimer's disease [116].

2.6.1 Radical Structures and Reactivity

One of the damages that oxidative stress causes at a molecular level is the hydrogen abstraction of the α-hydrogen atom of an amino acid either by hydroxy radicals (**IV**) or by other reactive oxygen species (ROS).

$$\underset{\diagup}{\overset{\diagdown}{=}}\text{C-H}\cdots\cdot\text{OH}$$

IV

It has been shown [120] that the hydrogen atom attached to the α-carbon is the most vulnerable part of proteinogenic residues for such a damaging attack. It has also been suspected that the radical center of a polypeptide chain would react differently as function of the main chain folding. Thus thermodynamic measures of the very same reaction would be different for a folded, unfolded and aggregated nano-system. Penta-glycine [121] and penta-alanine were studied in this respect. The hydrogen radical, (H•), recapturing by the Cα-radical, has been investigated [109]. The following Eq. (2.8) summarizes

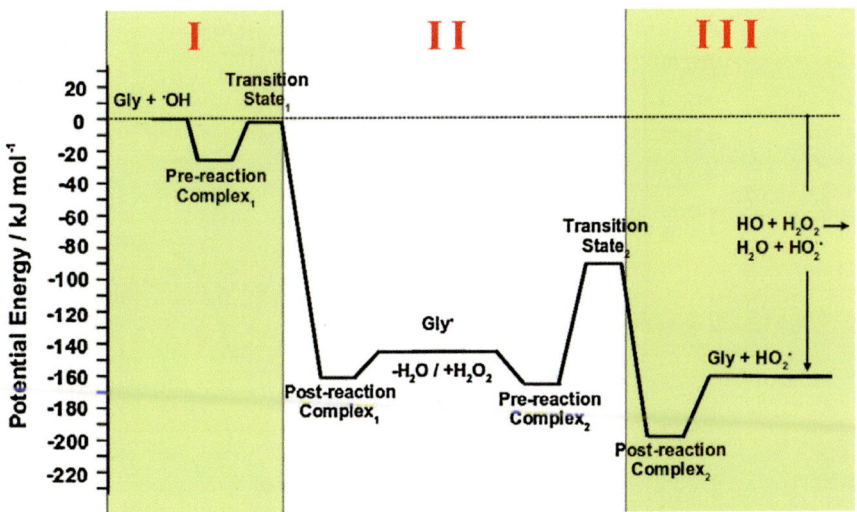

Fig. 2.29 H atoms are less likely to form α-radical followed radical unfolding and ended with H-atom capture of glycine–diamide. Potential Energy variation of the process is which the α-radical of glycine was formed by OH radical and the H-atom was donated the α-radical by H_2O_2. The *roman numerals* on the *top* indicating the initial, intermediate and final states

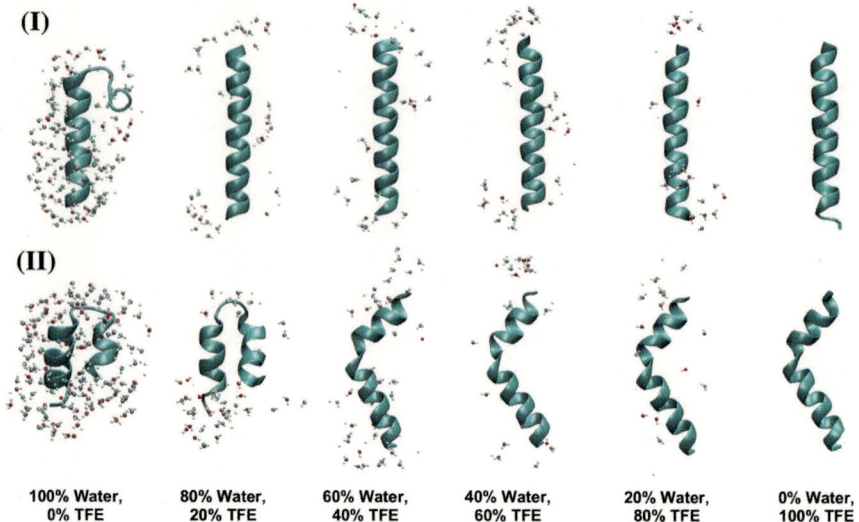

(I)

(II)

| 100% Water, | 80% Water, | 60% Water, | 40% Water, | 20% Water, | 0% Water, |
| 0% TFE | 20% TFE | 40% TFE | 60% TFE | 80% TFE | 100% TFE |

Fig. 2.30 **a** Change of polypeptide conformation from water to TFE solution. **b** Change of polypeptide radical conformation from water to TFE

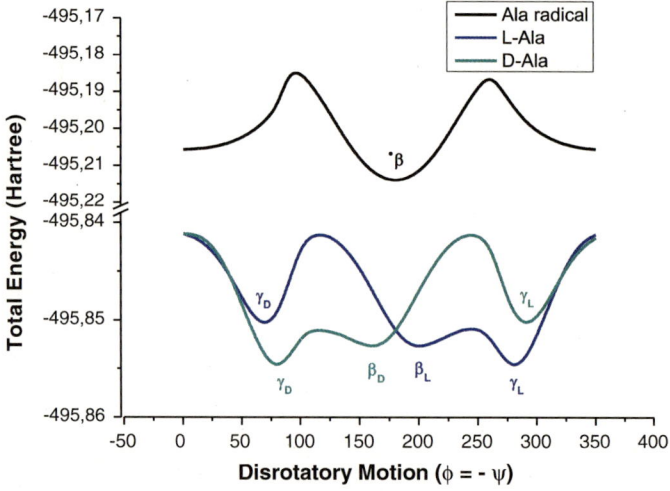

Fig. 2.31 Potential energy curves along the disrotatory ($\phi = -\psi$) cross section of the 2D Ramachandran potential energy surface, $E = f(\phi, \psi)$, for N- and C-protected Ala enantiomeric pairs (in *color*) and their "common" achiral α-radical (in *black*)

the H-abstraction followed by H-recapture and Fig. 2.29 illustrates the details of this process.

$$\text{>C—H + OH + H}_2\text{O}_2 \rightarrow \text{>C• + H——O—OH + H}_2\text{O}_2 \rightarrow \text{>C—H + •O—OH + H}_2\text{O}_2$$

$$(2.8)$$

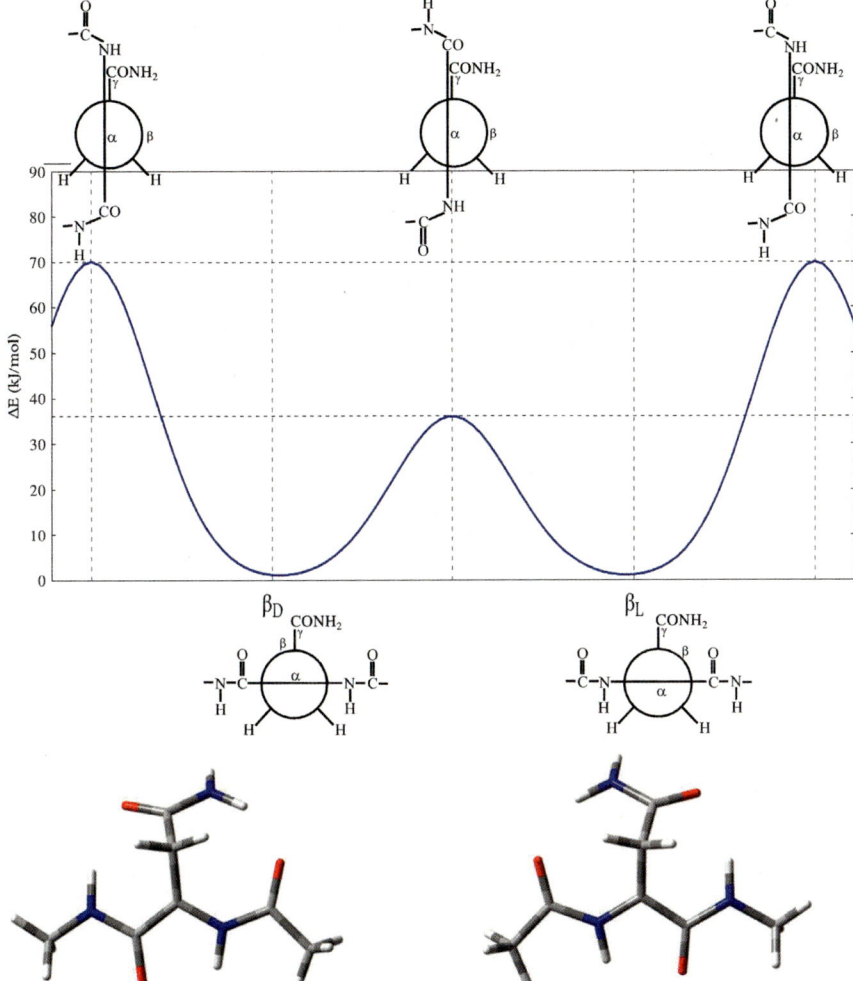

Fig. 2.32 Cross section of the full PES of Ac-Asn•-NHCH₃ and the schematic stereoisomers of Asn β_D and •Asn β_L, detailed path is show in Fig. 2.34

Unlike Gly, all the other amino acid residues are chiral and thus, the possibility of racemization when H. is recaptured is significant even for folded macromolecules, where the attack of the H. could have spatial preference.

The fact that an α-helix is not very stable as a foldamer is known from the literature. In fact if a helical subset of a protein is cut off and separated from the parent protein, it will almost always unfold in H₂O. The affinity of such a sequence to present nascent helices can be enhanced by adding TFE of other fluorinated cosolvents to water as shown for instance for penetratine (see below) [101, 122].

However, as the helical backbone gets more and more structured beside the equilibrium state of a folded and of an unfolded set of molecules the helix can also turn and/or bend (Fig. 2.30). The conformational change of a helix in water as well as in apolar environment, like it is in the case of trans-membrane proteins [109] is illustrated in Fig. 2.30.

Clearly, at the radical center (Fig. 2.30b) the folding of the helix is more pronounced in water than in a less polar environment (like in a membrane).

2.6.2 Atropisomerism of Radicals

The potential energy surface (PES) associated with Ac-Ala-NHMe clearly shows that regardless of whether the $C\alpha$-radical was generated from the L- or D-alanine, it has a single, fully extended β-like conformation, with ($\phi = \psi = 180°$). (Center of the black line in Fig. 2.31.)

However, if the side-chain is larger than that of a CH_3 as it is in Ala, then the PES gets more complex. For Asn which has a longer side-chain, the C_α-radical itself will present a pair of degenerate β-conformers, labeled as β_L and β_D as shown in Fig. 2.32. This type of potential energy curve is typical of axis chirality which leads to atropisomerisation. Atropisomerism occurs even in simple compounds, the rotation about C_2–C_3 of n-butane is a typical example; the g^+ and g^- conformers are atropisomers.

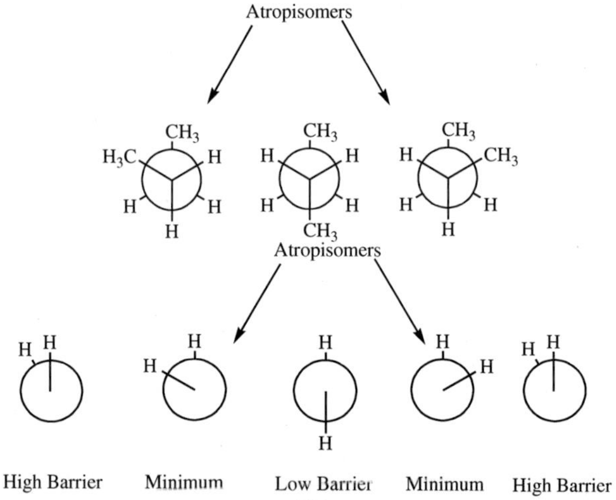

One of the simplest examples for axis chirality is hydrogen peroxide.

Fig. 2.33 Schematic illustration of enantiomeric topomerization paths A and B on the 2D-PES cross sections associated with backbone (*top*) and side-chain (*bottom*) conformational change

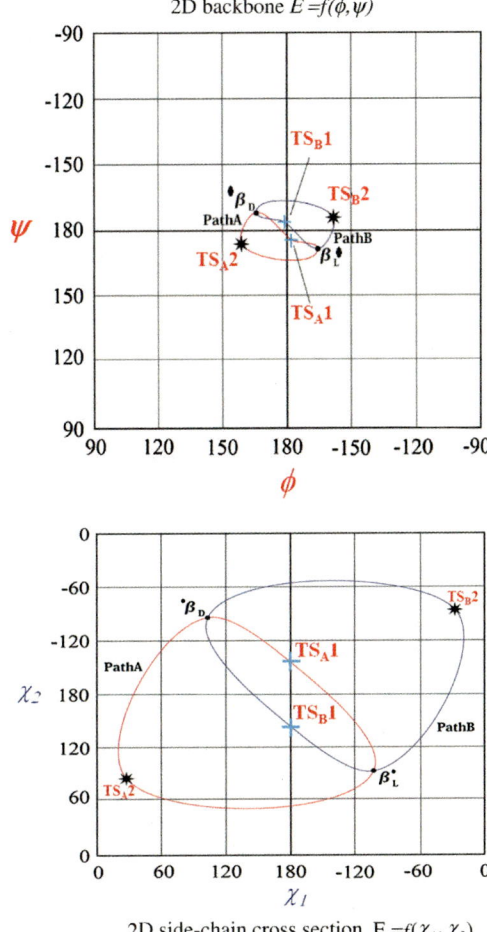

2D backbone $E = f(\phi, \psi)$

2D side-chain cross section $E = f(\chi_1, \chi_2)$

The above 1D representation of Asn radical (Fig. 2.32) is an over simplification of the molecular topomerization phenomenon [3] shown in Fig. 2.33. However, it is clear that the system may proceed along two directions. One of the two directions is along the path toward the low energy maximum point (TS1), and the other one toward the high energy maximum point (TS2). Clearly, when we are moving in a 2D PES we have two orthogonal paths that may be labeled as path A and path B. This situation will lead us to four transition states (TS$_A$1, TS$_A$2, TS$_B$1 and TS$_B$2). Figure 2.34 illustrates this phenomenon in terms of a pair of 2D cross-sections. Noting that Asn conformations are describable by a 4D-Ramachandran Potential Energy Surface, such a 4D problem may be represented by a pair of 2D energy

Quantum System

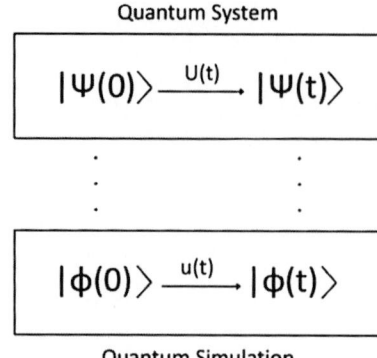

Quantum Simulation

Fig. 2.34 A schematic illustration of the relationship between the quantum system to be investigated and the quantum simulation actually studied. Note that instead of the exact initial state $\psi(0)$ actually the approximate $\phi(0)$ is used as initial state

Table 2.5 Free energy changes, ΔG^0 (kJ/mol) of radical formation and enantiomeric topomerization computed at several levels of theory

Species	DFT[a]				MP2[b]			
	ΔE_1	Corr[d]	ΔG_1^0	E(CC[c])	ΔE_2	Corr[d]	ΔG_2^0	E(CC[c])
TS A1	38.47	−5.9	32.57	43.29	42.71	−7.34	35.37	42.29
TS B1	38.47	−5.88	32.58	43.28	42.71	−7.34	35.38	42.29
TS A2	70.88	−5.46	65.42	67.5	72.8	−5.11	67.69	63.72
TS B2	70.87	−5.46	65.41	67.49	72.8	−5.11	67.69	63.72
Asn β_L[e]	−1676.3	31.03	−1645.2	−1604.9	−1694.6	32.22	−1662.4	−1602.4
Asn β_D[e]	−1676.3	31.03	−1645.2	−1604.9	−1694.6	32.22	−1662.4	−1602.4
•Asn β_L	0	0	0	0	0	0	0	0
•Asn β_D	0	0	0	0	0	0	0	0

[a]Geometry optimized at B3LYP/6-311 ++G(d,p) level of theory (calculated vibrational wavenumbers are scaled by 0.97)
[b]Values are obtained by MP2/cc-pVTZ//B3LYp/6-311 ++G(d,p) level of theory
[c]Values are obtained by CCSD(T)/cc-pVTZ//B3LYp/6-311 ++G(d,p) level of theory
[d]Correction for ΔE to ΔG^0
eFor dissociation free energy the ΔG^0 must be reduced by the hydrogen energy which may takes as 0.5 hartree (1312.75 kJ/mol)

surfaces, one in the backbone subspace, $E = f(\phi, \psi)$, and the other one in the side-chain subspace, $E = f(\chi_1, \chi_2)$ as shown in Fig. 2.33 (Table 2.5).

Clearly, the enantiotopic TS A1 and TS B1 represent the lower barrier (38.47 kJ/mol) while the enantiotopic TS A2 and TS B2 correspond to the higher barrier (70.87 kJ/mol).

2.7 Future Perspectives

According to a classical and old cliché "the future is no longer what it used to be". This poetic statement implies that we cannot predict the future by direct extrapolation from the events of the past. Nevertheless the final section of this chapter was constructed to envisage certain visions to highlight certain principles, knowing that they might not be true and thus may not turn into reality.

2.7.1 Time Dependent Quantum Simulations

To become an exact science for chemistry is an eschatological hope. In order to avoid any misunderstanding, we first need to give a careful definition of the term "exact". We mean by it not only something which is accurately computable, but rather a rigorous theory behind what is computed. The rigorous theory is quantum mechanics in principle and the equation to be solved is the time-dependent form of the Schrödinger equation:

$$ih\frac{d}{dt}[\Psi(t)] = \hat{H}[\Psi(t)] \tag{2.9}$$

For the sake of simplicity, $\hat{H}$ is the time independent Hamiltonian. Time dependent Schrödinger equations are needed to describe processes, and time will occure in the form of an evolutionary operator U shown below in Eq. (2.11). The time independent Schrödinger equation (at t $= 0$)

$$\hat{H}[\Psi(0)] = E[\Psi(0)] \tag{2.10}$$

where $\psi(0)$ is the initial state. The solution of the time dependent state is

$$\Psi(t) = \hat{U}\Psi(0) = e^{-\hbar\hat{H}t}\Psi(0) \tag{2.11}$$

where the unitary transformation is achieved by the following operator

$$\hat{U} = \exp(-i\hbar\hat{H}t) \tag{2.12}$$

With the advent of quantum computers it was hoped that such time dependent solution will be possible. Since all processes, including chemical changes are time dependent, the solution of the time dependent Schrödinger equation is the ultimate goal [123–129].

It turned out that most of the efforts today in chemistry and particularly in protein folding studies is to overcome the computational struggle on the time dependent aspect of the latter problem, by brushing aside the determination of the initial state $\psi(0)$. Furthermore, since the problem is not solvable rigorously, at this time, therefore a simple problem is solved instead and that is actually regarded as a simulation of the actual process. This simulation process is illustrated by Fig. 2.34.

Fig. 2.35 A schematic illustration for the computation of the initial state φ(0) from the one electron functions AO, $\chi(0)$, and MO, $\varphi(0)$. In between the *top* and the *bottom*, the various methods developed during the 20th century are shown

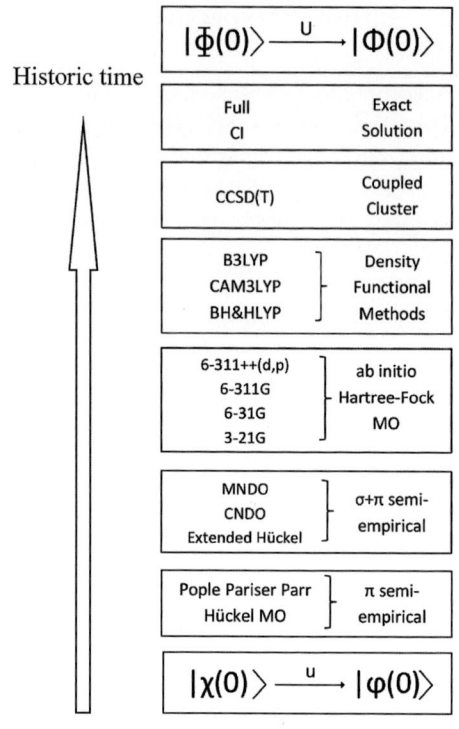

2.7.2 Time Independent Quantum Simulations

In order to be ever successful in the time dependent quantum simulation we have to make a serious effort to generate a reliable initial state wave function: $\psi(0)$. Since the time of Schrödinger's publication in 1927, which is getting closer and closer to a century, we made a considerable scientific effort to generate an accurate wave function for the initial state, $\psi(0)$, of a stable molecular system which undergoes the time dependent process. Figure 2.35 shows the hierarchy of the various attempts during the 20th century to state wave functions $\Phi(0)$ from orbitals (χ or φ).

We are today in the second box from the top: The transition to the top box is occurring in the 21st century.

2.7.3 The Protein Folding Problem

This is an old problem. It is a fact that the molecular conformation problem coupled with intramolecular interactions such as hydrogen bondings.

The ultimate rationale behind all purposeful structures and behavior of living things is embodied in the sequence of residues of nascent polypeptide chains – the precursors of the folded proteins which in biology play the role of Maxwell's demons. In a very real sense it is at this level of organization that the secret of life (if there is one) is to be found. If we could only determine these sequences but also pronounce the law by wich they fold, then the secret of life would be found – the ultimate rationale discovered!

(Jaques Monod (1970))

The social impact of the discovery how proteins fold exceeds the combined social impact of

the discovery of fire
the discovery of writing
the discovery of wheel

(Unknown authors)

For a while, it was assumed that the biologically active conformer or folded structure is thermodynamically the most stable one. Protein misfolding or even denaturation studies suggest that the biologically active structure is thermodynamically not the most stable one. At that time, the existence of chaperones were postulated in which case the protein-chaperone complex would find the appropriate folded structure. Since the folding itself, like all chemical process, is time dependent therefore the process of folding could be studied by Quantum Simulation (perhaps on a quantum computer), if the initial conditions are defined [130–143]. Figure 2.36 illustrates a four-amino acid peptide either without or with chaperone assistance. This figure also shows how such computations are treating the initial and find final quantum states for such a tetrapeptide.

Thus the Quantum Simulation that is jumping from the initial state to the final state is currently reducing protein folding to a lattice folding problem.

However, even if we ignore all complexity arising from the method applied we have to face the problem rising from molecular structure flexibility. For example if we start with a tetra-glycine to make the problem easy and simple (e.g. Ac-Gly$_4$-NHMe), the folding has to be represented on a Ramachandran conformational hypersurface of eight independent variables (if counting only the most important backbone torsional angles):

$$E = f(\phi_1, \psi_1, \phi_2, \psi_2, \phi_3, \psi_3, \phi_4, \psi_4) \qquad (2.13)$$

Fig. 2.36 Lattice folding simulation of a tetrapeptide folding, **a** alone **b** with chaperone assistance

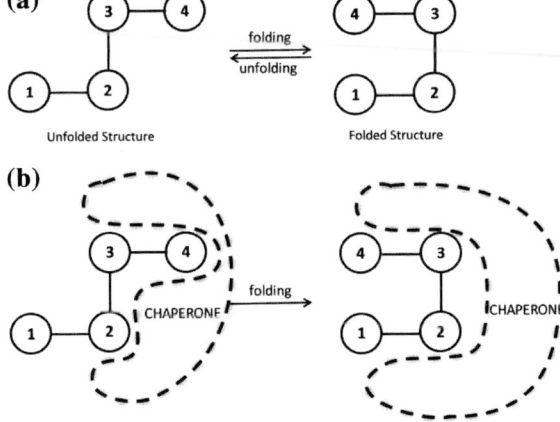

Table 2.6 N_0 (number of minima), N_1 (number of transitional states or TS) and N grid points (in case of 30° grid spacing) to be calculated for a 2D and a 4D amino acid PEHS

n	2nD-PEHS			4nD-PEHS		
	N_0 (min) backbone or side chain	N_1 (TS) backbone or side chain	N grid points	N_0 (min) backbone + side chain	N_1 (TS) backbone + side chain	N grid points
1	9	18	144	81	324	20,736
2	81	324	20,736	6,561	52,488	4.30×10^8
4	6,561	52,488	4.30×10^8	4.30×10^7	6.89×10^8	1.85×10^{17}
10	3.49×10^9	6.97×10^{10}	3.83×10^{21}	1.21×10^{19}	4.86×10^{20}	1.47×10^{43}
20	1.21×10^{19}	4.86×10^{20}	1.47×10^{43}	1.47×10^{38}	1.18×10^{40}	2.16×10^{86}

Therefore such a potential energy hypersurface (PEHS) is expected to have up to the following number of minima (N_0) and the following number of transition states (N_1).

$$N_0 = 3^8 = 6561 \tag{2.14}$$

$$N_1 = 8 \cdot 3^8 = 52488 \tag{2.15}$$

So the initial and final states need to be selected from something like 6,516 stable conformers and the paths interconnecting these minima must pass through a number of transition states (TS) to be picked from the 52,488 TS structures. Keeping track of such simple folding events seems already complicated enough, not talking about how to visualize such a phenomena.

To view the situation from a more general point of view we may consider the followings: (i) Backbone of an amino acid residue has two rotors (ϕ, ψ) and (ii) on average (see Table 2.6) side chain also has two rotors (χ_1, χ_2). Thus, both backbone (BB) 2D- and side chain (SC)—on average 2D—makes for each amino acid residues at least a 4D-problem.

For a 2D amino acid problem, where n is the number of amino acid residues and $2n$ is the number of rotors we have the following relationships.

$$N_0 = 3^{2n} \tag{2.16}$$

$$N_1 = 2n \cdot N_0 = 2n \cdot 3^{2n} \tag{2.17}$$

The reason we expect to have $2n$ times N_0 transitional states, N_1, is illustrated on Fig. 2.37, where it can be seen that each minimum, N_0, has two nearest transitional states along each variable.

One possible way to estimate all N_0 initial minimal energy structures is to code algorithmically our knowledge of conformational analysis. These predicted foldamer structures can then be optimized. However, the coded algorithm will be quite complicated, meaning that in certain cases, perhaps several times we may need to attempt a given optimization to find the rotamer or foldamer. Then the algorithm can be modified to look for transitional states. Another alternative would be to generate grid points and subject the fitted functions to the process of mathematical analysis.

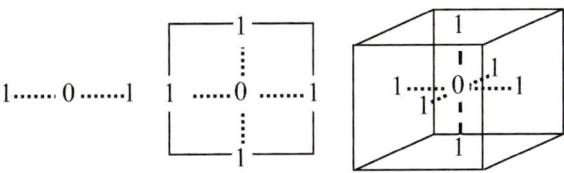

Fig. 2.37 Location of the transitional states (1) around a minima (0), in a one, two and three dimensional space. Critical points labelled by their index (λ) meaning the number of imaginary frequencies. Minima have 0, first order TS have 1 etc

If we generate a grid with 30° intervals i.e. 12 points in the range 0–360° then we need N grid points

$$N = 12^{2n} \tag{2.18}$$

for a 4D problem where n is again the numbers of the amino acid residues we have

$$N_0 = 3^{4n} \tag{2.19}$$

$$N_1 = 4n \cdot N_0 = 4n \cdot 3^{4n} \tag{2.20}$$

$$N = 12^{4n} \tag{2.21}$$

Figure 2.38 shows the variation of N_0 and N as a function of n on a logarithmic scale.

At this time of human history, we can handle 20,736 grid points, what is a 4D representation, either a single amino acid diamide with a 2D side chain, like Asn, or diglycine–diamide or dialanine–diamide backbone conformations. However, for a peptide of 20 amino acids, a 10^{82} times increase in the computing capabilities would be demanded.

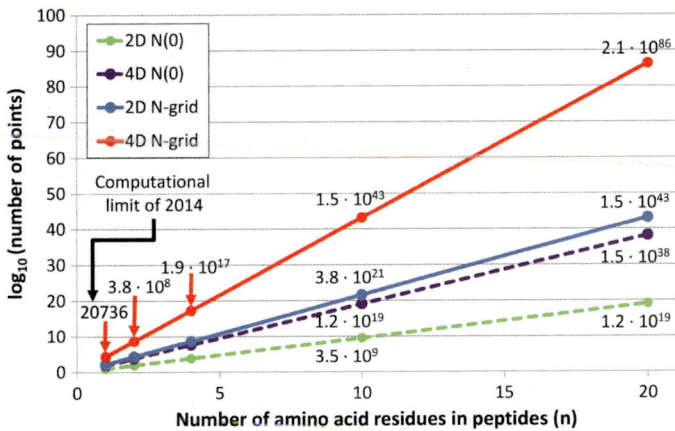

Fig. 2.38 Logarithmic scale of N_0 (number of minima) for a 2D and a 4D case (*broken lines*) and N (number of *grid points*) for a 2D and a 4D case (*solid lines*). Data is summarized in Table 2.6

Fig. 2.39 Schematic folding
pathways between foldamer
A and foldamer B

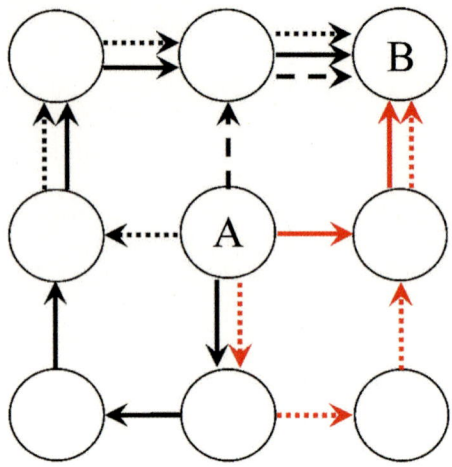

All in all we need to find the folding pathways and there are several ways to go from foldamer **A** to foldamer **B**. The following simple example shows 5 paths (Fig. 2.39).

Thus, we have some homework to do before a full-fledged time dependent Quantum Simulation could be performed on a future quantum computer that may have sufficient speed and accuracy needed for chemical and biological considerations.

2.7.4 *The Future of Protein Folding*

The protein folding problem is more than a century old. The real solution is expected to come after the chemical problem is converted to a mathematical problem. From the Born-Oppenheimer approximation emerged the concept of PEHS which represents the chemical problem. Thus, the chemical problem becomes a travel path from one minimum to another minimum. In the case of protein folding, the PEHS is a conformational PEHS that represents structure changes that involve neither making nor breaking any covalent bond. That conformation PEHS would need to be computed, with such a rigorous theory at hand, that the design of a meaningful experimental method would lead to realistic possibility.

As with all evolutionary process, such development takes time, but the development has already started. Force field methods emerged in the 1970s mostly due to the initial work of Norman Allinger [144]. This was a molecular mechanical (MM) simulation of the quantum mechanically (QM) computable potential energy hypersurface (PEHS). Since such an ab initio surface is not available, yet it would represent the ultimate solution, therefore the use of the currently available force field softwares make a remarkable service to chemistry and biology. Needless to say, the currently available force field softwares are far more sophisticated than their early versions. Clearly, many improvements were implemented during the past 40 years.

Further development represents the combination of QM and MM methods in which the chemically more important regions, like active sites of a proteins are computed quantum mechanically, and the less important regions are represented by molecular mechanics. This QM/MM method is a 'hybrid offspring', like a mule is, of the QM and MM methods, yet remarkable achievements can be made by this more recent development.

Nevertheless, force field potential energy hyper-surfaces are simulations of the ab initio PEHS. However, the time has arrived to aim for such a gigantic effort that could produce the ab initio surfaces. It has taken 50 years for about 20 researchers to a develop Gaussian and Schrödinger from the initial POLYATOM software. That is a 1,000 man–years. However, it could take considerably less to develop a rigorous mathematical equation that encompasses all possible structures of a given protein. Certainly this is doable in the 21st Century if we are really dedicated to achieve that.

Acknowledgments This work was supported by grants from the Hungarian Scientific Research Fund (grant numbers: OTKA NK101072 and TÁMOP-4.2.2.A-11/1/KONV-2012-0047. TÁMOP-4.2.1.B-09/1/KMR). This work was also supported by the following NDA grants: TÁMOP-4.2.2.C-11/1/KONV-2012-0010, TÁMOP-4.2.2.A-11/1/KONV-2012-0047. The authors thank M. Labádi and L. Müller for the administration of the computer clusters used for this work. Technical assistances of Anita Rágyanszki, Klára Gerlei, Kyungseop Lee and Csaba Hatvani are greatly appreciated.

References

1. Csizmadia IG, Harrison MC, Sutcliffe BT (1963) Non-empirical LCAO-SCF calculation of formyl fluoride with gaussian atomic orbitals. Q Prog Rep 50:1
2. Csizmadia IG, Harrison MC, Moskowitz JW, Sutcliffe BT (1966) Non-empirical LCAO-MO-SCF-CI calculations on organic molecules with gaussian type functions. Part I Theor Chim Acta 6:191
3. Gerlei KZ, Jákli I, Szőri M et al (2013) Atropisomerism of the Asn α radicals revealed by Ramachandran surface topology. J Phys Chem B 117:12402–12409. doi:10.1021/jp4070906
4. Robb MA, Csizmadia IG (1968) Non-empirical LCAO-MO-SCF-CI calculations on organic molecules with gaussian type functions. Theor Chim Acta 10:269
5. Mucsi Z, Tsai A, Szori M et al (2007) A quantitative scale for the extent of conjugation of the amide bond. Amidity percentage as a chemical driving force. J Phys Chem A 111:13245–13254. doi:10.1021/jp0759325
6. Weintraub SJ, Manson SR (2004) Asparagine deamidation: a regulatory hourglass. Mech Ageing Dev 125:255–257. doi:10.1016/j.mad.2004.03.002
7. Robinson NE, Robinson AB (2001) Deamidation of human proteins. Proc Natl Acad Sci USA 98:12409–12413. doi:10.1073/pnas.221463198
8. Ritz-Timme S, Laumeier I, Collins M (2003) Age estimation based on aspartic acid racemization in elastin from the yellow ligaments. Int J Legal Med 117:96–101. doi:10.1007/s00414-002-0355-2
9. Ohtani S, Yamamoto T (2010) Age estimation by amino acid racemization in human teeth. J Forensic Sci 55:1630–1633. doi:10.1111/j.1556-4029.2010.01472.x
10. Thorpe CT, Streeter I, Pinchbeck GL et al (2010) Aspartic acid racemization and collagen degradation markers reveal an accumulation of damage in tendon collagen that is enhanced with aging. J Biol Chem 285:15674–15681. doi:10.1074/jbc.M109.077503
11. Peters B, Trout BL (2006) Asparagine deamidation: pH-dependent mechanism from density functional theory. Biochemistry 45:5384–5392. doi:10.1021/bi052438n

12. Catak S, Monard G, Aviyente V, Ruiz-López MF (2008) Computational study on nonen-zymatic peptide bond cleavage at asparagine and aspartic acid. J Phys Chem A 112:8752–8761. doi:10.1021/jp8015497

13. Fujii N, Tajima S, Tanaka N et al (2002) The presence of D-beta-aspartic acid-containing peptides in elastic fibers of sun-damaged skin: a potent marker for ultraviolet-induced skin aging. Biochem Biophys Res Commun 294:1047–1051. doi:10.1016/S0006-291X(02)00597-1

14. Ramachandran GN, Ramakrishnan C, Sasisekharan V (1963) Stereochemistry of polypeptide chain configurations. J Mol Biol 7:95–99. doi:10.1016/S0022-2836(63)80023-6

15. Perczel A, Angyan JG, Kajtar M et al (1991) Peptide models. 1. Topology of selected peptide conformational potential energy surfaces (glycine and alanine derivatives). J Am Chem Soc 113:6256–6265. doi:10.1021/ja00016a049

16. Csizmadia IG (1997) Basic principles for introductory organic chemistry. Quirk Press, Toronto, pp 135–141

17. Perczel A, Kajtar M, Marcoccia JF, Csizmadia IG (1991) The utility of the four-dimensional Ramachandran map for the description of peptide conformations. J Mol Struct THEOCHEM 232:291–319. doi:http://dx.doi.org/10.1016/0166-1280(91)85261-5

18. Berman HM, Westbrook J, Feng Z et al (2000) The protein data bank. Nucleic Acids Res 28:235–242

19. Radom L, Hehre JW, Pople JA (1972) Molecular orbital theory of the electronic structure of organic compounds. XIII. Fourier component analysis of internal rotation potential functions in saturated molecules. J Am Chem Soc 94:2371–2381. doi:10.1021/ja00762a030

20. Radom L, Lathan WA, Hehre WJ, Pople JA (1973) Molecular orbital theory of the electronic structure of organic compounds. XVII. Internal rotation in 1,2-disubstituted ethanes. J Am Chem Soc 95:693–698. doi:10.1021/ja00784a008

21. Kehoe TAK, Peterson MR, Chass GA, et al (2003) The fitting and functional analysis of a double rotor potential energy surface for the R and S enantiomers of 1-chloro-3-fluoro-isobutane. J Mol Struct THEOCHEM 666–667:79–87. doi:http://dx.doi.org/10.1016/j.theochem.2003.08.015

22. Demaré GR, Peterson MR, Csizmadia IG, Strausz OP (1980) Conformational energy surfaces of triplet-state isomeric methyloxiranes. J Comp Chem 1:141–148. doi:10.1002/jcc.540010206

23. Peterson R, Mare RDE, Roosevelt AFD (1981) Conformations formaldehyde, acetone of triplet carbonyl compounds: and the reaction of 0 (3P) atoms with olefins proceeds via a C % Ci, triplet biradical [1], which can undergo intersystem crossing to the So state and be stabilized as an epoxide, 86:131–147

24. Rágyanszki A, Surányi A, Csizmadia IG, et al (2014) Fourier type potential energy function for conformational change of selected organic functional groups. Chem Phys Lett 599:169–174. doi:http://dx.doi.org/10.1016/j.cplett.2014.03.029

25. Pohl G, Perczel A, Vass E et al (2007) A matrix isolation study on Ac-Gly-NHMe and Ac-L-Ala-NHMe, the simplest chiral and achiral building blocks of peptides and proteins. Phys Chem Chem Phys 9:4698–4708

26. Pohl G, Perczel A, Vass E et al (2008) A matrix isolation study on Ac-L-Pro-NH2: a frequent structural element of beta- and gamma-turns of peptides and proteins. Tetrahedron 64(9):2126–2133

27. Jákli I, Knak Jensen SJ, Csizmadia IG, Perczel A (2012) Variation of conformational properties at a glance. True graphical visualization of the Ramachandran surface topology as a periodic potential energy surface. Chem Phys Lett 547:82–88. doi:10.1016/j.cplett.2012.08.002

28. Viviani W, Rivail J-L, Csizmadia I (1993) Peptide models II. Intramolecular interactions and stable conformations of glycine, alanine, and valine peptide analogues. Theor Chim Acta 85:189–197. doi:10.1007/BF01374587

29. McAllister MA, Perczel A, Császár P et al (1993) Peptide models 4. Topological features of molecular mechanics and ab initio 2D-Ramachandran maps. J Mol Struct THEOCHEM 288:161–179. doi:10.1016/0166-1280(93)87048-I

30. Endrédi G, Perczel A, Farkas O, et al (1997) Peptide models XV. The effect of basis set size increase and electron correlation on selected minima of the ab initio 2D-Ramachandran map of For-Gly-NH$_2$ and For-l-Ala-NH$_2$. J Mol Struct THEOCHEM 391:15–26. doi:http://dx. doi.org/10.1016/S0166-1280(96)04695-7

31. Viviani W, Rivail JL, Perczel A, Csizmadia IG (1993) Peptide models. 3. Conformational potential energy hypersurface of formyl-L-valinamide. J Am Chem Soc 115:8321–8329. doi:10.1021/ja00071a046

32. Perczel A, Farkas O, Csizmadia IG (1995) Peptide models. 17. The role of the water molecule in peptide folding. An ab initio study on the right-handed helical conformations of N-formylglycinamide and N-formyl-L-alaninamide monohydrates [H(CONH-CHR-CONH) H.H$_2$O; R=H or CH$_3$]. J Am Chem Soc 117:1653–1654. doi:10.1021/ja00110a028

33. Perczel A, Daudel R, Ångyán JG, Csizmadia IG (1990) A study on the backbone/side-chain interaction in N-formyl-(L)serineamide. Can J Chem 68:1882–1888. doi:10.1139/v90-291

34. Farkas Ö, Perczel A, Marcoccia J-F, et al (1995) Peptide models XIII. Side-chain conformational energy surface E=E(χ1, χ2) of N-formyl-l-serinamide (For-l-Ser-NH$_2$) in its γL or C7 eq backbone conformation. J Mol Struct THEOCHEM 331:27–36. doi:http://dx.doi.org/10.1016/0166-1280(94)03929-F

35. Perczel A, Farkas Ö, Csizmadia IG (1996) Peptide models XVI. The identification of selected HCO—L—SER—NH$_2$ conformers via a systematic grid search using ab initio potential energy surfaces. J Comput Chem 17:821–834. doi:10.1002/(SICI)1096-987X (199605)17:7<821:AID-JCC6>3.0.CO;2-U

36. Perczel A, Farkas Ö, Csizmadia IG (1996) Peptide models. 18. hydroxymethyl side-chain induced backbone conformational shifts of l-serine amide. All ab Initio conformers of for-l-Ser-NH$_2$. J Am Chem Soc 118:7809–7817. doi:10.1021/ja960464q

37. Perczel A, Farkas Ö, Jákli I, Csizmadia IG (1998) Peptide models XXI. Side-chain/backbone conformational interconversions in HCO-l-Ser-NH$_2$. Tracing relaxation paths by ab initio modeling. An exploratory study. J Mol Struct THEOCHEM 455:315–338

38. Imre Jákli AP (2000) Peptide models XXIII. Conformational model for polar side-chain containing amino acid residues: a comprehensive analysis of RHF, DFT, and MP2 properties of HCO-L-SER-NH$_2$. J Comput Chem 21:626–655. doi:10.1002/(SICI)1096-987X(200006)21:83.0.CO;2-P

39. Fang D-C, Fu X-Y, Tang T-H, Csizmadia IG (1998) Ab initio modelling of peptide biosynthesis. J Mol Struct THEOCHEM 427:243–252. doi:http://dx.doi.org/10.1016/ S0166-1280(97)00257-1

40. Galant NJ, Lee DR, Fiser B, et al (2012) Disulfidicity: a scale to characterize the disulfide bond strength via the hydrogenation thermodynamics. Chem Phys Lett 539–540:11–14. doi:http://dx.doi.org/10.1016/j.cplett.2012.05.017

41. Galant NJ, Song HC, Jákli I et al (2014) A theoretical study of the stability of disulfide bridges in various β-sheet structures of protein segment models. Chem Phys Lett 593:48–54

42. Vank JC, Sosa CP, Perczel A, Csizmadia IG (2000) Peptide models XXVII. An exploratory ab initio study on the 21st amino acid side-chain conformations of N-formyl-L-selenocysteinamide (For-L-Sec-NH$_2$) and N-acetyl-L-selenocysteine-N-methylamide (Ac-L-Sec-NHMe) in their γL backbone conformation. Can J Chem 78:395–408. doi:10.1139/v00-029

43. Fiser B, Mucsi Z, Viskolcz B et al (2013) Controlled antioxidative steps of the cell. The concept of chalcogenicity. Chem Phys Lett 590:83–86. doi:10.1016/j.cplett.2013.10.033

44. Láng A, György K, Csizmadia IG, Perczel A (2003) A conformational comparison of N- and C-protected methionine and N- and C-protected homocysteine. J Mol Struct THEOCHEM 666–667:219–241. doi:http://dx.doi.org/10.1016/j.theochem.2003.08.029

45. Láng A, Csizmadia IG, Perczel A (2005) Peptide models XLV: conformational properties of N-formyl-L-methioninamide and its relevance to methionine in proteins. Proteins 58:571–588. doi:10.1002/prot.20307

46. Sheraly AR, Chass GA, Csizmadia IG (2003) The multidimensional conformational analysis for the backbone across the disrotatory axis at selected side-chain conformers of

N-Ac-homocysteine-NHMe—an ab initio exploratory study. J Mol Struct THEOCHEM 666–667:243–249. doi:http://dx.doi.org/10.1016/j.theochem.2003.08.030

47. Sahai MA, Motiwala SS, Chass GA et al (2003) An ab initio exploratory study of the full conformational space of MeCO-l-threonine-NH-Me. J Mol Struct THEOCHEM 666–667:251–267. doi:10.1016/j.theochem.2003.08.031

48. Sahai MA, Fejer SN, Viskolcz B et al (2006) First-principle computational study on the full conformational space of L-threonine diamide, the energetic stability of cis and trans isomers. J Phys Chem A 110:11527–11536. doi:10.1021/jp0680488

49. Lam JSW, Koo JCP, Hudaky I, et al. Predicting the conformational preferences of N-acetyl-4-hydroxy-L-proline-N$'$-methylamide from the proline residue. J Mol Struct Theochem 666–667:285–289

50. Rassolian M, Chass GA, Setiadi DH, Csizmadia IG (2003) Asparagine—ab initio structural analyses. J Mol Struct THEOCHEM 666–667:273–278. doi:http://dx.doi.org/10.1016/j.theochem.2003.08.032

51. Berg M, Salpietro SJ, Perczel A et al (2000) Peptide models XXVI Side chain conformational analysis of N-formyl-L-asparagin amide and N-acetyl-L-asparagin N-methylamide in their gL backbone conformation. J Mol Struct 504:127–140

52. Masman MF, Zamora MA, Rodrıguez AM et al (2002) Exploration of the full conformational space of N-acetyl-L-glutamate-N-methylamide. Eur Phys J D-At Mol Opt Plasma Phys 20:531–542. doi:10.1140/epjd/e2002-00150-y

53. Salpietro SJ, Perczel A, Farkas Ö et al (2000) Peptide models XXV. Side-chain conformational potential energy surface, of N-formyl-l-aspartic acidamide and its conjugate base N-formyl-l-aspartatamide in their γl backbone conformations. J Mol Struct THEOCHEM 497:39–63. doi:10.1016/S0166-1280(99)00196-7

54. Koo JCP, Chass GA, Perczel A et al (2002) Exploration of the four-dimensional-conformational potential energy hypersurface of N-Acetyl-l-aspartic Acid N$'$-Methylamide with its internally hydrogen bonded side-chain orientation. J Phys Chem A 106:6999–7009. doi:10.1021/jp014514b

55. Koo JCP, Chass GA, Perczel A et al (2002) N-acetyl-L-aspartic acid-N$'$-methylamide with side-chain orientation capable of external hydrogen bonding. Eur Phys J D-At Mol Opt Plasma Phys 20:499–511. doi:10.1140/epjd/e2002-00148-5

56. Koo JCP, Lam JSW, Chass GA, et al. (2003) Ramachandran backbone potential energy surfaces of aspartic acid and aspartate residues: implications on allosteric sites in receptor–ligand complexations. J Mol Struct THEOCHEM 666–667:279–284. doi:http://dx.doi.org/10.1016/j.theochem.2003.08.055

57. Farkas Ö, McAllister MA, Ma JH et al (1996) Peptide models XIX: side-chain conformational energy surface and amide I vibrational frequencies of N-formyl-l-phenylalaninamide (For-Phe-NH$_2$) in its γL or γinv or C7 eq backbone conformation. J Mol Struct THEOCHEM 369:105–114. doi:10.1016/S0166-1280(96)04548-4

58. Perczel A, Farkas Ö, Csizmadia IG, Császár AG (1997) Peptide models XX. Aromatic side-chain–backbone interaction in phenylalanine-containing diamide model system. A systematic search for the identification of all the ab initio conformers of N-formyl-L-phenylalanine-amide. Can J Chem 75:1120–1130. doi:10.1139/v97-134

59. Jákli I, Perczel A, Farkas Ö et al (1998) Peptide models XXII. A conformational model for aromatic amino acid residues in proteins. A comprehensive analysis of all the RHF/6–31 + G* conformers of for-L-Phe–NH$_2$. J Mol Struct THEOCHEM 455:303–314

60. Chass GA, Mirasol RS, Setiadi DH et al (2005) Characterization of the conformational probability of N-acetyl-phenylalanyl-NH$_2$ by RHF, DFT, and MP2 computation and AIM analyses, confirmed by jet-cooled infrared data. J Phys Chem A 109:5289–5302. doi:10.1021/jp040720i

61. Chass GA, Lovas S, Murphy RF, Csizmadia IG (2002) The role of enhanced aromatic-electron donating aptitude of the tyrosyl sidechain with respect to that of phenylalanyl in intramolecular interactions. Eur Phys J D 20:481–497. doi:10.1140/epjd/e2002-00155-6

62. Sahai MA, Kehoe TAK, Koo JCP et al (2005) First principle computational study on the full conformational space of l-proline diamides. J Phys Chem A 109:2660–2679. doi:10.1021/jp040594i

63. Füzéry AK, Csizmadia IG (2000) An exploratory density functional study on N- and C-protected trans-α, β-didehydroalanine. J Mol Struct THEOCHEM 501–502:539–547. doi:10.1016/S0166-1280(99)00469-8

64. Baldoni HA, Zamarbide GN, Enriz RD, et al (2000) Peptide models XXIX. cis–trans Isomerism of peptide bonds: ab initio study on small peptide model compound; the 3D-Ramachandran map of formylglycinamide. J Mol Struct THEOCHEM 500:97–111. doi:http://dx.doi.org/10.1016/S0166-1280(00)00372-9

65. Sahai MA, Szöri M, Viskolcz B et al (2007) Transition state infrared spectra for the trans → Cis isomerization of a simple peptide model. J Phys Chem A 111:8384–8389. doi:10.1021/jp074991f

66. Van Alsenoy C, Cao M, Newton SQ et al (1993) Conformational analysis and structural study by ab initio gradient geometry optimizations of the model tripeptide N-formyl L-alanyl L-alanine amide. J Mol Struct THEOCHEM 286:149–163. doi:10.1016/0166-1280(93)87160-F

67. McAllister MA, Perczel A, Császár P, Csizmadia IG (1993) Peptide models 5. Topological features of molecular mechanics and ab initio 4D-Ramachandran maps. J Mol Struct THEOCHEM 288:181–198. doi:10.1016/0166-1280(93)87049-J

68. Perczel A, McAllister MA, Császár P, Csizmadia IG (1994) Peptide models. IX. A complete conformational set of For-Ala-Ala-NH$_2$ from ab inito computations. Can J Chem 72:2050–2070. doi:10.1139/v94-262

69. Perczel A, McAllister MA, Csaszar P, Csizmadia IG (1993) Peptide models 6. New.beta.-turn conformations from ab initio calculations confirmed by x-ray data of proteins. J Am Chem Soc 115:4849–4858. doi:10.1021/ja00064a053

70. Perczel A, Jákli I, McAllister MA, Csizmadia IG (2003) Relative stability of major types of beta-turns as a function of amino acid composition: a study based on Ab initio energetic and natural abundance data. Chemistry 9:2551–2566. doi:10.1002/chem.200204393

71. Sahai MA, Setiadi DH, Chass GA et al (2003) A model study of the IgA hinge region: an exploratory study of selected backbone conformations of MeCO-l-Pro-l-Thr-NH-Me. J Mol Struct THEOCHEM 666–667:311–319. doi:10.1016/j.theochem.2003.08.036

72. Sahai MA, Viskolcz B, Pai EF, Csizmadia IG (2007) Quantifying the intrinsic effects of two point mutation models of pro-pro-pro triamino acid diamide. A first-principle computational study. J Phys Chem B 111:13135–13142. doi:10.1021/jp074046r

73. Perczel A, Endrédi G, McAllister MA et al (1995) Peptide models VII the ending of the right-handed helices in oligopeptides [For-(Ala)n-NH$_2$ for $\leq$ n $\leq$ 4] and in proteins. J Mol Struct THEOCHEM 331:5–10. doi:10.1016/0166-1280(94)03972-N

74. Endrédi G, McAllister MA, Farkas Ö et al (1995) Peptide models XII Topological features of molecular mechanics and ab-initio 8D-Ramachandran maps. Conformational data for Ac-(l-Ala)4-NHMe and For-(l-Ala)4-NH$_2$. J Mol Struct THEOCHEM 331:11–26. doi:10.1016/0166-1280(94)03915-8

75. Sahai MA, Sahai MR, Chass GA et al (2003) An ab initio exploratory study on selected conformational features of MeCO-l-Ala-l-Ala-l-Ala-NH-Me as a XxxYyyZzz tripeptide motif within a protein structure. J Mol Struct THEOCHEM 666–667:327–336. doi:10.1016/j.theochem.2003.08.041

76. Masman MF, Rodríguez AM, Svetaz L et al (2006) Synthesis and conformational analysis of His-Phe-Arg-Trp-NH$_2$ and analogues with antifungal properties. Bioorg Med Chem 14:7604–7614. doi:10.1016/j.bmc.2006.07.007

77. Perczel A, Gáspári Z, Csizmadia IG (2005) Structure and stability of beta-pleated sheets. J Comput Chem 26:1155–1168. doi:10.1002/jcc.20255

78. Perczel A, Farkas Ö, Marcoccia JF, Csizmadia IG (1997) Peptide models. XIV. Ab initio study on the role of side-chain backbone interaction stabilizing the building unit of right- and left-handed helices in peptides and proteins. Int J Quantum Chem 61:797–814. doi:10.1002/(SICI)1097-461X(1997)61:5<797::AID-QUA6>3.0.CO;2-R

79. Topol IA, Burt SK, Deretey E et al (2001) alpha- and 3(10)-helix interconversion: a quantum-chemical study on polyalanine systems in the gas phase and in aqueous solvent. J Am Chem Soc 123:6054–6060
80. Viskolcz B, Fejer SN, Knak Jensen SJ et al (2007) Information accumulation in helical oligopeptide structures. Chem Phys Lett 450:123–126. doi:10.1016/j.cplett.2007.11.001
81. Jákli I, Csizmadia IG, Fejer SN et al (2013) Helix compactness and stability: Electron structure calculations of conformer dependent thermodynamic functions. Chem Phys Lett 563:80–87. doi:10.1016/j.cplett.2013.01.060
82. Galant NJ, Wang H, Lee DR et al (2009) Thermodynamic role of glutathione oxidation by peroxide and peroxybicarbonate in the prevention of Alzheimer's disease and cancer. J Phys Chem A 113:9138–9149. doi:10.1021/jp809116n
83. Ding VZY, Dawson SSH, Lau LWY et al (2011) A computational study of glutathione and its fragments: N-acetylcisteinylglycine and γ-glutamylmethylamide. Chem Phys Lett 507:168–173. doi:10.1016/j.cplett.2011.03.067
84. Fiser B, Szori M, Jójárt B et al (2011) Antioxidant potential of glutathione: a theoretical study. J Phys Chem B 115:11269–11277. doi:10.1021/jp2049525
85. Fiser B, Jójárt B, Csizmadia IG, Viskolcz B (2013) Glutathione–hydroxyl radical interaction: a theoretical study on radical recognition process. PLoS ONE 8:e73652. doi:10.1371/journal.pone.0073652
86. Veeravalli K, Boyd D, Iverson BL et al (2011) Laboratory evolution of glutathione biosynthesis reveals natural compensatory pathways. Nat Chem Biol 7:101–105. doi:10.1038/nchembio.499
87. Ganguly D, Srikanth CV, Kumar C et al (2003) Why is glutathione (a tripeptide) synthesized by specific enzymes while TSH releasing hormone (TRH or thyroliberin), also a tripeptide, is produced as part of a prohormone protein? IUBMB Life 55:553–554. doi:10.1080/15216540310001623064
88. Fahey RC, Sundquist AR (1991) Evolution of glutathione metabolism. Adv Enzymol Relat Areas Mol Biol 64:1–53
89. Yi H, Ravilious GE, Galant A et al (2010) From sulfur to homoglutathione: thiol metabolism in soybean. Amino Acids 39:963–978. doi:10.1007/s00726-010-0572-9
90. Hermes-Lima M (2004) Oxygen in biology and biochemistry: role of free radicals. Funct Metab 319–368
91. Meuwly P, Thibault P, Rauser WE (1993) Gamma-glutamylcysteinylglutamic acid—a new homologue of glutathione in maize seedlings exposed to cadmium. FEBS Lett 336:472–476
92. Pálfi V, Perczel A (2008) How stable is a collagen triple helix? An ab initio study on various collagen and beta-sheet forming sequences. J Comput Chem 29(9):1374–1386
93. Bella J, Berman HM (1996) Crystallographic evidence for Cα–H···O=C hydrogen bonds in a collagen triple helix. J Mol Biol 264:734–742. doi:10.1006/jmbi.1996.0673
94. Henkelman RM, Stanisz GJ, Kim JK, Bronskill MJ (1994) Anisotropy of NMR properties of tissues. Magn Reson Med 32:592–601
95. Pálfi V, Perczel A (2010) Stability of the hydration layer of tropocollagen: a QM study. J Comput Chem 31(4):764–777
96. Masman MF, Eisel ULM, Csizmadia IG et al (2009) Model peptides for binding studies on peptide—peptide interactions: the case of amyloid β(1-42) aggregates. J Phys Chem A 113:11710–11719
97. Eakin CM, Berman AJ, Miranker AD (2006) A native to amyloidogenic transition regulated by a backbone trigger. Nat Struct Mol Biol 13:202–208. doi:10.1038/nsmb1068
98. Nelson R, Sawaya MR, Balbirnie M et al (2005) Structure of the cross-beta spine of amyloid-like fibrils. Nature 435:773–778. doi:10.1038/nature03680
99. Wright CF, Teichmann SA, Clarke J, Dobson CM (2005) The importance of sequence diversity in the aggregation and evolution of proteins. Nature 438:878–881

100. Beke T, Csizmadia IG, Perczel A (2006) Theoretical study on tertiary structural elements of beta-peptides: nanotubes formed from parallel-sheet-derived assemblies of beta-peptides. J Am Chem Soc 128:5158–5167. doi:10.1021/ja0585127
101. Czajlik A, Perczel A (2004) Peptide models XXXII. Computed chemical shift analysis of penetratin fragments. J Mol Struct THEOCHEM 675:129–139. doi:10.1016/j.theochem.2003.12.036
102. Salpietro SJ, Viskolcz B, Csizmadia IG (2003) An exploratory ab initio study on the entropy of various backbone conformers for the HCO-Gly-Gly-Gly-NH$_2$ tripeptide motif. J Mol Struct THEOCHEM 666–667:89–94. doi:10.1016/j.theochem.2003.08.016
103. Viskolcz B, Fejer SN, Csizmadia IG (2006) J Phys Chem A 110:3808
104. Fejer SN, Csizmadia IG, Viskolcz B (2006) Thermodynamic functions of conformational changes: conformational network of glycine diamide folding, entropy lowering, and informational accumulation. J Phys Chem A 110:13325–13331. doi:10.1021/jp065595k
105. Pohl G, Jákli I, Csizmadia IG et al (2012) The role of entropy in initializing the aggregation of peptides: a first principle study on oligopeptide oligomerization. Phys Chem Chem Phys 14:1507
106. Hudáky I, Hudáky P, Perczel A (2004) Solvation model induced structural changes in peptides. A quantum chemical study on Ramachandran surfaces and conformers of alanine diamide using the polarizable continuum model. J Comput Chem 25:1522–1531. doi:10.1002/jcc.20073
107. Hudáky P, Perczel A (2004) Conformation dependence of pKa: Ab initio and DFT investigation of histidine. J Phys Chem A 108:6195–6205. doi:10.1021/jp048964q
108. Barnham KJ, Masters CL, Bush AI (2004) Neurodegenerative diseases and oxidative stress. Nat Rev Drug Discov 3:205–214. doi:10.1038/nrd1330
109. Owen MC, Szori M, Csizmadia IG, Viskolcz B (2012) Conformation-dependent ·OH/H2O2 hydrogen abstraction reaction cycles of Gly and Ala residues: a comparative theoretical study. J Phys Chem B 116:1143–1154. doi:10.1021/jp2089559
110. Büeler H (2009) Impaired mitochondrial dynamics and function in the pathogenesis of Parkinson's disease. Exp Neurol 218:235–246. doi:10.1016/j.expneurol.2009.03.006
111. Easton CJ (1997) Free-radical reactions in the synthesis of alpha-amino acids and derivatives. Chem Rev 97:53–82
112. Stadtman ER (2006) Protein oxidation and aging. Free Radic Res 40:1250–1258. doi:10.1080/10715760600918142
113. Berlett BS, Stadtman ER (1997) Protein oxidation in aging, disease, and oxidative stress. J Biol Chem 272:20313–20316
114. Gregersen N, Bolund L, Bross P (2005) Protein misfolding, aggregation, and degradation in disease. Mol Biotechnol 31:141–150. doi:10.1385/MB:31:2:141
115. Harris E (1992) Regulation of antioxidant enzymes. FASEB J 6:2675–2683
116. Shapira R, Chou CH (1987) Differential racemization of aspartate and serine in human myelin basic protein. Biochem Biophys Res Commun 146:1342–1349
117. Masters PM (1983) Stereochemically altered noncollagenous protein from human dentin. Calcif Tissue Int 35:43–47
118. Fujii N (2005) D-amino acid in elderly tissues. Biol Pharm Bull 28:1585–1589
119. Shapiro SD, Endicott SK, Province MA et al (1991) Marked longevity of human lung parenchymal elastic fibers deduced from prevalence of D-aspartate and nuclear weapons-related radiocarbon. J Clin Invest 87:1828–1834. doi:10.1172/JCI115204
120. Gerlei KZ, Élo L, Fiser B et al (2014) Impairment of a model peptide by oxidative stress: thermodynamic stabilities of asparagine diamide Cα-radical foldamers. Chem Phys Lett 593:104–108
121. Owen MC, Viskolcz B, Csizmadia IG (2011) Quantum chemical analysis of the unfolding of a penta-glycyl 3(10)-helix initiated by HO(•), HO2(•), and O2(-•). J Chem Phys 135:035101. doi:10.1063/1.3608168

122. Czajlik A, Meskó E, Penke B, Perczel A (2002) Investigation of penetratin peptides. Part 1. The environment dependent conformational properties of penetratin and two of its derivatives. J Pept Sci 8:151–171. doi:10.1002/psc.380
123. Smirnov AY, Savel'ev S, Mourokh LG, Nori F (2007) Modelling chemical reactions using semiconductor quantum dots. Europhys Lett 80:67008. doi:10.1209/0295-5075/80/67008
124. Aspuru-Guzik A, Dutoi AD, Love PJ, Head-Gordon M (2005) Simulated quantum computation of molecular energies. Science 309:1704–1707. doi:10.1126/science.1113479
125. Kassal I, Jordan SP, Love PJ et al (2008) Polynomial-time quantum algorithm for the simulation of chemical dynamics. Proc Natl Acad Sci USA 105:18681–18686. doi:10.1073/pnas.0808245105
126. Kassal I, Whitfield JD, Perdomo-Ortiz A et al (2011) Simulating chemistry using quantum computers. Annu Rev Phys Chem 62:185–207. doi:10.1146/annurev-physchem-032210-103512
127. Wang H, Ashhab S, Nori F (2009) Efficient quantum algorithm for preparing molecular-system-like states on a quantum computer. Phys Rev A 79:042335. doi:10.1103/PhysR evA.79.042335
128. Lu D, Xu N, Xu R et al (2011) Simulation of chemical isomerization reaction dynamics on a NMR quantum simulator. Phys Rev Lett 107:20501
129. Lidar DA, Wang H (1999) Calculating the thermal rate constant with exponential speedup on a quantum computer. Phys Rev E 59:2429–2438
130. Perdomo A, Truncik C, Tubert-Brohman I et al (2008) Construction of model Hamiltonians for adiabatic quantum computation and its application to finding low-energy conformations of lattice protein models. Phys Rev A 78:12320
131. Floudas CA, Pardalos PM (2000) Optimization in computational chemistry and molecular biology: local and global approaches. Kluwer Academic Publication, Berlin
132. Alexander K. Hartmann HR (2004) New optimization algorithms in physics
133. Kolinski A, Skolnick J (1996) Lattice models of protein folding. Dynamics and thermodynamics. Chapman & Hall, Austin
134. Mirny L, Shakhnovich E (2001) Protein folding theory: from lattice to all-atom models. Annu Rev Biophys Biomol Struct 30:361–396. doi:10.1146/annurev.biophys.30.1.361
135. Pande VS, Grosberg AY, Tanaka T (2000) Heteropolymer freezing and design: towards physical models of protein folding. Rev Mod Phys 72:259–314
136. Hart WE, Istrail S (1997) Robust proofs of NP-hardness for protein folding: general lattices and energy potentials. J Comput Biol 4:1–22
137. Dill KA, Ozkan SB, Shell MS, Weikl TR (2008) The protein folding problem. Annu Rev Biophys 37:289–316. doi:10.1146/annurev.biophys.37.092707.153558
138. Lau KF, Dill KA (1989) A lattice statistical mechanics model of the conformational and sequence spaces of proteins. Macromolecules 22:3986–3997. doi:10.1021/ma00200a030
139. Epstein CJ, Goldberger RF, Anfinsen CB (1963) The genetic control of tertiary protein structure: studies with model systems. Cold Spring Harb Symp Quant Biol 28:439–449. doi:10.1101/SQB.1963.028.01.060
140. Crescenzi P, Goldman D, Papadimitriou C et al (1998) On the complexity of protein folding. J Comput Biol 5:423–465
141. Amin MHS, Choi V (2009) First-order quantum phase transition in adiabatic quantum computation. Phys Rev A 80:062326. doi:10.1103/PhysRevA.80.062326
142. Farhi E, Goldstone J (2009) Quantum adiabatic algorithms, small gaps, and different paths. Comput Res Repos—CORR
143. Perdomo-Ortiz A, Venegas-Andraca SE, Aspuru-Guzik A (2010) A study of heuristic guesses for adiabatic quantum computation. Quantum Inf Process 10:33–52. doi:10.1007/s11128-010-0168-z
144. Allinger NL (1977) Conformational analysis. 130. MM2. A hydrocarbon force field utilizing V1 and V2 torsional terms. J Am Chem Soc 99:8127–8134. doi:10.1021/ja004 67a001

Chapter 3
Car-Parrinello Simulations of Chemical Reactions in Proteins

Carme Rovira

3.1 Introduction

It is nowadays accepted that proteins should not be pictured as static objects, but rather as dynamic systems [1, 2]. It is also recognized that there must be a detailed balance, encoded in the amino acids sequence, between stability and flexibility since protein function generally requires some degree of conformational motion [1]. These requirements are unified under the concept of the free energy landscape (FEL), i.e., the free energy as a function of the atomic coordinates of the system [3, 4]. Its knowledge would tell us about the stable states of the system (wells on the FEL), their population (Boltzmann probabilities) and the paths by which they are interconnected (likely overcoming energy barriers), guiding us towards an understanding of protein folding [5–7], signal transmission [8], ligand binding [9], transportation [10] and chemical catalysis [11]. It is thus not surprising that much effort has been and is dedicated to the development of experimental and computational techniques for its characterization. Yet, even for the most studied system, large parts of the surface are still unknown. Much of today's understanding of the topology of a protein FEL has been modelled on extensive studies of the heme protein myoglobin [3, 12]. X-ray crystallography, spectroscopy and molecular dynamics simulations contributed to picture the FEL as a rugged surface organized

C. Rovira (✉)
Departament de Química Orgànica and Institut de Química Teòrica i Computacional
(IQTCUB), Universitat de Barcelona, Martí i Franquès 1, 08028 Barcelona, Spain
e-mail: c.rovira@ub.edu

C. Rovira
Institució Catalana de Recerca i Estudis Avançats (ICREA),
Passeig Lluís Companys 23, 08010 Barcelona, Spain

© Springer International Publishing Switzerland 2014
G. Náray-Szabó (ed.), *Protein Modelling*, DOI 10.1007/978-3-319-09976-7_3

in a hierarchy of tiers depending on the magnitude of the barriers that separate the wells. Higher tiers comprise few functional states separated one another by large barriers, such that transitions from one to another are rare events, with each one comprising a large number of conformational substates separated by smaller energy barriers [3, 13]. This classifications turns out also to be an operational one, as the magnitude of the barriers determine the time scale by which (sub)states interconvert and thus dictate the choice of the technique to be used to investigate transitions [1].

In the context of enzymatic catalysis, the mechanisms by which protein fluctuations are coupled to the reactive events are still under debate [14–16], but it is becoming increasingly evident that characteristic motions of the protein, present in the native state, preferentially follow the pathways that create the configuration optimum for catalysis [11]. Some of these pathways can be analysed by classical molecular dynamics (MD) simulations [13], whereas subtle motions of active site residues, often coupled to electronic rearrangements, can be captured by quantum chemical and ab initio MD approaches [17] such as the Car-Parrinello (CP) method.

Following a brief description of the CP method, I summarize the results of almost two decades of CP simulations of electronic processes in proteins [18], with particular emphasis on enzymatic reactions. The current status of the applications of the CPMD methodology is illustrated by a recent study on heme enzyme reactivity. Further examples can be found on recent reviews to which this work is complementary (see e.g. [17–21]).

3.2 Ab Initio Molecular Dynamics. The Car-Parrinello Approach

Ab initio molecular dynamics (AIMD) is a powerful technique for the study of biological processes at an atomic-electronic level [22, 23]. It can be viewed as a series of density functional theory (DFT) calculations for a different set of atomic positions $\left\{ \vec{R}_N \right\}$ at successive instants of time. These atomic positions are related by the Newton's equations of motion (e.o.m.),

$$M_N \ddot{\vec{R}}_N = -\frac{\partial E_{el}}{\partial \vec{R}_N} \tag{3.1}$$

which can be derived from the Lagrangian:

$$\mathcal{L} = E_N^{kin} - E_{el} \tag{3.2}$$

where $E_N^{kin} = \sum_N \frac{1}{2} M_N \dot{\vec{R}}_N^2$ is the kinetic energy of the nuclei, M_N and $\vec{R}_N$ are nuclear masses and positions, respectively, and the electronic energy E_{el} is their potential energy, i.e., E^{DFT} (we are assuming throughout this section that the Born-Oppenheimer approximation holds, i.e., the electrons are moving in the field of fixed nuclei).

Fig. 3.1 Schematic diagram of an ab initio molecular dynamics simulation (AIMD)

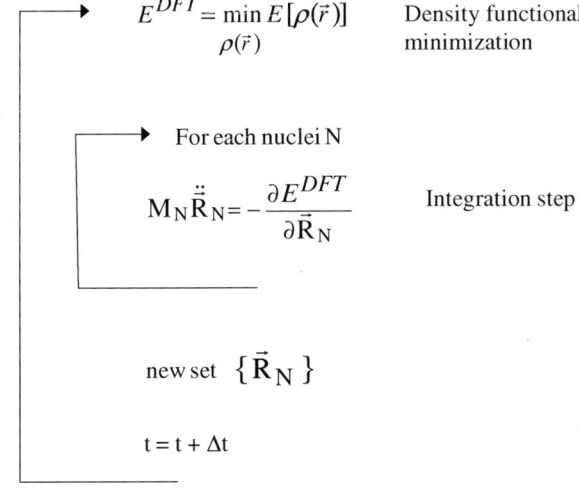

$$E^{DFT} = \min_{\rho(\vec{r})} E[\rho(\vec{r})]$$ Density functional minimization

For each nuclei N

$$M_N \ddot{\vec{R}}_N = -\frac{\partial E^{DFT}}{\partial \vec{R}_N}$$ Integration step

new set $\{\vec{R}_N\}$

$t = t + \Delta t$

The basic AIMD procedure consists in repeating two main steps: (i) For a given set of atomic co-ordinates $\{\vec{R}_N\}$, find the total energy E^{DFT}. (ii) Solve Newton's equations of motion, Eq. (3.1). This procedure is illustrated in Fig. 3.1.

The basic difference between AIMD and classical MD lies in the way the interatomic energy is obtained. In classical MD, the potential energy (E_{el}) is computed from a parameterized energy expression that depends on the structural properties of our system (atomic positions, bond distances, angles,...) as variables [24]. Instead, in AIMD the interatomic energy (i.e., E^{DFT}) is obtained from quantum mechanics and depends on the atomic positions and the electron density. Nevertheless, the integration of Newton equations of motion to update the atomic positions at each time instant is performed using similar techniques [25] as in standard MD.

A very elegant and efficient approach to perform AIMD was introduced by Car and Parrinello in 1985 [26]. Rather than minimizing the density functional and integrating Newton's equations separately, the authors introduced a generalized fully classical Lagrangian for both electrons and nuclei,

$$\mathcal{L} = E_N^{kin} + E_{el}^{kin} - E_{el}^{pot} + \sum_{ij} \Lambda_{ij} \left(\int d\vec{r} \, \psi_i^*(\vec{r}) \psi_j(\vec{r}) - \delta_{ij} \right) \quad (3.3)$$

where $E_{el}^{kin} = \sum_i \mu \int d\vec{r} |\dot{\psi}_i(\vec{r})|^2$ is a "fictitious" classical kinetic energy term associated with the electronic subsystem $\{\psi_i(\vec{r})\}$, μ is a parameter that controls the timescale of the electronic motion, Λ_{ij} are Lagrangian multipliers that impose the orthonormality constraints between the orbitals and E_{el}^{pot} is the electronic energy (i.e., the Kohn-Sham energy, E^{KS} [27]). The total energy of the CP Lagrangian is given by $E_{tot}^{CP} = E_{el}^{kin} + E_N^{kin} + E^{KS}$ and it is a constant of motion. The corresponding equations of motion are,

$$\mu \ddot{\psi}_i = -\frac{\delta E^{KS}}{\delta \psi_i^*} + \sum_j \Lambda_{ij} \psi_j(\vec{r}) \quad (3.4a)$$

$$M_N \ddot{\vec{R}}_N = -\frac{\partial E^{KS}}{\partial \vec{R}_N} \tag{3.4b}$$

The integration of the coupled Eqs. (3.4a) and (3.4b) provides the time evolution of not only the atomic positions $\left\{\vec{R}_N(t)\right\}$ but also the KS orbitals $\{\psi_i(\vec{r}, t)\}$. In practice, the orbitals are expanded in a basis set (for efficiency reasons, a plane-wave basis set is commonly used) and what is obtained from the integration is the value of the expansion coefficients at each time instant.

Therefore, in a CP simulation both electrons and nuclei are evolved simultaneously. It can be demonstrated that, provided that the electrons are initially in the ground state, they will follow adiabatically the nuclear motion, remaining very close to the instantaneous ground state [22]. From this point of view, the Car-Parrinello method is a procedure to describe computationally what occurs in reality which is that electrons follow the nuclear motion [28, 29]. In a Car-Parrinello simulation, the electronic energy only needs to be calculated at the beginning of the simulation, and the KS orbitals evolve, following the nuclear motion, as the simulation proceeds.

The electronic energy obtained at a given instantaneous structure $\left\{\vec{R}_N\right\}$ generally differs slightly from the exact DFT energy. However, if the energy exchange between the electronic and nuclear subsystems is small, the trajectory generated will be identical to the one obtained in a standard AIMD simulation [23]. This decoupling of the two subsystems can be achieved by a suitable choice of the fictitious electronic mass μ. As discussed in several reviews, the choice of μ affects the efficiency of the calculation: the higher the electronic mass, the lower the integration time step [22, 30]. As the time needed for energy equipartition between electrons and nuclei is larger than physical nuclear relaxation times, meaningful statistical averages can be obtained from the trajectories (see Refs. [23, 28, 29] for reviews of the Car-Parrinello method).

In recent times, the Car-Parrinello method has been used in conjunction to methods to enhance the sampling of the free energy landscape, such as thermodynamic integration (the *blue moon* approach), umbrella sampling and metadynamics, as well as methods that enable the description of large systems, such as the combined quantum mechanical molecular mechanical (QM/MM) approach [31].

3.3 Car-Parrinello Simulations in Biology

Ab initio (Car-Parrinello) MD simulations have been used in the last two decades to study electronic processes in proteins. Of course, the description of a full protein with complete sampling of the configuration space is unaffordable for ab initio methods. Some electronic processes, however, take place in a relatively small region of the protein (e.g. the active site in the case of enzymatic reactions) and involve a small group of atoms. These atoms, whose electronic structure is expected to change during the reactive process, are the only ones for which one needs first-principles accuracy, whereas a more approximate method (force-field)

Fig. 3.2 Setup for an enzymatic QM/MM molecular dynamics calculation. The particular example corresponds to the complex of the glycoside hydrolase 1,3–1,4-β-glucanase with a polysaccharide substrate (1,3–1,4-β-glucan tetrasaccharide). The *QM region* involves part of the polysaccharide and three protein residues (see further details in Ref. [32])

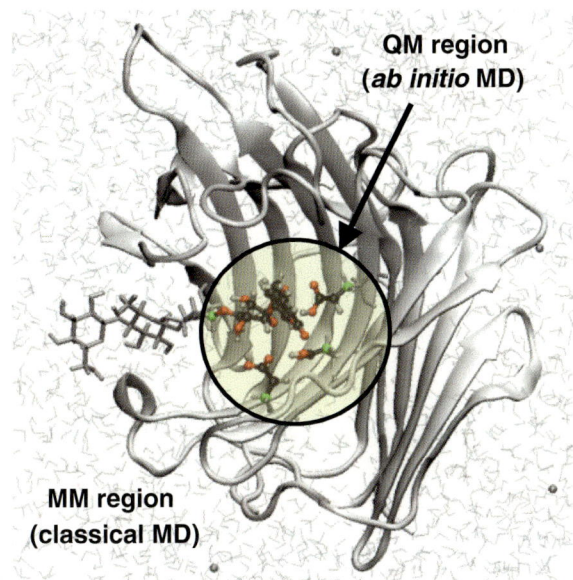

can be used for the rest of the system. This is the rationale behind the QM/MM approach [31], which can also be used in conjunction with molecular dynamics, either in the context of Car-Parrinello or standard AIMD (also named Born-Oppenheimer MD) (Fig. 3.2).

In a QM/MM calculation, the computational load is dictated by the number of atoms included in the "QM region" (and, in the case of using plane-waves as basis set functions, by its distribution in space). The use of QM/MM (or any QM/MM method in general) can efficiently solve the so-called *size issue* to a great extent whenever the importance of the quantum description is confined in space at least on the time scale of the simulations. To overcome (or minimize) the second issue (protein flexibility or dynamics, requiring proper sampling of the phase space) we should rely on a starting structure that corresponds to the portion of the protein landscape we are interested in and perform an extensive classical MD prior to the QM/MM simulation. The accuracy and predictive power of QM/MM MD simulations for enzymatic reactions is very dependent of the initial structure taken as starting point (because one is limited in the sampling of the phase space, it is necessary to make sure we start in the appropriate local minimum of the complex protein landscape, or at least close to it). A high resolution crystal structure is a good starting point. However, subtle motions of active site residues, essential for catalysis, may not, or only partially, be evidenced by analysis of the crystal structures, especially when not all reactive species are present [33], underscoring the need to complement structural studies with classical and AIMD simulations to obtain a more complete picture of the biological process [34].

The way we handle the above issues (system size and adequate sampling of the relevant phase space) [35] as well as the reliability of the QM method to describe the process of interest (DFT in CP simulations) will determine the overall accuracy and predictive capability of our computational model to study electronic processes in proteins. The first attempt to implement a QM/MM approach based on the CP method was the coupling the Car-Parrinello CPMD [36] with a classical MD code based on the CHARMM/FAMUSAMM force field [37, 38]. The QM-MM interface was modelled with the scaled position link atom method, [38] an approach that was later successfully adapted to other AIMD QM-MM codes [39]. The sensitivity of the vibrations of the CO ligand in myoglobin with respect to the active site environment (in particular, with the protonation state of the distal histidine) was investigated with this methodology. It was shown [40] that the distal histidine is protonated at $N_{\$\$}$, confirming the proposals of spectroscopic measurements [41, 42]. Later on, Laio, Vandevondele and Röthlisberger developed a much more efficient CP QM/MM code by coupling CPMD with the GROMOS code [43]. The corresponding interface was adapted to both GROMOS96 and AMBER force-fields [44]. In this CP QM/MM code, electrostatic interactions are taken into account within an efficient multilayer approach within a fully Hamiltonian electrostatic coupling (i.e., electrostatic interactions among QM and MM atoms take explicitly into account the electronic density of the quantum system):

$$H_{nonbonded} = \sum_{i \in MM} q_i \int dr \frac{\rho(r)}{r - r_i} + \sum_{\substack{i \in MM \\ j \in QM}} \upsilon_{vdw}(r_{ij})$$

where r_i is the position of the MM atom i, with charge q_i, r is the total (electronic plus ionic) charge of the quantum system, and $\upsilon_{vdw}(r_{ij})$ is the van der Waals interaction between atom i and atom j. The QM-MM interface is handled via monovalent pseudopotentials [45] or link-atoms [46].

There have been numerous applications of CP QM/MM dynamics to study electronic processes in proteins, such as the description of the electronic structure of protein active sites, charge transfer in ion channels, ligand-target interactions and enzymatic reaction pathways. The method has been particularly successful in describing fast dynamic processes such as concerted proton transfer processes in proteins and enzymes, as they normally take place in a short (i.e., picoseconds) time scale. For instance, de Vivo et al. found that proton shuttles are concomitant with the enzymatic phosphoryl transfer reaction catalyzed by soluble epoxide hydrolase [47, 48]. Derat et al. elucidated the key role of a water solvent molecule in the mechanism of formation of the principal reaction intermediate of the peroxidase reaction in horseradish peroxidase [49]. Brunk et al. [50] showed the involvement of water molecules via concerted proton transfer during the main catalytic step (cleavage of the adenine N-glycosidic bond) in the DNA repair enzyme MutY, potentially enabling the enzyme to lower the energetic barrier of a proton transfer. But the clearest examples of a concerted proton transfer is probably the migration of protons in water wires via the Grottus mechanism in ion channels [51, 52] (see also the exclusion of protons in aquaporins [53]) or the proton delocalization upon formation of wire-like water structures $[H^+ \cdot (H_2O)_3$ and $H^+ \cdot (H_2O)_4]$ in barteriorhodopsin [54, 55]. These processes, with a high entropic component, in which fast concerted atomic motions takes place at room temperature, are often very difficult capture using structural searches, underscoring the usefulness of the dynamical treatment to obtain full insight.

3.4 Car-Parrinello Studies of Enzymatic Reactions

Among biological processes with a high electronic component, enzymatic reactions have always fascinated scientists, because the understanding of enzymatic mechanisms can help in the design of molecules that can affect the protein function (i.e., inhibitors), as well as guide the design of point mutations to make desired changes in the enzyme (rational design). Theoretical analysis of enzymatic reactions is routinely used in combination with kinetic, structural or spectroscopic experiments to disentangle the mechanisms of enzymes. Computer simulations have also the advantage that can be performed in situations that experimentally would be unreachable (e.g. the native enzyme in complex with its natural substrates [56]) or occur on very short time scales and very small space scales.

In general, an enzymatic reaction starts from a local minimum (reactants state) and goes into another minimum (product state). Both states are separated by a free energy barrier of ~15–25 kcal/mol, which is much higher than the thermal energy available for the system. Because the typical time scale to cross an energy barrier grows exponentially with the barrier, the probability to overcome it and reach the products state during the timescale of a "standard" CP QM/MM simulation (tens of picoseconds) is practically zero. Therefore, the mechanism of enzymatic reactions cannot be modelled with "standard" CP QM/MM simulations. This problem can be overcame with the use of techniques aimed to enhance the sampling of the energy landscape (thus accelerating rare events), such as the Blue Moon method and metadynamics [20]. The latter has proved to be a flexible tool that can be used not only for efficiently computing the free energy, but also for finding reaction pathways in chemical transformations (see also Refs. [57, 58]). Metadynamics is particularly suited for cases in which there is little information a priori about the reaction coordinate (i.e., search for reaction pathways) [59] whereas the use of the Blue Moon approach is limited to one-dimensional energy profiles in which the reaction coordinate is already known to a great extent (extensions to multi-dimensional problems have been developed [60] but remain of difficult application) [50, 61]. As an example, Boero [61] recently studied the possible roles of the $3'$-OH group of the ribose of tRNA in LeuRS, a RNA-binding protein responsible for the translation of genetic code. Both the Blue Moon ensemble and the metadynamics approach were used, although only metadynamics captured the two pathways operative in the enzyme. Another popular approach to compute free energy profiles is the umbrella sampling (US) method. Here, a biasing potential is added to the system during the MD simulation, which enhances sampling of regions with low statistical weights. One dimensional US simulations are comparable to the Blue Moon approach. Ensing showed [62], for a simple S_N2 reaction in a small system, that metadynamics can be efficiently combined with the umbrella sampling method [63] (US simulations were performed by using the lowest free energy path obtained by metadynamics as reaction coordinate) to improve the accuracy of calculated free-energy profiles. However, the US/MTD simulations are generally recognized as being rather costly to perform and this is probably the reason that this combination is not yet routinely used for complex biological systems.

The CP QM/MM approach has also been applied to non-reactive processes, such as electron transfer (ET), ion transport and ligand conformational changes. Many essential biological functions, such as photosynthesis and respiration, depend on electron transfer processes. Recent works have been successful in computing reorganization free energies and redox potentials in an efficient way [64, 65]. The free energy of ET is not only determined by the redox active cofactors but to a large extent by the protein and surrounding solvent. Therefore, sufficient sampling of the energy for the time and length scales of the ET process becomes a critical aspect (see discussion in Ref. [65]).

The study of conformational changes of protein ligands in general do not require the use of a first-principles approach, as electrostatic and vdW interactions are well described by classical force-fields. An exception is provided by the problem of substrate conformational changes in the enzymes that catalyze the cleavage of the glycosidic bond in carbohydrates (glycoside hydrolases, GHs). A fascinating thread of the research on GHs, and one with major impact on the design of enzyme inhibitors, is the conformational analysis of reaction pathways within the diverse enzyme families [66, 67]. Over the preceding decade, research has highlighted the harnessing of non-covalent interactions to aid this distortion of the sugar substrates from their lowest energy chair conformation to a variety of different boat, skew boat, and half-chair forms, each of which is adopted for a given enzyme and substrate. Using CP/MM MD, Biarnés et al. showed that distortion of the substrate (Fig. 3.3) is required to bind to *Bacillus* 1,3-1,4-β-glucanase and that

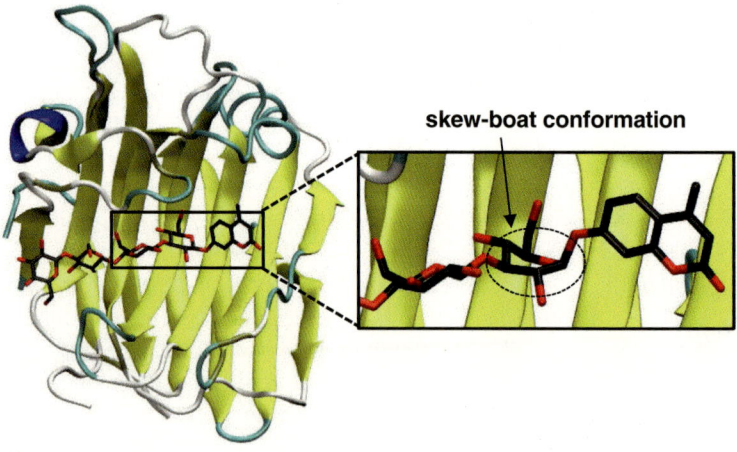

Fig. 3.3 Computed complex of the 1,3–1,4-β-glucanase enzyme with a 1,3–1,4-β-glucan tetrasaccharide. The saccharide ring located at the active site was predicted to be distorted in *a skew-boat conformation* [56]

this distortion, not captured by current force-fields, results in electronic changes in the substrate that favour cleavage of the glycosidic linkage (*substrate preactivation*) [56]. In recent years, CP/MM metadynamics simulations have thrown light upon the catalytic mechanisms of several GHs [32, 68–71]. Protein residues implicated in catalysis have been identified and the conformation of the carbohydrate during the complete reaction obtained, providing a probe for structural biology predictions.

3.4.1 Example: The Catalase Reaction

Heme catalases are one of the most efficient enzymes known: they are able to decompose up to one million molecules of hydrogen peroxide per second [72]. Structurally, they are tetrameric proteins with each subunit containing a heme group [73, 74] (Fig. 3.4). The active site contains two residues that are essential for catalysis: the proximal Tyr and the distal His (Tyr339 and His56, respectively, in *Helicobacter pylori* catalase, HPC, Fig. 3.4). The former is coordinated to the heme iron and is negatively charged, contributing to the +3 formal oxidation state of the iron atom. The latter is essential for the formation of the main reaction intermediate, compound I (Cpd I). A proximal arginine (hydrogen-bonded to tyrosine), together with a distal asparagine and serine complete the list of conserved residues in the active site [73].

The active species responsible for the decomposition of H_2O_2 is a high-valent iron intermediate, known as Compound I (Cpd I) [75], obtained by reaction of catalase with hydrogen peroxide.

$$\text{Enz}\left(\text{Por}-\text{Fe}^{III}\right) + H_2O_2 \rightarrow \text{Cpd I}\left(\text{Por}^{\bullet+}-\text{Fe}^{IV}=O\right) + H_2O \qquad (R1)$$

Cpd I is characterized to be an oxoferryl porphyrin cation radical ($\text{Por}^{\bullet+}-\text{Fe}^{IV}=O$), [75] and it is formed not only in catalases but also by other heme proteins, such as peroxidases or cytochrome P450. The subsequent Cpd I reactivity is determined by the protein frame in which the heme is buried and research aimed to grasp the origin of this functional diversity is an extremely active field [76–78].

Kinetics studies [79] have shown that once catalase Cpd I forms (reaction 1, R1) [80], it rapidly reacts with another molecule of H_2O_2 to generate a water molecule and O_2 (R2, Fig. 3.5). Isotope labeling studies demonstrated that both oxygen atoms of the O_2 molecule originate from the same H_2O_2 molecule [81, 82].

$$\text{Cpd I}\left(\text{Por}^{\bullet+}-\text{Fe}^{IV}=O\right) + H_2O_2 \rightarrow \text{Enz}\left(\text{Por}-\text{Fe}^{III}\right) + H_2O + O_2 \qquad (R2)$$

Reaction 2 (R2), extremely efficient in catalases, occurs at a much slower pace in a few other heme enzymes (e.g. chloroperoxidase (CPO), catalase-peroxidase (KatG) and myoglobin (Mb)) [83]. The origin of this disparity has long been sought and, even though the catalase reaction has been known since 1940s [84, 85],

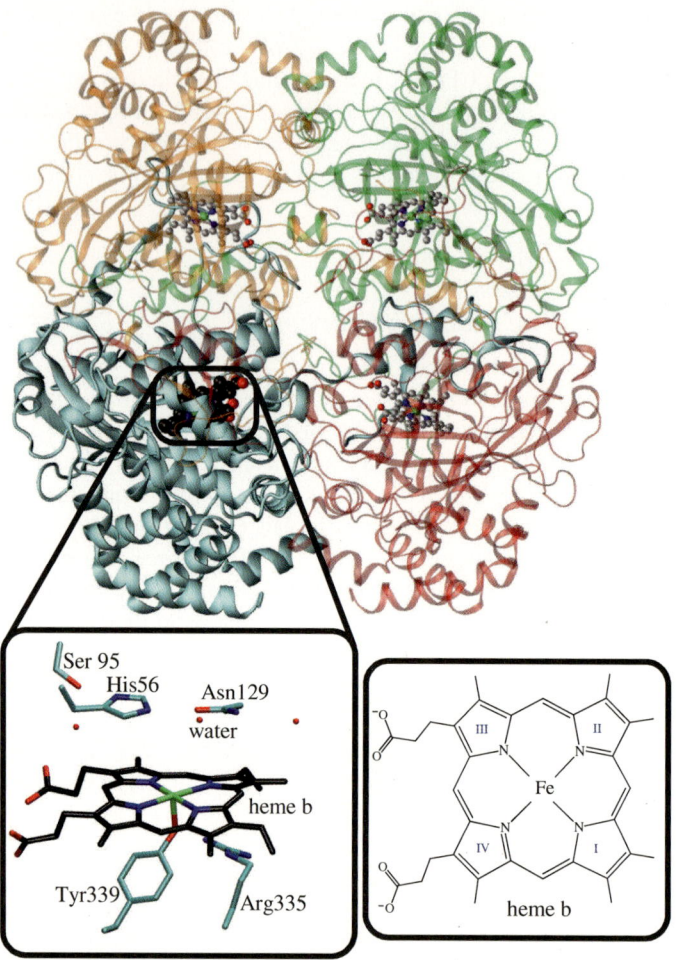

Fig. 3.4 Structure of native *Helicobacter pylori* catalase (HPC, PDB entry 2IQF). *Top* Cartoon picture of the protein, with the four subunits colored in *blue*, *red*, *yellow* and *green*, respectively. *Bottom* Heme binding pocket of one of the subunits (*blue*) and molecular structure of heme b

Fig. 3.5 Schematic representation of Cpd I reduction by H_2O_2 in catalase

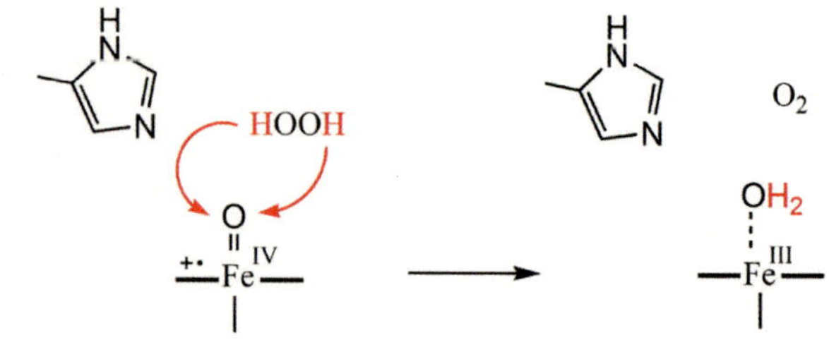

Fig. 3.6 a The His-mediated mechanism of Cpd I reduction (Fita–Rossmann) of Cpd I reduction in catalase. **b** The direct mechanism (Watanabe)

the detailed mechanism of Cpd I reduction by H_2O_2 remained a challenge until recently [59, 86].

Although there had been preliminary suggestions from kinetic data [87], Cpd I reduction was first considered at a molecular level in 1985 by Fita and Rossmann [88]. On the basis of the crystal structure of beef liver catalase, Fita and Rossmann proposed that the two hydrogens of H_2O_2 are sequentially transferred to the oxoferryl unit of Cpd I, with the distal His playing an active role in the reaction (Fig. 3.6a). Nevertheless, the precise mechanism how the two protons and two electrons of H_2O_2 are transferred to Cpd I was not discussed. Recently, the group of Watanabe, by means of a detailed kinetic study, was able to disentangle the rate constants of Cpd I formation and reduction for *Micrococcus lysodeikticus* catalase (MLC) and a series of Mb mutants [89, 90]. Two different kinetic behaviours were observed in H_2O and D_2O for reaction 2, which were interpreted as two different mechanisms. Namely, it was proposed that the reduction of Cpd I by H_2O_2 in native catalase, as well as in the F43H/H64L Mb mutant, involves the transfer of a hydride ion from H_2O_2 to Cpd I and the transfer of a proton mediated by the distal His (Fig. 3.6a, hereafter named as the His-mediated mechanism). This mechanism thus follows the Fita-Rossmann model [88], with the distal His acting as an acid-base catalyst. Besides, for certain Mb mutants lacking a distal residue that could act as acid-base catalyst, an alternative mechanism was proposed [89, 90], in which two hydrogen atoms of H_2O_2 are directly transferred to the oxoferryl group (Fig. 3.6b, from now on referred to as the direct mechanism).

Car-Parrinello QM/MM studies of catalases [91] showed that the H_2O_2 spontaneously transfers one hydrogen atom to the oxoferryl of Cpd I, forming a Cpd II-like species (Fig. 3.7).

(a)   **(b)**

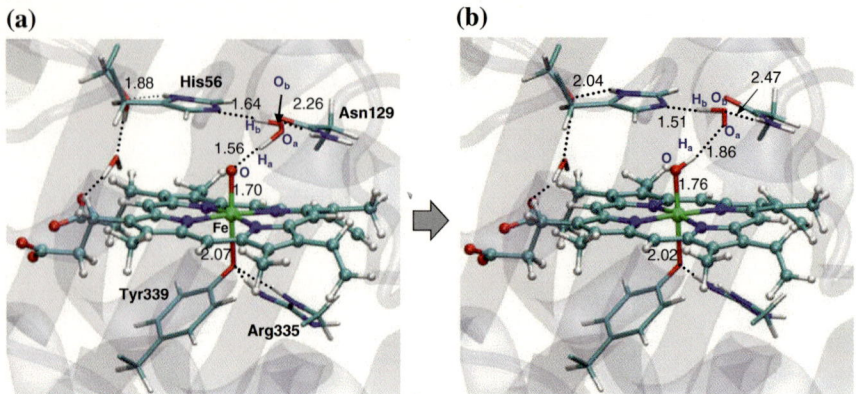

Fig. 3.7 a Optimized structure of the Cpd I···H_2O_2 complex of *Helicobacter pylori* catalase using CP QM/MM. The BP functional and a plane-wave basis set, with a kinetic cut-off of 70 Ry, was used [59]. The intramolecular distances of the H_2O_2 molecule are H_a–O_a = 1.04 Å, H_b–O_b = 1.05 Å, O_a–O_b = 1.47 Å. **b** Average structure of the Cpd II-like species obtained after the CP QM/MM simulation: H_a–O_a = 0.99 Å, O_a–O_b = 1.31 Å, O–H_b = 1.08 Å

To complete the reaction, two mechanisms may be operative: a His-mediated or a direct mechanism. These two pathways were clearly differentiated by combining CP QM/M dynamics with metadynamics [92], using two collective variables. The first collective variable (CV_1) was taken as the coordination number between the two oxygen atoms of the H_2O_2 molecule and their two hydrogens (see definition of the coordination number in the supporting information of reference [59]). Therefore, this CV gives an idea of the degree of detachment of the two hydrogens from the two hydrogen peroxide oxygens. An interesting feature of CV_1 is that it does not dictate which hydrogen is bonded to which oxygen during the simulation. Any hydrogen is allowed to coordinate to any oxygen (i.e., the two hydrogen atoms and the two oxygen atoms of H_2O_2 are treated in an equivalent way). The second collective variable (CV_2) was taken as the coordination number between the oxoferryl oxygen and the two peroxide hydrogens. Therefore this CV indicates the degree of formation of the product water molecule. The values of the two collective variables at the initial (CV_1, CV_2 ≈ 0.5, 1.0) and final states (CV_1, CV_2 ≈ 0.0, 2.0) of the process are different enough to ensure that the two states will appear in different regions of the free energy surface, a necessary condition for a suitable characterization of the reaction path in a metadynamics simulation. It is important to note that this choice of the collective variables does not force nor restrict that any of the two peroxide hydrogens bind to the distal His during the reaction.

The free energy surface (FES) reconstructed from the metadynamics simulation of HPC is shown as a contour plot in the central panel of Fig. 3.8. Two competing pathways (A and B) joining the Cpd II-like-HOO˙ complex and products valleys appear clearly differentiated. Each pathway contains several local minima of

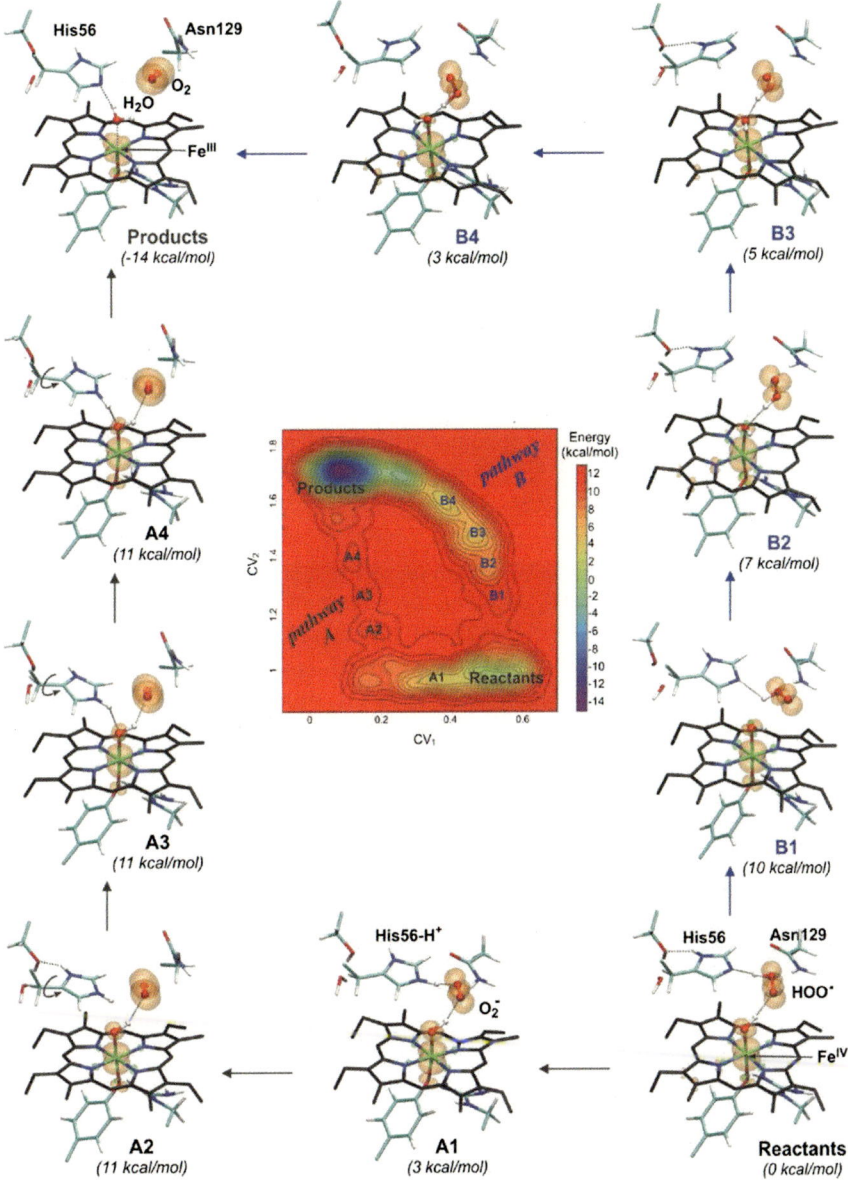

Fig. 3.8 Electronic/structural rearrangement during the conversion of Cpd II···OOH· (reactants state, *bottom-right picture*) to the resting state in HPC (products state, *top-left picture*). Contour lines of the free energy surface (*center of the picture*) are spaced by 1 kcal/mol. The energy of the different points along the pathway, relative to the reactants, is given below each structure label

different well depth, separated by transition states. The atomic and spin reorganization along each pathways are also shown in Fig. 3.8.

Pathway A Following proton transfer (PT) to the distal His, the $HisH^+\text{-}O_2^-$ complex is formed (basin A1 in Fig. 3.8). Afterward, the distal His rotates such that it breaks the H-bond with the superoxide anion (A2). At the same time, the hydroxoferryl group rotates with respect to the Fe–O bond, breaking the hydrogen bond with O_2^-, and positioning one oxygen lone pair in a suitable orientation to interact with the histidine proton (A3). A4 differs from A3 in the degree of rotation of the distal His around the C_β–C_γ bond and the absence of hydrogen bond between the hydroxoferryl hydrogen and O_2^-. In the products state, H_b has been transferred to the hydroxoferryl oxygen, the Fe–O distance has increased (from 1.83 ± 0.06 Å in A4 to 2.08 ± 0.07 Å in the products), and a water molecule has been formed. The decrease of the O–O distance (from 1.33 ± 0.02 Å in A1 to 1.25 ± 0.02 Å in the products, see Table 3.1), together with the change of the spin density distribution (Fig. 3.5 and Table 3.1), signals the change from O_2^- to O_2. Altogether, *pathway A consists of an electron transfer from O_2 to reduce Fe(IV) to Fe(III) early in the path, followed by a proton transfer from the distal His to Fe–OH.* Interestingly, the distal Asn changes conformation gradually (from the Cpd II-like configuration to the products), allowing the release of the product oxygen toward the main channel. Structures A2, A3, and A4 are at similar energies, 8 kcal/mol over A1 (Fig. 3.8). Thus, the transition state along this pathway would correspond to all the process through A2, A3, and A4, that is, rupture of the H-bond between the superoxide and the distal $HisH^+$, rotation of the latter to form an H-bond

Table 3.1 Distances and number of unpaired electrons of relevant fragments for stationary points along pathways **A** and **B** of *Helicobacter Pylori* catalase

Structure	Distance (Å)		Number of unpaired electrons[a]			
	O–H_b	O_a–O_b	Fe=O	O_a–O_b	Porph	Tyr
Cpd I: H_2O_2	(3.19)	(1.47)	2.15	0.23	0.52	0.06
Cpd II-like	(3.68) 3.45 ± 0.23	(1.34) 1.35 ± 0.03	1.87	0.99	0.10	0.02
A1	3.51 ± 0.13	1.33 ± 0.02	1.83	1.05	0.03	0.07
A2	2.41 ± 0.10	1.29 ± 0.02	1.65	1.45	0.01	0.07
A3	2.01 ± 0.06	1.28 ± 0.01	1.56	1.54	0.02	0.06
A4	1.73 ± 0.02	1.27 ± 0.02	1.52	1.65	0.01	0.10
B1	2.08 ± 0.08	1.34 ± 0.02	1.83	0.98	0.08	0.09
B2	1.79 ± 0.04	1.32 ± 0.00	1.72	1.00	0.22	0.03
B3	1.53 ± 0.05	1.33 ± 0.03	1.79	1.01	0.12	0.04
B4	1.29 ± 0.05	1.33 ± 0.03	1.66	1.07	0.22	0.01
Products	(1.02) 1.01 ± 0.04	(1.25) 1.26 ± 0.02	1.12	1.77	0.02	0.06

Distances are given as averages along the metadynamics simulation, except for values in parentheses that refer to optimized structures. Spin densities correspond to representative snap-shots along the path

[a]The spin density was integrated using Bader's Atoms-In-Molecules theory [93]

with the hydroxoferryl and transfer of H_b toward it. The highest states along this sequence of events are 9 kcal/mol over A1 and 12 kcal/mol over the initial Cpd II-like state.

Pathway B The hydroxoferryl unit first rotates around the Fe–O bond, breaking the hydrogen bond with O_a (B1). Afterward, the peroxyl radical flips orientation with Hb changing hydrogen bond partner from the distal His to the hydroxoferryl oxygen (B2). The distal His, not involved in any hydrogen bond interaction, moves upward (B1 f B2) to facilitate the rotation of the peroxyl radical. Structures B2, B3, and B4 mainly differ for the $O \cdots H_b$ distance and the degree of rotation of the hydroxoferryl unit. Finally, transfer of H_b to the hydroxoferryl oxygen leads to the product water and oxygen molecules. The change in O–O distances (Table 3.1) from B4 (1.33 ± 0.03 Å) to the products (1.25 ± 0.02 Å), together with the changes in spin density distribution (Fig. 3.8), evidences that H_b transfers as a hydrogen atom. Once the oxygen molecule forms, the distal Asn rotates to facilitate its escape toward the main channel, as it was also observed for pathway **A**. This illustrates the interplay of the His and Asn active site residues in the catalytic mechanism and highlights the importance of taking into account their dynamics on the modelling of the reaction. Along pathway **B**, the highest barrier that the system needs to overcome is in going from the Cpd II-like intermediate to the B1 basin (12 kcal/mol), corresponding to the rupture of the H-bond between the hydroxoferryl and the peroxyl radical. From B1 onward, the energetic profile is downhill, with each intermediate lower in energy than the preceding one.

In summary, two alternative pathways were identified for reduction of Cpd I in catalase. One pathway (A in Fig. 3.8) involves the distal His as an acid-base catalyst, mediating the transfer of a proton associated with an electron transfer. Another pathway (B in Fig. 3.8) requires two hydrogen atom transfers and does not involve the distal His. Independently of the pathway, the reaction proceeds by two one-electron transfers, rather than one two-electron transfer, as has long been assumed. It is also worth noting that the steps with the highest energy barrier along each pathway do not correspond to hydrogen atom transfer processes, but rather to changes of the hydrogen bond pattern. This is consistent with the small kinetic isotope effect (KIE) determined for catalase [89]. Interestingly, the distal Asn changes conformation gradually (from the Cpd II-like configuration to the products), allowing the release of the product oxygen toward the main channel. Finally, the results confirmed that the oxygen molecule released by catalase is in the triplet state [59].

Additional calculations on an in silico mutant of the distal histidine (H56G) showed that the hydrogen-bond network at the distal site plays a key role in positioning the peroxide such that the reaction can proceed with a low barrier. For the H56G HPC mutant, the first hydrogen atom transfer becomes rate-limiting, explaining the large KIE observed for the Mb mutants lacking the distal His [89, 90]. Therefore, in line with previous investigations [83, 94], Cpd I reactivity depends on the shape and the nature of the distal pocket (i.e., the interactions between H_2O_2 and the distal residues).

3.5 Final Remarks

The study electronic processes in proteins such as enzymatic reactions, charge-transfer and substrate preactivation processes in enzymes, has benefited in the last years from the development of very efficient QM/MM interfaces and enhanced sampling techniques that can be coupled to Car-Parrinello MD. Here we have overviewed the "evolution" of CPMD computations, from the description of the complex electronic structures of protein active centres to the modelling of reaction cycles in which the dynamics of all active species is described in detail using full models of the enzyme. Enhanced sampling techniques such as metadynamics allows for the treatment of chemical reactions, even including competing reaction channels with separate saddle points. This approach of exploring the multidimensional free energy surface is a powerful method to treat intrinsically concerted processes that require inclusion of the entropy contribution. The use of CP dynamics for the study of such processes gives, within the limits of DFT, a higher insight with respect to more traditional quasi-static approaches based on structure optimization along predefined reaction coordinates. As a result, CPMD simulations (and AIMD in general) are pushing the frontiers of first-principles based computer simulation of biological systems. The continuous development of ab initio MD techniques, in close interplay with classical approaches [20, 22, 95], will most likely lead to the cracking of more complex biological processes the next years. The modelling of processes involving several time and length scales (e.g. large conformational protein motions are coupled with electron and proton transfers), requiring a multi-scale type of modelling approach, is one of the areas in which significant progress will probably be made.

Acknowledgments I acknowledge the Spanish Ministry of Economy and Competitiveness (MINECO) (grant CTQ2011-25871) and the Generalitat de Catalunya (GENCAT) (grant 2009SGR-1309) for its financial assistance. I would like to thank my previous group members Xevi Biarnés, Mercedes Alfonso-Prieto, Pietro Vidossich and Albert Ardèvol with whom I carried out some of the work presented here.

References

1. Henzler-Wildman K, Kern D (2007) Dynamic personalities of proteins. Nature 450:964–972
2. McCammon JA, Karplus M (1983) The dynamic picture of protein-structure. Acc Chem Res 16:187–193
3. Frauenfelder H, Sligar SG, Wolynes PG (1991) The energy landscapes and motions of proteins. Science 254:1598–1603
4. Dill KA, Chan HS (1997) From Levinthal to pathways to funnels. Nat Struct Biol 4:10–19
5. Baker D (2000) Surprising simplicity to protein folding. Nature 405:39–42
6. Kalbitzer HR, Spoerner M, Ganser P, Hozsa C, Kremer W (2009) Fundamental link between folding states and functional states of proteins. J Am Chem Soc 131:16714–16719
7. Granata D, Camilloni C, Vendruscolo M, Laio A (2013) Characterization of the free-energy landscapes of proteins by NMR-guided metadynamics. Proc Natl Acad Sci USA 110:6817–6822

8. Smock RG, Gierasch LM (2009) Sending signals dynamically. Science 324:198–203
9. Lee GM, Craik CS (2009) Trapping moving targets with small molecules. Science 324:213–215
10. Noskov SY, Berneche S, Roux B (2004) Control of ion selectivity in potassium channels by electrostatic and dynamic properties of carbonyl ligands. Nature 431:830–834
11. Henzler-Wildman KA, Thai V, Lei M, Ott M, Wolf-Watz M, Fenn T, Pozharski E, Wilson MA, Petsko GA, Karplus M, Hubner CG, Kern D (2007) Intrinsic motions along an enzymatic reaction trajectory. Nature 450:838–844
12. Ansari A, Berendzen J, Bowne SF, Frauenfelder H, Iben IE, Sauke TB, Shyamsunder E, Young RD (1985) Protein states and proteinquakes. Proc Natl Acad Sci USA 82:5000–5004
13. Karplus M, McCammon JA (2002) Molecular dynamics simulations of biomolecules. Nat Struct Biol 9:646–652
14. Olsson MH, Parson WW, Warshel A (2006) Dynamical contributions to enzyme catalysis: critical tests of a popular hypothesis. Chem Rev 106:1737–1756
15. Moliner V (2011) Eppur si muove (Yet it moves). Proc Natl Acad Sci USA 108:15013–15014
16. Garcia-Meseguer R, Marti S, Ruiz-Pernia JJ, Moliner V, Tuñon I (2013) Studying the role of protein dynamics in an S_N2 enzyme reaction using free-energy surfaces and solvent coordinates. Nat Chem 5:566–571
17. Carloni P, Rothlisberger U, Parrinello M (2002) The role and perspective of a initio molecular dynamics in the study of biological systems. Acc Chem Res 35:455–464
18. Rovira C (2013) The description of electronic processes inside proteins from Car-Parrinello molecular dynamics: chemical transformations. WIREs Comput Mol Sci 3:393–407
19. Dal Peraro M, Ruggerone P, Raugei S, Gervasio FL, Carloni P (2007) Investigating biological systems using first principles Car-Parrinello molecular dynamics simulations. Curr Opin Struct Biol 17:149–156
20. Barducci A, Bonomi M, Parrinello M (2011) Metadynamics. WIREs Comput Mol Sci 1:826–843
21. Colombo MC, Guidoni L, Laio A, Magistrato A, Maurer P, Piana S, Rohrig U, Spiegel K, Sulpizi M, VandeVondele J, Zumstein M, Rothlisberger U (2002) Hybrid QM/MM Car-Parrinello simulations of catalytic and enzymatic reactions. Chimia 56:13–19
22. Hutter J (2012) Car-Parrinello molecular dynamics. WIREs Comput Mol Sci 2:604–612
23. Marx D, Hutter J (2000) In: Grotendorst J (ed) Modern methods and algorithms of quantum chemistry. John von Neumann Institute for Computing, Jülich, pp 301–449
24. Stote RH, Dejaegere A, Karplus M (2002) In: Náray-Szabó G, Warshel A (eds) Computational approaches to biochemical reactivity. Springer, New York, pp 153–198
25. Verlet L (1967) Computer experiments on classical fluids I thermodynamical properties of Lennard-Jones molecules. Phys Rev 159:98–103
26. Car R, Parrinello M (1985) Unified approach for molecular dynamics and density-functional theory. Phys Rev Lett 55:2471–2474
27. Kohn W, Sham LJ (1965) Self consistent equations including exchange and correlation effects. Phys Rev 140:1133–1138
28. Remler DK, Madden PA (1990) Molecular-dynamics without effective potentials via the Car–Parrinello approach. Mol Phys 70:921–966
29. Galli G, Pasquarello A (1993) In: Allen MP, Tildesley DJ (eds) Computer simulation in chemical physics. Kluwer, Dordrecht, pp 261–313
30. Rovira C (2005) Study of ligand-protein interactions by means of density functional theory and first-principles molecular dynamics. Methods Mol Biol 305:517–554
31. Warshel A, Levitt M (1976) Theoretical studies of enzymic reactions: dielectric, electrostatic and steric stabilization of the carbonium ion in the reaction of lysozyme. J Mol Biol 103:227–249
32. Biarnés X, Ardèvol A, Iglesias-Fernández J, Planas A, Rovira C (2011) Catalytic itinerary in 1,3-1,4-β-glucanase unraveled by QM/MM metadynamics. Charge is not yet fully developed at the oxocarbenium ion-like transition state. J Am Chem Soc 133:20301–20309
33. Rojas-Cervellera V, Ardevol A, Boero M, Planas A, Rovira C (2013) Formation of a covalent glycosyl-enzyme species in a retaining glycosyltransferase. Chemistry 19:14018–14023

34. Vidossich P, Alfonso-Prieto M, Carpena X, Fita I, Loewen PC, Rovira C (2010) The dynamic role of distal side residues in heme hydroperoxidase catalysis. Interplay between X-ray crystallography and ab initio MD simulations. Arch Biochem Biophys 500:37–44
35. Parrinello M (2008) In: Zewail AH (ed) Physical biology: from atoms to medicine. Imperial College Press, London, pp 247–266
36. CPMD program, Copyright MPI für Festkörperforschung, Stuttgart 1997–2001. URL: http://www.cpmd.org
37. Eichinger M, Grubmuller H, Heller H, Tavan P (1997) An algorithm for rapid evaluation of electrostatic interactions in molecular dynamics simulations. J Comput Chem 18:1729–1749
38. Eichinger M, Tavan P, Hutter J, Parrinello M (1999) A hybrid method for solutes in complex solvents: density functional theory combined with empirical force fields. J Chem Phys 110:10452–10467
39. Crespo A, Scherlis DA, Marti MA, Ordejon P, Roitberg AE, Estrin DA (2003) A DFT-based QM-MM approach designed for the treatment of large molecular systems: application to chorismate mutase. J Phys Chem B 107:13728–13736
40. Rovira C, Schulze B, Eichinger M, Evanseck JD, Parrinello M (2001) Influence of the heme pocket conformation on the structure and vibrations of the Fe-CO bond in myoglobin: a QM/MM density functional study. Biophys J 81:435–445
41. Li T, Quillin ML, Phillips GN Jr, Olson JS (1994) Structural determinants of the stretching frequency of CO bound to myoglobin. Biochemistry 33:1433–1446
42. Braunstein DP, Chu K, Egeberg KD, Frauenfelder H, Mourant JR, Nienhaus GU, Ormos P, Sligar SG, Springer BA, Young RD (1993) Ligand binding to heme proteins: III. FTIR studies of his-E7 and val-E11 mutants of carbonmonoxymyoglobin. Biophys J 65:2447–2454
43. Christen M, Hünenberger PH, Bakowies D, Baron R, Bürgi R, Geerke DP, Heinz TN, Kastenholz MA, Kräutler V, Oostenbrink C, Peter C, Trzesniak D, van Gunsteren WF (2005) The GROMOS software for biomolecular simulation: GROMOS05. J Comput Chem 26:1719–1751
44. Laio A, VandeVondele J, Rothlisberger U (2002) A Hamiltonian electrostatic coupling scheme for hybrid Car-Parrinello molecular dynamics simulations. J Chem Phys 116:6941–6947
45. Zhang YK, Lee TS, Yang WT (1999) A pseudobond approach to combining quantum mechanical and molecular mechanical methods. J Chem Phys 110:46–54
46. Reuter N, Dejaegere A, Maigret B, Karplus M (2000) Frontier Bonds in QM/MM Methods: a comparison of different approaches. J Phys Chem A 104:1720–1735
47. De Vivo M, Ensing B, Klein ML (2005) Computational study of phosphatase activity in soluble epoxide hydrolase: high efficiency through a water bridge mediated proton shuttle. J Am Chem Soc 127:11226–11227
48. De Vivo M, Ensing B, Dal M, Peraro G, Gomez A, Christianson DW, Klein ML (2007) Proton shuttles and phosphatase activity in soluble epoxide hydrolase. J Am Chem Soc 129:387–394
49. Derat E, Shaik S, Rovira C, Vidossich P, Alfonso-Prieto M (2007) The effect of a water molecule on the mechanism of formation of compound 0 in horseradish peroxidase. J Am Chem Soc 129:6346–6347
50. Brunk E, Arey JS, Rothlisberger U (2012) Role of environment for catalysis of the DNA repair enzyme MutY. J Am Chem Soc 134:8608–8616
51. Mei HS, Tuckerman ME, Sagnella DE, Klein ML (1998) Quantum nuclear ab initio molecular dynamics study of water wires. J Phys Chem B 102:10446–10458
52. Sagnella DE, Laasonen K, Klein ML (1996) Ab initio molecular dynamics study of proton transfer in a polyglycine analog of the ion channel gramicidin A. Biophys J 71:1172–1178
53. Jensen MO, Rothlisberger U, Rovira C (2005) Hydroxide and proton migration in aquaporins. Biophys J 89:1744–1759
54. Rousseau R, Kleinschmidt V, Schmitt UW, Marx D (2004) Assigning protonation patterns in water networks in bacteriorhodopsin based on computed IR spectra. Angew Chem Int Ed Engl 43:4804–4807

55. Mathias G, Marx D (2007) Structures and spectral signatures of protonated water networks in bacteriorhodopsin. Proc Natl Acad Sci USA 104:6980–6985
56. Biarnes X, Nieto J, Planas A, Rovira C (2006) Substrate distortion in the Michaelis complex of Bacillus 1,3-1,4-beta-glucanase. Insight from first principles molecular dynamics simulations. J Biol Chem 281:1432–1441
57. Leone V, Marinelli F, Carloni P, Parrinello M (2010) Targeting biomolecular flexibility with metadynamics. Curr Opin Struct Biol 20:148–154
58. Boero M, Ikeda T, Ito E, Terakura K (2006) K. Hsc70 ATPase: an insight into water dissociation and joint catalytic role of K^+ and Mg^{2+} metal cations in the hydrolysis reaction. J Am Chem Soc 128:16798–16807
59. Alfonso-Prieto M, Biarnes X, Vidossich P, Rovira C (2009) The molecular mechanism of the catalase reaction. J Am Chem Soc 131:11751–11761
60. Coluzza I, Sprik M, Ciccotti G (2003) Constrained reaction coordinate dynamics for systems with constraints. Mol Phys 101:1927–2894
61. Boero M (2011) LeuRS synthetase: a first-principles investigation of the water-mediated editing reaction. J Phys Chem B 115:12276–12286
62. Ensing B, Klein ML (2005) Perspective on the reactions between F⁻ and CH_3CH_2F: the free energy landscape of the E_2 and SN_2 reaction channels. Proc Natl Acad Sci USA 102:6755–6759
63. Torrie GM, Valleau JP (1977) Nonphysical sampling distributions in Monte Carlo freeenergy estimation: Umbrella sampling. J Comput Phys 23:187–199
64. Cascella M, Magistrato A, Tavernelli I, Carloni P, Rothlisberger U (2006) Role of protein frame and solvent for the redox properties of azurin from Pseudomonas aeruginosa. Proc Natl Acad Sci USA 103:19641–19646
65. Blumberger J (2008) Free energies for biological electron transfer from QM/MM calculation: method, application and critical assessment. Phys Chem Chem Phys 10:5651–5667
66. Vocadlo DJ, Davies GJ (2008) Mechanistic insights into glycosidase chemistry. Curr Opin Chem Biol 12:539–555
67. Davies GJ, Planas A, Rovira C (2012) Conformational analyses of the reaction coordinate of glycosidases. Acc Chem Res 45:308–316
68. Petersen L, Ardevol A, Rovira C, Reilly PJ (2009) Mechanism of cellulose hydrolysis by inverting GH8 endoglucanases: a QM/MM metadynamics study. J Phys Chem B 113:7331–7339
69. Petersen L, Ardevol A, Rovira C, Reilly PJ (2010) Molecular mechanism of the glycosylation step catalyzed by Golgi alpha-mannosidase II: a QM/MM metadynamics investigation. J Am Chem Soc 132:8291–8300
70. Barker IJ, Petersen L, Reilly PJ (2010) Mechanism of xylobiose hydrolysis by GH43 beta-xylosidase. J Phys Chem B 114:15389–15393
71. Greig IR, Zahariev F, Withers SG (2008) Elucidating the nature of the Streptomyces plicatus beta-hexosaminidase-bound intermediate using ab initio molecular dynamics simulations. J Am Chem Soc 130:17620–17628
72. Nichols P (2012) Classical catalase: ancient and modern. Arch Biochem Biophys 525:95–101
73. Chelikani P, Fita I, Loewen PC (2004) Diversity of structures and properties among catalases. Cell Mol Life Sci 61:192–208
74. Díaz A, Loewen PC, Fita I, Carpena X (2012) Thirty years of heme catalases structural biology. Arch Biochem Biophys 525:102–110
75. Groves JT, Haushalter RC, Nakamura M, Nemo TE, Evans BJ (1981) High-valent iron-porphyrin complexes related to peroxidase and cytochrome P450. J Am Chem Soc 103:2884–2886
76. Dawson JH (1988) Probing structure-function relations in heme-containing oxygenases and peroxidases. Science 240:433–439
77. Messerschmidt A, Huber R, Wieghardt K, Poulos T (2001) Handbook of metalloproteins. Wiley, Chichester
78. de Visser SP, Shaik S, Sharma PK, Kumar D, Thiel W (2003) Active species of horseradish peroxidase (HRP) and cytochrome P450: two electronic chameleons. J Am Chem Soc 125:15779–15788

79. Schonbaum GR, Chance B (1976) Catalase. Academic Press, New York
80. Jones P, Dunford HB (2005) The mechanism of compound I formation revisited. J Inorg Biochem 99:2292–2298
81. Jarnagin RC, Wang JH (1958) Investigation of the catalytic mechanism of catalase and other ferric compounds with doubly o18-labeled hydrogen peroxide. J Am Chem Soc 80:786–787
82. Vlasits J, Jakopitsch C, Schwanninger M, Holubar P, Obinger C (2007) Hydrogen peroxide oxidation by catalase-peroxidase follows a non-scrambling mechanism. FEBS Lett 581:320–324
83. Matsui T, Ozaki S, Liong E, Phillips GN Jr, Watanabe Y (1999) Effects of the location of distal histidine in the reaction of myoglobin with hydrogen peroxide. J Biol Chem 274:2838–2844
84. Nichols IFP, Loewen PC (2001) In: Sykes GMAG (ed) Advances in inorganic chemistry, Academic Press, Waltham, p 51–106
85. Nichols P (2012) The reaction mechanisms of heme catalases: an atomistic view by ab initio molecular dynamics. Arch Biochem Biophys 525:95–101
86. Alfonso-Prieto M, Vidossich P, Rovira C (2012) A. The catalase-hydrogen peroxide system. A theoretical appraisal of the mechanism of catalase action. Arch Biochem Biophys 525:121–130
87. Jones P, Suggett A (1968) The catalase-hydrogen peroxide system. A theoretical appraisal of the mechanism of catalase action. Biochem J 110:621–629
88. Fita I, Rossmann MG (1985) The active center of catalase. J Mol Biol 185:21–37
89. Kato S, Ueno T, Fukuzumi S, Watanabe Y (2004) Catalase reaction by myoglobin mutants and native catalase: mechanistic investigation by kinetic isotope effect. J Biol Chem 279:52376–52381, doi:10.1074/jbc.M403532200
90. Watanabe Y, Nakajima H, Ueno T (2007) The structures and electronic configuration of compound I intermediates of Helicobacter pylori and Penicillium vitale catalases determined by X-ray crystallography and QM/MM density functional theory calculations. Acc Chem Res 40:554–562
91. Alfonso-Prieto M, Borovik A, Carpena X, Murshudov G, Melik-Adamyan W, Fita I, Rovira C, Loewen PC (2007) Efficient exploration of reactive potential energy surfaces using Car-Parrinello molecular dynamics. J Am Chem Soc 129:4193–4205
92. Iannuzzi M, Laio A, Parrinello M (2003) Efficient exploration of reactive potential energy surfaces using Car-Parrinello molecular dynamics. Phys Rev Lett 90:238302
93. Bader RFW (1994) Atoms in molecules: a quantum theory. Clarendon Press, Oxford
94. Poulos TL (1996) Real-world predictions from ab initio molecular dynamics simulations. J Biol Inorg Chem 1:356–359
95. Kirchner B, di Dio PJ, Hutter J (2012) Real-world predictions from ab initio molecular dynamics simulations. Top Curr Chem 307:109–153

Chapter 4
Strictly Localised Molecular Orbitals in QM/MM Methods

György G. Ferenczy and Gábor Náray-Szabó

4.1 Introduction

Combined quantum mechanical-molecular mechanical methods can be applied to the treatment of large molecular systems, where a local change (e.g., bond breaking or fission, electronic excitation) should be considered, which is under the influence of the effect of far-lying atoms. While the quantum region should be treated on the basis of high-level methods, consideration of the environment is possible using some method of lower sophistication. A typical example is the study of enzyme reactions, where the size of the whole system does not allow a high-level quantum mechanical description, while the chemical reaction cannot be treated by conventional force fields. In the combined QM/MM approach the total system is divided into two interacting parts, the central one (QM) described by quantum mechanics and its environment described by molecular mechanics (MM). The central part may contain the ligand and the amino acid residues directly involved in the reaction, while other regions of the protein, as well as the solvent, may belong to the MM region. A critical question is the separation of the subsystem, since it is bound to the environment by covalent bonds. Borderline

G.G. Ferenczy (✉)
Research Centre for Natural Sciences of the Hungarian Academy of Sciences,
Magyar tudósok körútja 2, Budapest 1117, Hungary
e-mail: ferenczy.gyorgy@ttk.mta.hu

G.G. Ferenczy
Department of Biophysics and Radiation Biology, Semmelweis University,
Tűzoltó u. 37-47, Budapest 1094, Hungary

G. Náray-Szabó
Laboratory of Structural Chemistry and Biology, Institute of Chemistry,
Eötvös Loránd University, Pázmány Péter St. 1A, 1117, Budapest, Hungary

© Springer International Publishing Switzerland 2014
G. Náray-Szabó (ed.), *Protein Modelling*, DOI 10.1007/978-3-319-09976-7_4

atoms and bonds need a special treatment in the calculations. Beside QM/MM methods QM/QM methods are also available, where both the central part and the environment are described by quantum mechanical methods, but of different sophistication. For the reactive region a higher, while for the environment a lower level method can be applied. In the following we focus on QM/MM methods, though in some cases the QM/QM approach will be also mentioned. For extensive, recent reviews see Refs. [1, 2].

Some approaches, like the fragment molecular orbital [3, 4], the "divide and conquer" [5, 6] and the frozen density functional methods [7, 8] inherently consist the mode of division of the subsystem from its environment. In other cases the link atom method is often applied. This means that the bond connecting the two systems will be cut and the dangling bonds of the subsystem will be saturated by hydrogen or dummy atoms. Though this is a straightforward solution, the presence of dummy atoms in the close vicinity of others may lead to artefacts in the calculations. However, careful application of this model may still lead to valuable results [9]. Another possibility for the fragmentation of the subsystems is the boundary atom approach, which does not introduce any new atom but handles the boundary atom by using a specially parameterised potential that interacts with the QM region and considers the atom as a regular MM atom within the MM region [10, 11].

4.2 Strictly Localised Molecular Orbitals

Another way for a partition is using strictly localised molecular orbitals (SLMO) for the bonds connecting the boundary atom to the quantum region, while other bonds, belonging to the MM region will be modelled by simple strings (see Fig. 4.1).

We will call atom A at the boundary a frontier atom. SLMOs were first applied by Warshel and Levitt in their semi-empirical method for the treatment of the enzyme reaction in lysozyme [12]. Their early applications can be related to QM/QM methods [13, 14]. These apply SLMOs for the description of the environment, while the central quantum mechanical subsystem is described by delocalised wave functions within the frame of the given approach. Extension of these

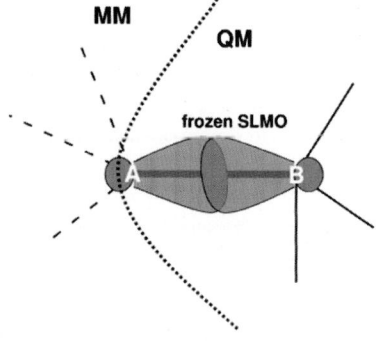

Fig. 4.1 Division of QM and MM subsystems by using strictly localised molecular orbitals (taken with permission from Ref. [42])

for QM/MM systems involves connection of the subsystem and its environment by rigid SLMOs with pre-calculated, fixed parameters. The local self-consistent field (LSCF) method, originally developed for semiempirical parameterisations, was extended to ab initio wave functions, too [15, 16].

Fixed orbitals are used also in similar QM/MM methods. Friesner and co-workers introduced parameterised interaction terms and obtained accurate energies as well as structural parameters for test systems of medium size [17, 18]. Use of the so-called general hybrid orbitals (GHO) places also appropriate hybrid orbitals on the frontier atoms. Hybrids pointing towards the QM subsystem, are optimised, while those oriented towards the MM environment, are kept rigid [19–21]. This model has been extended to post-Hartree-Fock methods [18, 22, 23].

A basic problem if using SLMOs in quantum chemical calculations is that their overlap is nonzero, therefore the conventional Hartree-Fock-Roothaan equations [24] cannot be used. A potential solution is to use the Adams-Gilbert equation [25–28]. An approximate form and the corresponding energy expression has been also proposed to obtain the orbitals [29, 30]. This latter is based on the series expansion of the inverse overlap matrix and neglect of higher order terms. This expression may replace that based on orthogonal orbitals. The method was used to optimize valence orbitals in the field of core electrons, as well as to calculate the energy, but the method can be applied quite generally [31–33].

4.3 Semiempirical Methods

An early combined QM/QM method has been proposed by Náray-Szabó and Surján [13]. The wave function of the total system is described in terms of SLMOs, which are typically centred at two neighbouring atoms, but they can be also lone pairs and delocalised π-orbitals. The corresponding Hartree-Fock equation within the complete neglect of differential overlap (CNDO) model [34] leads to low-dimension coupled equations, which can be solved by the conventional matrix diagonalisation technique. The Fockian breaks down to coupled blocks, which makes further iteration necessary. The wave function obtained by this procedure is modified in the central region, where e.g., a bond breaking reaction takes place. A conventional optimisation of the wave function has to be performed in the field of the environment represented by a set of SLMOs. An essential feature of the method is that two-centre SLMOs are obtained as linear combinations of hybrid orbitals oriented towards each other. Since hybrid orbitals are orthogonal within the frame of the CNDO approximation, simple secular equations are obtained, which can be solved easily.

This method was applied in an early work [14] within the frame of the neglect of diatomic differential overlap approximation [34]. In contrast to the CNDO approximation this does not neglect integrals computed from products of orbitals located on the same atom, thus new terms appear in the Fockian and hybrid orbitals centred on the same atom become non-orthogonal. Orthogonality can be secured by application of Löwdin's method [35]. Thus, we arrive to the Fragment Self-Consistent Field (FSCF) method, which is appropriate for the treatment of extended systems by describing the subsystem by delocalised, while the environment by strictly localised MOs. The method can be combined with various parameterisations [14, 36, 37] and it has been applied, among others, to the treatment of an enzyme mechanism [38].

Separation of subsystems using the above QM/QM scheme can be adapted to the QM/MM method, as well [39]. An SLMO appears at the boundary of the two subsystems, as depicted in Fig. 4.1. Other hybrid orbitals belonging to atom B are part of the basis set used for the expansion of the quantum mechanical wave function. There are no further orbitals located on atom A. Orientation of the hybrids, i.e., the relative weights of p-orbitals constructing the SLMOs is determined by atomic positions. Coefficients of SLMO hybrids and polarities are taken from calculations on model systems and kept fixed when calculating the quantum mechanical wave function. An essential feature of this separation scheme is that fixed SLMOs at the boundary are orthogonal to all other orbitals. This is secured partly by explicit orthogonalisation, partly by the neglect of diatomic differential overlap, as done within the NDDO scheme. This means that the approach cannot be extended directly to ab initio wave functions.

4.4 Orthogonal Molecular Orbitals

The conventional Hartree-Fock Roothaan equations [24] optimise all molecular orbitals on a common basis and provides orthogonal orbitals. However, the QM/MM wave function discussed above contains localised orbitals at the boundary. Orthogonality can be secured at the semiempirical level, but it is not straightforward in ab initio calculations. A potential solution of the problem is that we orthogonalise the basis functions to the fixed molecular orbitals [15, 40]. Orthogonal orbitals can be also obtained by solving a modified equation with a non-hermitic Fockian [17] or by neglecting the overlap [40].

Another way to obtain orthogonal molecular orbitals is to solve Huzinaga's equation (4.1) [41] and use an appropriate basis set [42]. The purpose of this is to determine orbitals interacting with fixed ones and making the energy stationary.

$$\left(\widehat{F} - \widehat{\rho}^f\widehat{F} - \widehat{F}\widehat{\rho}^f\right)\phi_i^a = \varepsilon_i^a\phi_i^a \tag{4.1}$$

here $\widehat{F}$ is the Fockian, $\widehat{\rho}^f$ projects it to the set of fixed orbitals, ϕ_i^a is the ith active orbital with eigenvalue ε_i^a. A typical application of the Huzinaga equation is the determination of valence electronic orbitals within the field of fixed core orbitals. Besides, it may serve as a starting point for further approximations, like representation of core electrons by a model potential [43]. In the following we give an overview on some important properties of Eq. (4.1), related to its application.

Equation (4.1) can be obtained if we suppose the orthogonality of active and fixed orbitals. This is automatically fulfilled if the same basis functions are assigned to both groups. If group-specific basis functions are used, they have to be orthogonal. Since fixed localised orbitals use only basis functions at the link atoms, these also have to be included in the basis set on which the active orbitals are expanded.

Fixed orbitals do not have to be eigenfunctions of the Fockian if orthogonality to active ones has to be secured. Idempotency of the operator $\hat{\rho}^f$ constructed from fixed orbitals allows that it can be commuted with the $\hat{F} - \hat{\rho}^f\hat{F} - \hat{F}\hat{\rho}^f$ Huzinaga operator of Eq. (4.1). Thus, these operators can be diagonalised on a common basis set, which means that appropriate linear combinations of fixed orbitals will be eigenfunctions of the Huzinaga operator. Since this is Hermitic, fixed and active orbitals are orthogonal if they are non-degenerated. Degeneracy can be avoided if non-zero eigenvalues of $\hat{\rho}^f\hat{F}\hat{\rho}^f$ are negative. Then, eigenvalues of Eq. (4.1) belonging to fixed orbitals are positive and those, belonging to occupied active orbitals, are negative. Non-zero eigenvalues of $\hat{\rho}^f\hat{F}\hat{\rho}^f$ are negative if the fixed orbitals are appropriate (i.e., occupied) eigenfunctions of F. Most probably they are negative also, if the fixed orbitals are good approximations to the eigenfunctions. On the other hand, if fixed orbitals are eigenfunctions of $\hat{\rho}^f\hat{F}\hat{\rho}^f$ with positive eigenvalues, then these appear among the solutions of Eq. (4.1) with negative eigenvalues. In this case the usual self-consistent iteration procedure, selecting orbitals with the lowest eigenvalues to construct the operator of the next step, does not provide active orbitals, which are orthogonal to the fixed ones.

Introducing the basis functions of the Huzinaga equation (4.1) will have the following form

$$\left[F - SR^f F - FR^f S\right]C^a = SC^a E^a \tag{4.2}$$

where F is the Fockian, S is the overlap matrix of the basis functions, C^a consists the coefficients of the active orbitals, E^a is the diagonal matrix of the corresponding eigenvalues and R^f projects to the space of the fixed orbitals. In case of our QM/MM wave function more fixed groups are allowed, thus for a single group of active functions Eq. (4.2) becomes

$$\left[F - SR^{\bar{a}}F - FR^{\bar{a}}S\right]C^a = SC^a E^a \tag{4.3}$$

where $R^{\bar{a}}$ projects to all orbitals not in the active group. If the QM and MM subsystems are linked by more than one chemical bonds, more fixed orbitals appear expanded on different basis sets. These are typically non-orthogonal, which, however, does not appear in Eq. (4.3) explicitly. However, if building up the matrix $R^{\bar{a}}$ then overlap has to be considered: $R^{\bar{a}} = C^{\bar{a}}\left(\sigma^{\bar{a}}\right)^{-1}\left(C^{\bar{a}}\right)^+$, where $\sigma^{\bar{a}}$ is the overlap matrix of all molecular orbitals, not belonging to the active group. Effective optimisation of the molecular structure requires calculation of forces acting at nuclei. The corresponding expression is similar to that referring to canonical orbitals that can be adapted easily to our case.

Fixed orbitals can be derived from calculations on model molecules. Thus, e.g., a C–C bond orbital within an apolar environment can be obtained from a calculation on the ethane molecule as follows. We perform a standard Hartree-Fock calculation with the actual basis set, then localise the orbitals obtained with e.g., the Pipek-Mezey procedure [44]. Cutting the tails of the localised orbitals and normalising the coefficients we obtain the SLMO belonging to the C–C bond, which has to be kept rigid when solving Eq. (4.3). Other electrons of atom B on Fig. 4.1 are part of the QM system. In case of the core orbitals of atom A more options are available. They can be optimised, can be kept rigid, or can be replaced by a point charge located at nuclei.

Model calculations have been performed, where the environment was modelled by point charges. This is appropriate for the correct reproduction of electrostatic (and potentially polarisation) interactions. Point charges were calculated according to Ferenczy et al. [45, 46]. These point charges reproduce the electrostatic potential well and can be easily calculated. Accordingly, our approximation can be considered as a model of QM/QM systems and harmonise with a QM/MM model using an AMBER force field, since the latter uses potential derived point charges [47]. In all calculations atom A of Fig. 4.1 was chosen to be a sp^3 hybridised carbon atom bound by a single, localised bond to the QM system (atom B of Fig. 4.1) and its further bonds point towards the MM system. Thus, explicit electrons on atom A are the core electron pair and the valence electron contributing to the fixed orbital. It was found advantageous to increase the core charge of atom A and to introduce compensating negative bond charges placed at the midpoints of the bonds connecting atom A to MM atoms. The use of increased core charges on atom A facilitates the convergence of the self-consistent determination of the QM wave-function. The magnitude of the charges are calculated by requiring that the sum of bond charges and the core charge of atom A is equal to the effective charge obtained for atom A according to Refs. [45, 46] (see Fig. 4.2). The introduction of extra charges has been proposed by others, too. Philipp and Friesner put point charges on the centroid of the bonds and compensate them by charges at close MM atoms [48]. Ferré and co-workers increase charges at boundary C-atoms and reduce those at

Fig. 4.2 Bond charges at the QM/MM boundary (taken with permission from Ref. [42])

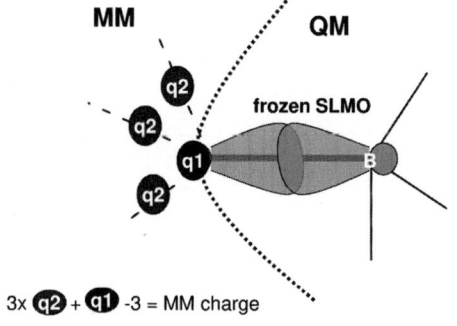

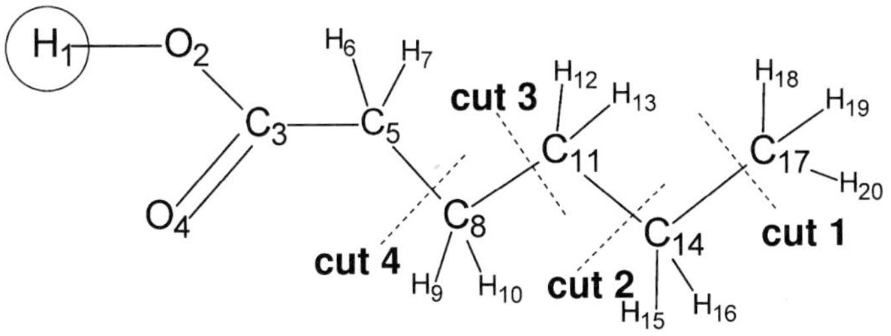

Fig. 4.3 Fragmentation schemes for the $C_5H_{11}COOH$ molecule. The QM subsystem is situated left to the *dashed line* (taken with permission from Ref. [42])

connecting H atoms [49], while Lin and Truhlar put charges on the bonds connecting QM and MM subsystems [50].

In the following we present results of some pilot calculations as obtained by the standard Hartree-Fock-Roothaan method with a 6-31G* basis set [51] and with the QM/MM scheme modelling the MM region by simple point charges. We used a modified version of the GAMESS-US software [52]. In order to obtain effective charges we made first a distributed multipole analysis [53], then used the MULFIT software [54]. For the $C_5H_{11}COOH$ molecule we obtained the following results with various choices of the boundary between the QM and MM regions as show on Fig. 4.3.

4.5 Deprotonation Energies

As it was discussed above, though the number of explicit electrons on atom A is 3, to solve Eq. (4.3) self-consistently, larger positive charge on atom A and compensating negative bond charges are necessary (cf. Fig. 4.2). The core charge was chosen to allow a good reproduction of the reference deprotonation energy by the QM/MM calculation in case of the largest QM system (see Fig. 4.3). This determines bond charges in such a way, that the sum of core and bond charges, as well as those of the three explicit electrons (-3) equals to the effective atomic charge. Choosing the core charge to be $+5.6$, deprotonation energy calculated for boundary 1 on Fig. 4.3 reproduces the QM/MM value within a 0.2 kcal/mol difference from the reference HFR value. We used this core charge in all subsequent calculations, while bond charges were calculated from effective charges obtained for the given system. Calculated deprotonation energies are displayed in Table 4.1. All structural parameters, referring to all QM atoms including atom B, were optimised both for the protonated and unprotonated species. All atoms belonging to subsystem MM and atom A were kept fixed (see Fig. 4.1).

Deprotonation energy is smallest for the largest QM system (boundary 1 in Fig. 4.3). Absolute values of the deviation increase with shrinking the size of the QM system. Accuracy of the structural parameters is high indicating that our

Table 4.1 Errors for QM/MM deprotonation energies and structural parameters for the $C_5H_{11}COOH$ molecule in case of various selections of the boundary (see Fig. 4.3)

System	d^a (Å)	n^b	ΔE (kcal/mol)	Δd^c (Å)	$\Delta \alpha^d$ (°)	$\Delta \tau^e$ (°)
Boundary 1	6.8	5	−0.17	0.000	0.5	0.1
Boundary 2	5.6	4	−1.43	0.001	0.4	0.3
Boundary 3	4.3	3	−1.58	0.003	0.4	0.7
Boundary 4	3.0	2	−3.78	0.003	0.6	0.2

[a]Distance between the O_2 atom and the centroid of the fixed SLMO
[b]Number of bonds between the O_2 atom and the centroid of the fixed SLMO
[c]RMS error of the interatomic distance
[d]RMS error of the bond angle
[e]RMS error of the torsion angle

approximations do not have practically any influence on the reproduction of these. It has to be stressed that errors of the computed deprotonation energies are sensitive indicators of the performance of the method, since these consist of the energy differences of differently charged systems, therefore they depend considerably on electrostatic interactions diminishing slowly with increasing distance.

4.6 Conformational Energies

Conformational energies of the $C_5H_{11}COOH$ molecule were calculated as a function of the torsional angle O_2–C_3–C_5–C_8 (see Fig. 4.4). Boundaries were selected identically as for deprotonation. Results are depicted in Fig. 4.4. As it is seen for boundaries 1 and 2 the reference curve is perfectly reproduced, the energy

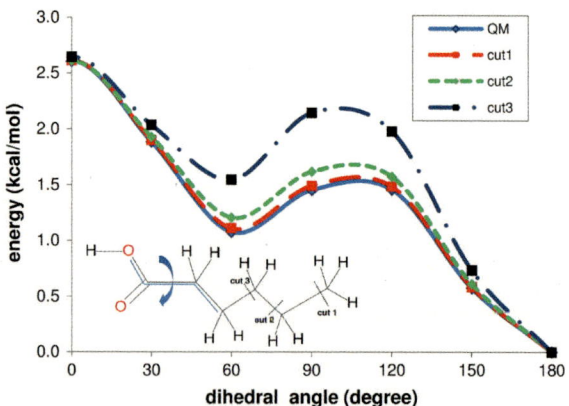

Fig. 4.4 Energy of the $C_5H_{11}COOH$ molecule as a function of the rotation of the –COOH group. Dihedral angle of rotation is indicated by lines parallel to bonds. System separations are shown by *dashed lines* (taken with permission from Ref. [42])

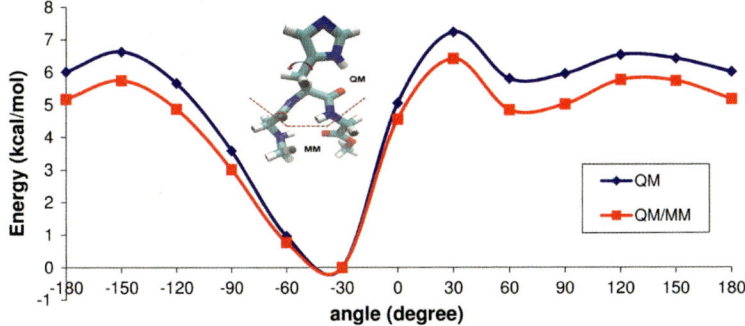

Fig. 4.5 Energy of the Gly-His-Gly tripeptide as a function of the rotation of the imidazole group. The QM/MM boundary is indicated by a *dashed line*. Curves are shifted to have the same minima near $-30°$ (taken with permission from Ref. [42])

difference is less than 0.1 kcal/mol for all values of the torsional angle. Even for boundary 3, i.e., the smallest QM subsystem, the shape of the curve is close to the reference one. The largest error is near 0.5 kcal/mol. The above results indicate that a precise description of the conformation energy change is obtained if we select a small QM subsystem. At least two bonds should separate the fixed bond at the QM/MM boundary from the one around which rotation takes place.

In another example the torsional energy of the Gly-His-Gly tripeptide is calculated as a function of the rotation of the imidazole group. The QM subsystem includes the central imidazole group, as well as the amide groups (cf. Fig. 4.5). It is reasonable to select the core charge of the atom at the boundary to be identical to that used for the $C_5H_{11}COOH$ molecule.

4.7 Proton Transfer

We calculated the energy curve of the proton transfer from the –COOH group of an Asp side chain to the neighbouring His side chain (for the molecular arrangement see Fig. 4.6). The boundary of the subsystems was chosen to have a fixed bond between atoms C_α and C_β. These are the farthest possible boundaries from the site of proton transfer, which ensure to have a single SLMO at the boundary. The energy curve is shown on Fig. 4.6. Both the reference and the QM/MM curves have their minima at 1.0 Å this is the location where both curves are shifted. The overlap between the two curves is very good, the difference between the energy at the second minimum is only 1 kcal/mol, less than 4 %. The size of the QM subsystem is small, which allows considerable reduction of the computational work. Such a selection of the boundary may be appropriate for calculations on proteins, too.

Fig. 4.6 Energy of the
Asp-His dyad as a function
of the distance between the
O atom of the Asp side chain
and the transferring proton.
The subsystem boundary is
indicated by a *dashed line*
(taken with permission from
Ref. [42])

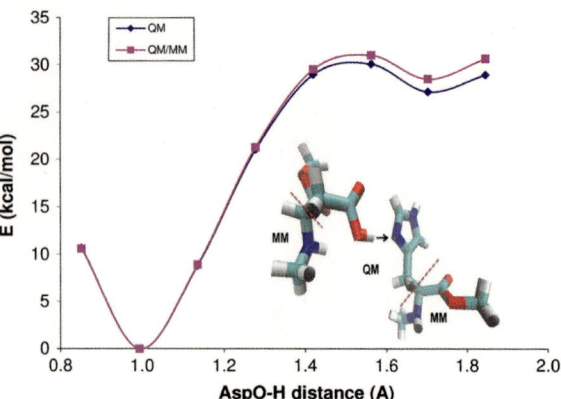

4.8 Non-orthogonal Molecular Orbitals

In the previous section we discussed application of the Huzinaga equation for
the calculation of wave functions of QM/MM systems with fixed SLMO-s. The
Huzinaga equation allows determination of molecular orbitals, which are orthogonal to the SLMO-s. As an alternative, we describe the calculation of orbitals, too,
which are non-orthogonal to the SLMO-s [55, 56]. First, we introduce the local
basis equation for the determination of orbitals in cases, when group functions are
expanded on different basis sets, thus orthogonality cannot be generally fulfilled.
Then, we will show, how the local basis equation can be used for the calculation of
QM/MM wave functions with fixed SLMOs. Molecular orbitals obtained this way
are, in general, not orthogonal to SLMOs.

We construct the wave function as a Slater determinant of non-orthogonal orbitals, combined in groups, orbitals in different groups can be expanded on different
basis sets. The electronic energy can be written as follows

$$E = \frac{1}{2} \sum_{\alpha\beta} P_{\alpha\beta} (h_{\beta\alpha} + F_{\beta\alpha}) \qquad (4.4)$$

where $P_{\alpha\beta}$, $h_{\beta\alpha}$ and $F_{\beta\alpha}$ are elements of the density matrix, the Hamiltonian and
the Fockian, respectively. Since orthogonality of orbitals belonging to different
groups is not required, $P = 2R$, where $R = C\sigma^{-1}C^+$ [35]. Matrix C consists the
orbital coefficients, thus according to the specific assignment of the basis set, it is
block diagonal. σ is the overlap matrix, which is, in general, non-diagonal, since
group orbitals overlap.

Putting the first derivatives of the energy as function of the coefficients of the
active orbitals we obtain the following equation [55]

$$(I_a. - S_a.R)FC\left(\sigma^{-1}\right)_A = 0_{aA} \qquad (4.5)$$

where I, 0 and S are the unit, zero and basis overlap matrices, respectively. Index a refers to basis functions of active orbitals, while index A refers to occupied active orbitals. The missing lower index or a dot as a lower index refers to the full dimension of the matrix. Thus, $S_{a.}$ denotes a block of the overlap matrix of the basis set, where the number of rows equals to the number of basis orbitals of the active orbitals, while the number of columns is equal to the total number of basis functions. Equation (4.5) is equivalent to that derived by Stoll et al. [57]. Derivatives appearing on the left side of the equation allow, in principle, localisation of a stationary point of the energy, however, this is not practicable because of the high computational demand. It can be shown [55], that solutions of the following eigenvalue equation satisfy Eq. (4.5)

$$\left(I_{a.} - S_{a.}R^{\bar{a}}\right)F\left(I_{.a} - R^{\bar{a}}S_{.a}\right)C_{aA} = \left(I_{a.} - S_{a.}R^{\bar{a}}\right)S_{.a}C_{aA}E_{AA} \qquad (4.6)$$

Matrix $R^{\bar{a}}$ projects on the subspace of the fixed orbitals: $R^{\bar{a}} = C^{\bar{a}}\left(\sigma^{\bar{a}}\right)^{-1}\left(C^{\bar{a}}\right)^{+}$. Equation (4.6) allows to determine the active orbitals with the usual iteration process, while keeping others fixed. The method is appropriate for the determination of valence orbitals with fixed core electron orbitals. A further possibility for application is the determination of a priori localised orbitals [55, 56]. In this case assignment of the proper basis function reduces the extension of the orbitals and allows the use of non-orthogonal localised orbitals. The calculated energy is in, general, higher than that obtained by the full basis set, but this still acceptable if reasonable basis functions are assigned.

In order to get solutions of Eq. (4.5) Stoll et al. proposed an eigenvalue equation, which is different from Eq. (4.6) [57]

$$\left(I_{a.} - S_{a.}R + S_{aa}\left(\tilde{R}^{a}_{a.}\right)^{+}\right)F\left(I_{.a} - RS_{.a} + \tilde{R}^{a}_{.a}S_{aa}\right)C_{aA} = S_{aa}C_{aA}E_{AA} \quad (4.7)$$

where $\tilde{R}^{a}_{.a} = C\sigma_{A}^{-1}C_{Aa}^{+}$. Matrices C and E of the latter equation are not equivalent to those of Eq. (4.6). On the other hand, solutions of both Eqs. (4.6) and (4.7) span the same space and both make the energy stationary. The convergence of Eq. (4.7) is unfortunately very slow [56], thus it is not appropriate for the determination of the orbitals. The reason for the slow convergence may be that the equation consists of the sixth power of the coefficients to be optimised. In contrast, Eq. (4.6) includes coefficients to be optimised only in the Fockian, their highest power is thus two, similarly as in standard HFR equations. If solving Eq. (4.6) we did not have convergence problems.

An equation, similar to Eq. (4.6) was proposed for the elimination of the basis set superposition error for such groups, which are expanded on basis sets not consisting common functions [58]. Later, the equation was generalised for groups, where a common basis set is also allowed [59, 60]. This is similar to Eq. (4.7), i.e., it contains the sixth power of the coefficients to be optimised, and convergence was found to be slow. The equation was applied to determine SLMOs [60, 61], which were used later in the Local Self-Consistent Field (LSCF) method [15, 16, 39]. In QM/MM calculations gradient optimisation was used for the determination of orbitals [60], which is much less effective than the iterative self-consistent solution. Optimisation using the first, in some cases the second, derivatives of the energy with respect to coefficients was proposed by other authors, too, in order to avoid slow convergence of the self-consistent iteration [57, 62, 63]. To our knowledge, Eq. (4.6) is the only one by now, which is appropriate for the determination of non-orthogonal orbitals by an iterative, self-consistent procedure.

Solutions of the local basis equation (4.6) are, in general, non-orthogonal molecular orbitals. However, the full wave function may be the same as that from the Huzinaga equation (4.3). If assignment of the basis functions to groups allows orthogonality, then solutions of Eq. (4.3) are simultaneously solutions of Eq. (4.6), too. In spite of this, if solving Eq. (4.6) iteratively, we do not necessarily obtain orthogonal orbitals.

As discussed above, if fixed orbitals are poor approximations of the occupied ones, solutions of Eq. (4.3) with negative eigenvalues will appear in the space of the fixed orbitals and this hinders determination of the active orbitals by the usual iterative self-consistent manner. Equation (4.6) is less sensitive to the shape of the fixed orbitals, since for solutions within the space of the fixed orbitals the eigenvalues are zero and not negative. In the following we will present some examples for cases, where iterative self-consistent solution of Eq. (4.3) is unsuccessful, however, Eq. (4.6) can be solved on the usual way.

4.9 QM/MM Calculations with the Local Basis Equation [56]

QM/MM-type calculations were done by solving Eq. (4.6), similarly, as with Eq. (4.3) providing orthogonal molecular orbitals as shown above. The full system was divided into QM and MM parts, connected by fixed SLMOs. Solution of Eq. (4.6) was built in the GAMESS-US program [52], the 6-31G* basis set was used [50]. For the determination of MM charges we performed a distributed multipole analysis [53, 64, 65] of the wave function by the GDMA program [53] then the MULFIT program [54] was used to calculate effective charges [45, 46].

As discussed above, the wave function from the local basis equation (4.6) is equivalent to the solution of the Huzinaga equation (4.3), if assignment of basis functions allows determination of orthogonal orbitals. All calculations, performed by solving Eq. (4.3), were repeated by solving Eq. (4.6), too. While calculations precisely reproduced results obtained from Eq. (4.3), no convergence problem arose. Number of iterations was about the same as in the standard HFR method. Deprotonation energy is well reproduced if the fixed orbital is separated from the deprotonation site by two or three bonds. Structural parameters, obtained while optimising the structure, are very close to reference values. Conformational energy curves and proton transfer energy values are also fine.

In the above calculations we retained the core charge and compensating bond charges. These latter are necessary for the iterative self-consistent solution of Eq. (4.3) (see above). Iterative solution of the local basis equation (4.6) converges well even if small atomic core charges are used. Therefore deprotonation energy of the $C_5H_{11}COOH$ molecule was repeated by smaller atomic core charges and without bond charges, separating the subsystem by boundary 1 of Fig. 4.3. The charge on the boundary atom C_{17} was chosen to be $+3$ (the number of explicit electrons) adding the effective charge of this atom. Effective charges were assigned to MM atoms, too, which were calculated for the neutral molecule, but used for the deprotonated molecule, as well. Calculated deprotonation energy differs by 6.1 kcal/mol from reference. (Note that with appropriately selected bond charges this reduces to 0.2 kcal/mol.) Analysing the charge distribution we found that near the boundary of the subsystems Mulliken charges considerably differ from those obtained in reference HFR calculations. The charge on C_{14} increases (-0.075 vs. -0.304), and decreases on atoms linked to it (C_{11} -0.326 vs. -0.307, H-atoms bound to C_{14} $+0.074$ vs. $+0.158$, for atom numbering see Fig. 4.3). On the other hand, Mulliken charges obtained by bond charges are close to reference HFR ones. This indicates that the fixed orbital itself cannot hinder distortion of the wave function at the boundary. This is why the error in deprotonation energy is quite large.

We further investigated the possibility for neglecting the bond charges with alteration of the basis functions of the fixed molecular orbitals. While in case of Eq. (4.3) orthogonality requires inclusion of the basis functions of the fixed orbitals in the basis set of the active group, this is not necessary in case of Eq. (4.6) allowing a more flexible assignment of the basis. If dropping the basis functions located on the boundary atom (*A* on Fig. 4.1) deprotonation energy practically does not change. Calculations were repeated by the application of more extended fixed orbitals. The basis set of the fixed localised orbital linking the subsystems contains the terminal C_{11}–C_{14}–C_{17} unit and hydrogen atoms bound to C_{14} (see Fig. 4.7). The basis set of the active orbitals does not contain basis functions of the C_{14}–C_{17} moiety in order to reduce the perturbation of the QM system due to the QM/MM boundary. Though the error of the deprotonation energy became smaller, still it remained near 5 kcal/mol.

Analysis of Mulliken charges has shown that the charge of the methylene group consisting of the C_{14} atom is closer to the reference (C_{14} -0.389 vs. -0.304,

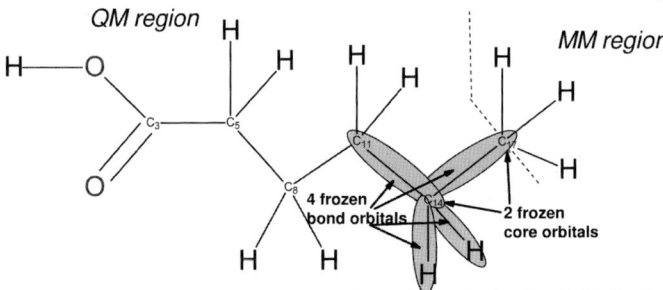

Fig. 4.7 Subsystems defined for the $C_5H_{11}COOH$ molecule and localised orbitals linking them (taken with permission from Ref. [56])

Fig. 4.8 Energy of the C$_5$H$_{11}$COOH molecule as a function of the torsional angle of the –COOH group for two subsystem boundaries (taken with permission from Ref. [56])

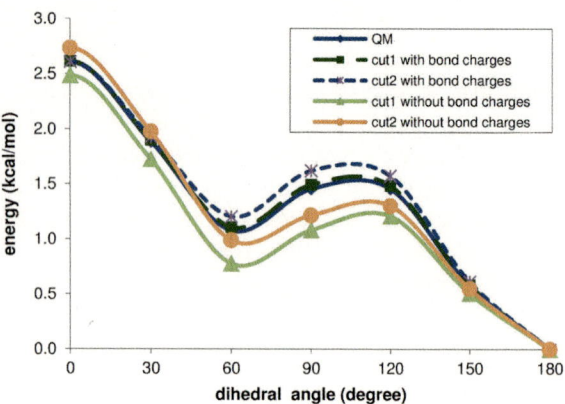

linked H-atoms 0.163 vs. 0.158), in agreement with the fact that coefficients of orbitals containing C$_{14}$ basis functions are fixed. On the other hand, the charge on the neighbouring atom, C$_{11}$ (−0.078), lying closer to the deprotonation site, shows a more pronounced difference from the reference (−0.307 vs. −0.326).

It is anticipated that the deformation of the wave function on the boundary of the subsystems has different effect on calculated properties. Though, as seen above, calculated deprotonation energies are considerably biased, energy differences between systems of the same charge are less influenced. In order to investigate this supposition we repeated calculation of conformational energy changes and proton transfer energies without bond charges. Charge of the atom at the boundary was chosen to be +3 (number of explicit electrons) added to the effective charge of the atom. Energy of the C$_5$H$_{11}$COOH molecule as function of the torsional angle of the –COOH group (C$_5$–C$_3$ bond, see Fig. 4.3) was calculated. Two subsystem separations were considered, one with boundary 1, another with boundary 2 (cf. Fig. 4.3). Former calculations with bond charges provided good results. Calculated energy curves without consideration of bond charges closely coincide with the reference curve (cf. Fig. 4.8). Deviation of energies as obtained without bond charges is larger than in case if using bond charges, however, still less than 0.5 kcal/mol.

Torsional energy curves for rotation of the imidazole group of the Gly-His-Gly tripeptide with and without bond charges are shown in Fig. 4.9. The shape of the reference curve is well reproduced by the calculations. Similarly as above, curves without bond charges differ more from the reference than those with them. The largest deviation is within 1 kcal/mol with and 2 kcal/mol without bond charges.

Energy curve for the proton transfer between Asp and His molecules was also calculated with and without bond charges (see Fig. 4.10). The shape of both curves follows well the reference. The other minimum, belonging to the protonated His is higher than the reference in both QM/MM calculations. The deviation is 1.3 kcal/mol with and 3.5 kcal/mol without bond charges.

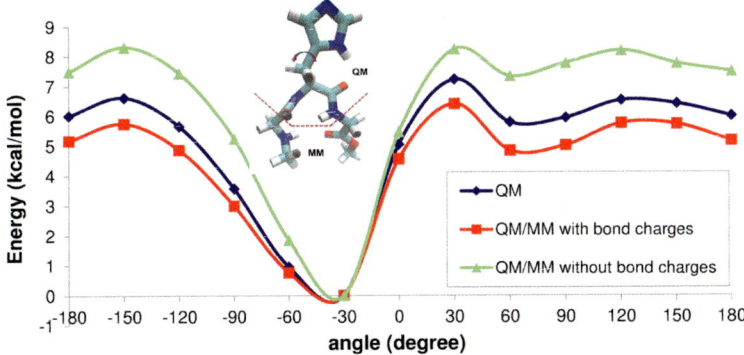

Fig. 4.9 Energy of the Gly-His-Gly molecule as a function of the rotation of the histidine group calculated with and without bond charges (taken with permission from Ref. [56])

Fig. 4.10 Energy of the Asp-His system as a function of the Asp(O) atom and the attached hydrogen atom, with and without bond charges (taken with permission from Ref. [56])

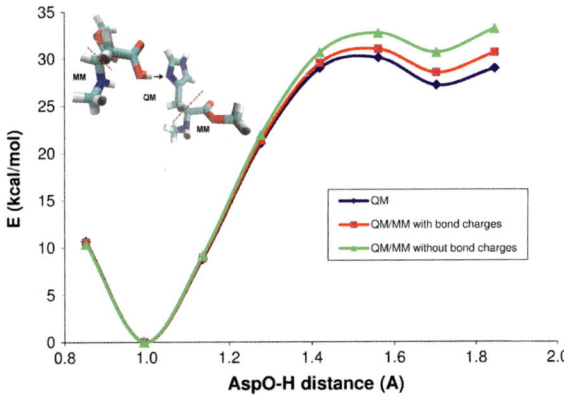

The above results show that application of increased core and bond charges allows better reproduction of reference results. It has to be noticed that atomic core charges were selected by a trial-and-error method to allow a good reproduction of the deprotonation of the $C_5H_{11}COOH$ molecule within a given QM/MM separation. Bond charges are determined by the core charge and the effective charge of the atom. Charges are well transferable among various subsystem separations and different system, since in all calculations the same core charge was used in spite of the fact that the chemical environment of the boundary atom was different. The advantage of calculations without bond charges is that they use less parameters, but this pays with the lower accuracy. This can be probably increased by increasing the size of the QM subsystem in a way that the site of the chemical or physical change gets further from the perturbed electron density at the boundary. However, using a QM subsystem of larger size involves larger computation times.

4.10 Localised Non-orthogonal Orbitals in QM/QM Systems [56]

As it was mentioned above, Eq. (4.6) is appropriate for the calculation of a priori localised orbitals. In contrast to the more generally applied a posteriori localisation, when the canonical orbitals of the HFR wave function are subjected to a transformation, which results in localised orbitals, Eq. (4.6) makes it is possible to calculate localised orbitals directly. The Adams-Gilbert equation, allowing a priori localisation [25–28], exploits the invariance of the HF wave function to a non-singular linear transformation of the occupied orbitals, similarly to a posteriori localisation [66]. On the other hand, Eq. (4.6) is appropriate for the calculation of group functions on a local basis set, where localisation is secured by the assignment of basis functions to molecular orbitals. This equation can be used for the determination of the wave function of QM/QM systems, where the central, delocalised QM subsystem is surrounded by a QM subsystem of localised orbitals. This latter may be composed of more than one groups. Molecular orbitals of both systems can be optimised by solving Eq. (4.6) for the central QM subsystem, then for all groups of the localised subsystems. These equations are coupled through $R^{\bar{a}}$ matrix (see Eq. 4.6), projecting to the space of other orbitals not containing those to be optimised. Consequently, the equations can be solved iteratively. As an alternative, we may assign appropriately selected orbitals to the subsystem containing localised orbitals, which are kept fixed and only the orbitals of the central QM system will be optimised. Similar solutions were proposed for density functional methods [66–68].

A QM/QM system, composed of systems with delocalised and localised orbitals mentioned above, can be expanded by an MM subsystem, too. Such a three-layer QM/QM/MM system is shown for the $C_5H_{11}COOH$ molecule in Fig. 4.11. Calculated deprotonation energy reproduced the HFR results within 1 kcal/mol.

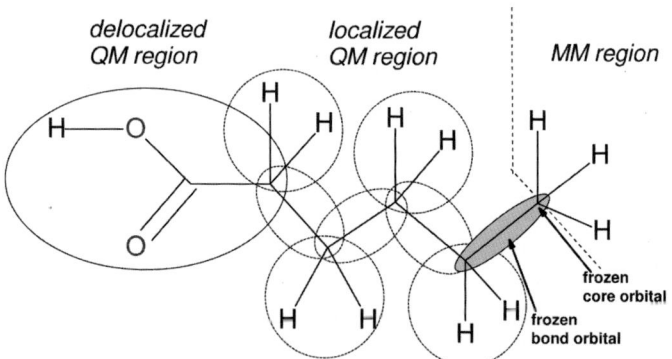

Fig. 4.11 Subsystems for the $C_5H_{11}COOH$ molecule in the QM/QM/MM model. The central delocalised QM subsystem contains the C–COOH moiety with 12 doubly occupied molecular orbitals. The localised QM subsystem contains 17 localised orbitals including 4 core orbitals. Fixed orbitals are a C–C bond and the core orbital at the QM/MM boundary. H atoms of the terminal CH_3 group are modelled by point charges (taken with permission from Ref. [56])

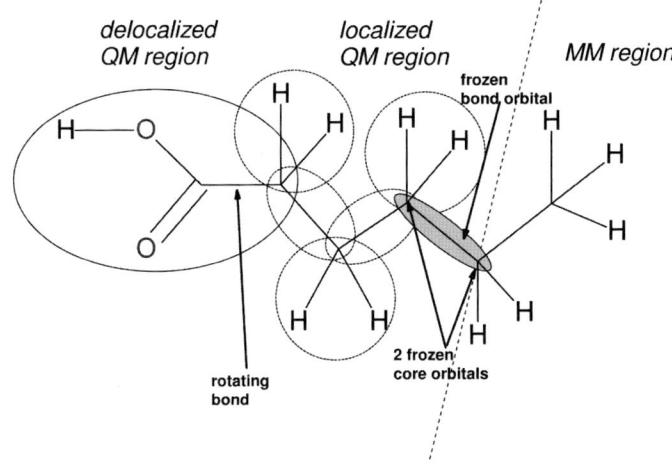

Fig. 4.12 Subsystems defined for the $C_5H_{11}COOH$ molecule in the QM/QM/MM model. The central delocalised QM subsystem contains the C–COOH moiety with 12 doubly occupied molecular orbitals. The localised QM subsystem contains 13 localised orbitals, among them 3 core orbitals. Fixed orbitals are a C–C bond and 2 core orbitals at the boundary. H atoms of the terminal CH_3 group and the connected CH_2 group are modelled by point charges (taken with permission from Ref. [56])

Fig. 4.13 Energy of the $C_5H_{11}COOH$ molecule as the function of the torsion of the –COOH group. The two-layer curve belongs to the respective separation scheme on Fig. 4.4, while the three-layer curve corresponds to the subsystem separation of Fig. 4.12 (taken with permission from Ref. [56])

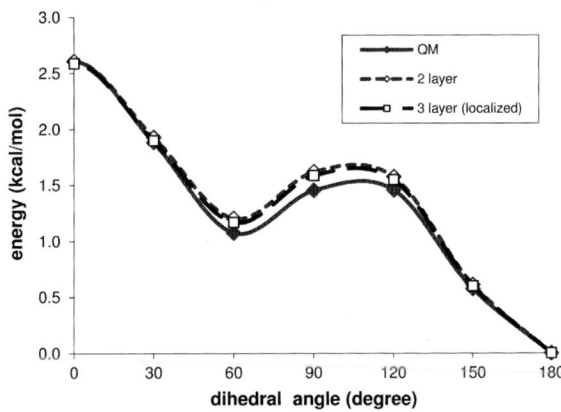

Another three-layer system is shown on Fig. 4.12. Selection of this partition scheme is motivated by the fact that a similar subsystem definition with a delocalised QM system provided very good results for the rotation of the –COOH group (see Fig. 4.8).

Figure 4.13 displays results obtained for the two-layer QM/MM and the three-layer QM/QM/MM partitions. The energy curve for the three-layer model overlaps almost perfectly with that of the two-layer curve. This indicates that the introduction of localised orbitals does not lead to the decrease of accuracy. This is especially interesting if we note that the extension of localised orbitals is small, and they are close to the rotation axis.

References

1. Lin H, Truhlar DG (2007) Theor Chem Acc 117:185–199
2. Senn HM, Thiel W (2009) Angew Chem Int Ed 48:1198–1229
3. Fedorov DG, Kitaura K (2007) J Phys Chem A 111:6904–6914
4. Yang W (1991) Phys Rev Lett 66:1438–1441
5. Yang W, Lee T-S (1995) J Chem Phys 103:5674–5678
6. He X, Merz KM (2010) J Chem Theory Comput 6:405–411
7. Wesołowski TA, Warshel A (1993) J Phys Chem 97:8050–8053
8. Wesołowski TA (2008) Phys Rev A77:012504
9. Reuter N, Dejaegere A, Maigret B, Karplus M (2000) J Phys Chem A 104:1720–1735
10. Antes I, Thiel W (1999) J Phys Chem A 103:9290–9295
11. Zhang Y, Lee T-S, Yang W (1999) J Chem Phys 110:46–54
12. Warshel A, Levitt M (1976) J Mol Biol 103:227–249
13. Náray-Szabó G, Surján P (1983) Chem Phys Lett 96:499–501
14. Ferenczy GG, Rivail J-L, Surján PR, Náray-Szabó G (1992) J Comput Chem 13:830–837
15. Assfeld X, Rivail JL (1996) Chem Phys Lett 263:100–106
16. Ferré N, Assfeld X, Rivail J-L (2002) J Comput Chem 23:610–624
17. Philipp DM, Friesner RA (1999) J Comput Chem 20:1468–1494
18. Murphy RB, Philipp DM, Friesner RA (2000) J Comput Chem 21:1442–1457
19. Gao J, Amara P, Alhambra C, Field MJ (1998) J Phys Chem 102:4714–4721
20. Pu J, Gao J, Truhlar DG (2004) J Phys Chem A 108:632–650
21. Jung J, Choi CH, Sugita Y, Ten-no S (2007) J Chem Phys 127:204102
22. Jung J, Sugita Y, Ten-no S (2010) J Chem Phys 132:084106
23. Kitagawa Y, Akinaga Y, Kawashime Y, Jung J, Ten-no S (2012) Chem Phys 401:95–102
24. Roothaan CCJ (1951) Rev Mod Phys 23:69–89
25. Adams WH (1961) J Chem Phys 34:89–102
26. Adams WH (1962) J Chem Phys 37:2009–2018
27. Adams WH (1965) J Chem Phys 42:4030–4038
28. Gilbert TL (1964) In: Löwdin P-O, Pullman B (eds) Molecular orbitals in chemistry, physics and biology. Academic Press, New York, p 409
29. Ferenczy GG (1995) Int J Quant Chem 53:485
30. Ferenczy GG (1996) Int J Quant Chem: Quant Chem. Symp 57: 361
31. Tchougréeff AL (1999) Phys Chem Chem Phys 1:1051–1060
32. Ángyán JG (2000) Theor Chem Acc 103:238–241
33. Tokmachev AM, Dronskowski R (2006) J Comput Chem 27:296–308
34. Pople JA, Beveridge DL (1970) Approximate molecular orbital theory. McGraw-Hill, New York
35. Löwdin PO (1950) J Chem Phys 18:365–375
36. Náray-Szabó G, Ferenczy GG (1992) J Mol Struct (Theochem) 261:55
37. Náray-Szabó G, Tóth G, Ferenczy GG, Csonka G, (1994) Int J Quant Chem: Quant Biol. Symp 21: 227
38. Ferenczy GG, Náray-Szabó G, Várnai P (1999) Int J Quant Chem 75:215
39. Théry V, Rinaldi D, Rivail J-L, Maigret B, Ferenczy GG (1994) J Comput Chem 15:269
40. Pu J, Gao J, Truhlar DG (2004) J Phys Chem A 108:632–650
41. Huzinaga A, Cantu AA (1971) J Chem Phys 55:5543
42. Ferenczy GG (2013) J Comput Chem 34:854–861
43. Sakai Y, Miyoshi E, Klobukowski M, Huzinaga S (1987) J Comput Chem 8:256–264
44. Pipek J, Mezey PG (1989) J Chem Phys 90:4916–4926
45. Ferenczy GG (1991) J Comput Chem 12:913
46. Ferenczy GG, Winn PJ, Reynolds CA (1997) J Phys Chem A 101:5446
47. Cornell WD, Cieplak P, Bayly CI, Gould IR, Merz KM Jr, Ferguson DM, Spellmeyer DC, Fox T, Caldwell JW, Kollman PA (1995) J Am Chem Soc 111:5179–5197
48. Philipp DM, Friesner RA (1999) J Comput Chem 20:1468–1494

49. Ferré N, Assfeld X, Rivail J-L (2002) J Comput Chem 23:610–624
50. Lin H, Truhlar DG (2005) J Phys Chem A 109:3991–4004
51. Hehre WJ, Ditchfield R, Pople JA (1972) J Chem Phys 56:2257–2261
52. Schmidt MW, Baldridge KK, Boatz JA, Elbert ST, Gordon MS, Jensen JJ, Koseki S, Matsunaga N, Nguyen KA, Su S, Windus TL, Dupuis M, Montgomery JA (1993) J Comput Chem 14:1347–1363
53. Stone AJ (2005) J Chem Theory Comput 1:1128–1132
54. Ferenczy GG, Reynolds CA, Winn PJ, Stone AJ Mulfit: http://www-stone.ch.cam.ac.uk/programs/gdma.html
55. Ferenczy GG, Adams WH (2009) J Chem Phys 130:134108
56. Ferenczy GG (2013) J Comput Chem 34:862–869
57. Stoll H, Wagenblast G, Preuss H (1980) Theor Chim Acta 57:169–178
58. Gianinetti E, Raimondi M, Tornaghi E (1996) Int J Quant Chem 60:157–166
59. Sironi M, Famulari A (2000) Theor Chem Acc 103:417–422
60. Fornili A, Sironi M, Raimondi M (2003) J Mol Struct (Theochem) 632:157–172
61. Fornili A, Moreau Y, Sironi M, Assfeld X (2006) J Comput Chem 27:515–523
62. Smits GF, Altona C (1985) Theor Chem Acc 67:461–475
63. Couty M, Bayse CA, Hall MB (1997) Theor Chem Acc 97:96–109
64. Stone AJ (1981) Chem Phys Lett 83:233–239
65. Stone AJ, Price SL (1988) J Phys Chem 92:3325–3335
66. Lee T-S, Yang W (1998) Int J Quantum Chem 69:397–404
67. Hong G, Strajbl M, Wesolowski TA, Warshel A (2000) J Comput Chem 21:1554–1561
68. Wesołowski TA, Warshel A (1993) J Phys Chem 97:8050–8053

Chapter 5
Polarizable Force Fields for Proteins

Oleg Khoruzhii, Oleg Butin, Alexey Illarionov, Igor Leontyev,
Mikhail Olevanov, Vladimir Ozrin, Leonid Pereyaslavets and Boris Fain

5.1 Introduction

Computer simulations of molecular systems have become an integral part of scientific investigations across a wide range of biophysical and biochemical problems [1]. Today computational methods offer detailed atomic-level information about physiological processes that is frequently unobtainable by experimental approaches [2, 3]. Computer modeling complements bench science by providing interpretation of experimental results, guiding further experiments and, perhaps in the near future, itself becoming the experiment [4, 5].

High-level ab initio calculations are the most reliable and consistent computational tool capable of reproducing many molecular properties [6]. For intermolecular interactions quantum mechanics (QM) can provide not only the absolute value of the interaction energy, but it also allows the partitioning of the energy into several physically meaningful components, thus helping to analyze and to understand the mechanism of the interaction [7–11]. Despite continuous progress in first-principles methods, calculations of sufficient accuracy are still very computationally demanding and rapidly become intractable as the system size increases.

Phenomenological classical potentials known as force fields (FFs) are much more useful in practical applications that involve large molecular systems consisting of thousands of atoms. In this approach, known as molecular mechanics (MM) simulation, atoms or groups of atoms are represented as particles that interact through relatively simple potential functions based on physical models.

O. Khoruzhii (✉) · O. Butin · A. Illarionov · I. Leontyev · M. Olevanov ·
V. Ozrin · L. Pereyaslavets · B. Fain
InterX Inc, 811 Carleton Street, Berkeley, CA 94710, USA
e-mail: bokonon@yandex.ru

© Springer International Publishing Switzerland 2014
G. Náray-Szabó (ed.), *Protein Modelling*, DOI 10.1007/978-3-319-09976-7_5

FF potentials are constructed as functions of the nuclear coordinates only, while the electronic subsystem in the Born-Oppenheimer approximation is assumed relaxed in some sense and is buried within the electrostatic part of the model.

A specific FF is characterized by a potential functional form as well as by a set of associated parameters obtained by matching of FF predictions to experimental data and/or ab initio calculated properties. This representation is then used to sample the conformational phase space of the molecules via simulation techniques such as Monte Carlo (MC) and molecular dynamics (MD). This sampling, in turn, permits the characterization of the time evolution of the molecular system, its fluctuations and interactions, and therefore, the investigation of the system's structural, kinetic and thermodynamic properties.

A FF can be considered as a classical approximation of the quantum interaction and should ideally include all the energy components revealed by quantum mechanics. This, however, encounters two basic difficulties. First, interactions of chemically bonded atoms are inherently quantum and cannot be described classically without crude simplifications. For this reason most FFs describe the interaction of bonded atoms by phenomenological two-body, three-body, and four-body terms that depend on bond distances, angles and dihedral angles and provide reasonable molecular geometry along with conformational energetics. The second difficulty is the intrinsically many-body character of some quantum effects, of which the most prominent representative is the electronic polarization (induction). The inclusion of such terms in a FF makes it much less computationally efficient.

The significance of the polarization effect was recognized long ago [12–14] and polarizable FFs have been developed since the very early stage of the simulation era [15–18], primarily for water modeling [19, 20]. Later, the pressing demand of large scale simulations of biological interest resulted in domination of simplified non-polarizable FFs (although, about a half of specialized water potentials was still polarizable [21]). In non-polarizable potentials, which are also called pairwise or additive, polarization is included in some averaged way by empirically tuning model parameters so that simulations reproduce the target properties of a given molecule in an environment of particular polarity.

Empirical parameterization is an unavoidable element of additive FFs, and it allows one to achieve sufficient accuracy provided by partial compensation of errors in standard solvation conditions. Several decades of testing and refinement have resulted in impressive progress in additive FFs such as AMBER [22–24], MMFF [24, 25], OPLS [25–27], GROMOS [28, 29], CHARMM [30–32], etc. Today these FFs have become very popular and valuable tools for biomolecular MM simulations.

Though based on several drastic simplifications, existing pairwise FFs have been remarkably successful in modeling many complex molecular systems. They have produced a great number of useful insights concerning biological function, as well as a demonstrated ability of obtaining quantitative results for structures and energetics [33, 34].

Despite some success and the wide usage of additive FFs, implicit description of the polarization is frequently considered as their main deficiency [35]. As the state of a molecule in these models is insensitive to its environment, problems arise when the environment changes. Simulations of gas-liquid interfaces and phase transitions, polar solutes in nonpolar solvents, properties of polar liquids at high temperature and low density, solvation and affinity energies by non-polarizable FFs are often unreliable. The special issue of Journal of Chemical Theory and Computations on polarization has illuminated a number of such cases [35].

The implicit treatment of polarization, however, is just one of many other deficiencies and oversimplifications of modern additive FFs.

The much less recognized problem is that parameterization of empirical potentials is typically limited by insufficient experimental data available for the target solvation condition. Moreover, due to integral character of experimentally observable properties, the empirical fitting generally provides little insight into how the model's inadequacies can be removed. For example, though the inability of additive FFs in reproducing ion absorption [36] and conductivity of ionic liquids [37, 38] being well recognized, only recently has it been resolved with implementation of a new theory [39].

Clearly high level QM calculations are superior in this regard because they can provide much more accurate, diverse, focused and representative data sets for the FF parameterization, training and testing. QM based parameterization strategy benefits not only from existing extensive and physically well-grounded theoretical methods, but it also reaps the rewards of continuous advances in ab initio techniques where rapid increase in computer power will eventually deliver results of any required precision. Transferable polarizable force fields derive the most benefit from QM parameterization.

Thus, aided by recent and ongoing great progress in computer technologies as well as by a fundamental increase in accuracy and speed of QM techniques, the development of explicitly polarizable FFs is finally starting to fully blossom.

The polarization "Renaissance" began at the very end of the last century. In the past 15 years a great deal of effort has been put into developing accurate and efficient ways to introduce polarization into classical FF paradigm, elaborating specific parameterization techniques, and producing a new generation of general purpose FFs. The amount of work and the interest in the subject can be characterized by the statistics of the relevant reviews: starting with the first review of Halgren and Damm in 2001[40] new reviews have appeared every year (more than 3 per year after 2006, with the most recent in 2014 [41]) referring to over a thousand pages and reviewing development of polarizable FFs in almost every possible detail.

A very clear description of the different methods to make the FF polarizable can be found in Rick and Stuart [42]. Specific features of protein FFs have been described by Ponder and Case [43]. Warshel [44, 45] has discussed a broad range of problems related to proteins that require accounting for polarization, along with a historical perspective. Patel and Brooks [46, 47] reviewed many aspects concerning construction and parameterization of polarizable FF using the fluctuating charges approach, while Yu and van Gunsteren [48] had concentrated on Drude oscillator approach with most recent related results in Lopes et al. [49].

Cisneros et al. [50] outlined a more sophisticated approach to the description of the electronic component, which they implemented in SIBFA and GEM models. Various polarizable models in the CHARMM framework have been reviewed by MacKerell and coworkers [51]. Stone and Misquitta [9] and Wang with colleagues [52] have discussed quantum mechanical counterparts of energy components of classical FFs including polarization, and possibilities for their parameterization from the first principles. Simplified QM approaches to model interactions in large molecular systems as well as QM/MM approaches were reviewed by Gordon et al. [53, 54]. Gong [55] thoroughly described development and applications of the sophisticated ABEEM fluctuating charge molecular force field. Cisneros and coauthors [56] have given an exhaustive picture of all electrostatic aspects of MM simulations. Marshall [57] provided arguments to decide whether accurate molecular polarizability is the most limiting aspect in MM electrostatic models.

With such extensive and detailed bibliographic support we shall not attempt to cover in our review all aspects of a very broad and actively developing subject of polarizable force fields. Instead we shall take a more educational approach and try to discuss in more detail several specific and previously poorly elucidated points, and to conclude with our perspective as to where we are in the subject and what we can expect in the nearest future.

The latter is not a trivial question. In their prominent review in 2001 Halgren and Damm stated a hope that: "Given the accelerated progress made in the past five years, the next few years bode well to establish the limitations of standard, non-polarizable fixed-charge force fields and to make the case for routinely including polarizability in bio-molecular calculations" [40]. Unfortunately their hope has not been realized yet. Despite clear physical arguments for the necessity of explicit inclusion of molecular polarizability into MM models, success of current polarizable FFs is not widely accepted, and their superiority over simpler additive models is debatable. Thus to move forward one first needs to find a reason why more than 15 years of active development have not achieved the clear superiority and prevalence of a physically more accurate model.

5.2 Polarization: Basic Notions and Mechanisms

In classical electrodynamics [58] polarization density (or electric polarization, or simply polarization) is the vector field that expresses the density of induced electric dipole moments in a dielectric sample V:

$$\vec{P} = \left\langle \vec{d}_V \right\rangle \Big/ V. \tag{5.1}$$

In moderate electric fields polarization is proportional to the field

$$\vec{P} = \chi \vec{E}, \tag{5.2}$$

where χ is electrical susceptibility and is a tensor in a general case.

Polarization can be induced by different mechanisms that include shift of ionic species, orientation of dipolar elements, change in electronic distribution, etc. As MM models characterize a system on an atomic level, the majority of polarization factors are automatically included except for electronic polarization. For this reason when one says that a FF is polarizable one means that the FF includes special additional terms to describe electronic polarization explicitly.

The response of each polarization mechanism to the applied field may be characterized by a specific time scale that results in χ frequency dependence. In the simplest case of the unimodal polarization mechanism this dependency can be approximated by

$$\chi(\omega) = \chi_\infty + \frac{\chi_s - \chi_\infty}{1 + \omega^2 \tau^2}. \tag{5.3}$$

Here χ_∞ and χ_s are high frequency and low frequency (stationary) asymptotic values, τ is the relaxation time.

For water in this approximation, the high frequency response is determined by electronic polarization and the low frequency response is mainly provided by orientational polarization of water molecules. For liquid water at 25 °C τ is 8.27 ps and for hexagonal ice at 0 °C $\tau \sim 20$ μs both characterizing the time required for the reorientation of a water molecule in the electric field [59].

Additionally, we should mention two specific polarization mechanisms: geometric [60, 61] and mechanical [42] (steric) polarization. Actually they could be considered as special cases of electronic polarization.

Geometric polarization arises as a result of change in the molecular geometry. In addition to the trivial contribution due to nuclear displacement it includes a much more complex effect of electron density redistribution. The naive description of this effect uses an idea of an "effective" atom in a molecule: the assumption that the change in nuclear positions results in similar displacement of related electron clouds without change of their partitioning between nucleolus (atoms).

The physical invalidity of such an idea can be easily understood by considering the result of increase of bond length in a simple ionic diatomic molecule like NaCl. In equilibrium state it consists approximately of two bonded ions Na^+ and Cl^- but at the dissociation limit it will be a pair of neutral atoms—Na and Cl, with some continuous change of the electronic state in the intermediate.

Both experiment [62] and ab initio calculations [63] demonstrate that a change in molecular geometry results in complex and nonlocal changes in the electron density. For example, in water molecule, the angle between the vectors $\partial \vec{\mu}_{H_2O}/\partial l_{OH}$ describing change in the molecular dipole with change in either bond length is about 150° (Fig. 5.1) [64] while under the effective atom hypothesis it should be equal to the molecular bending angle ($\approx 104°$ for gas phase geometry).

Mechanical polarization arises from the exchange-repulsion of electrons at close contacts between molecules. Basic features of this phenomenon can be understood by considering the interaction of a pair of noble gas atoms placed at a small distance from each other (the scales in this case are unrealistic but help to understand the effects qualitatively). In the simplest but qualitatively correct

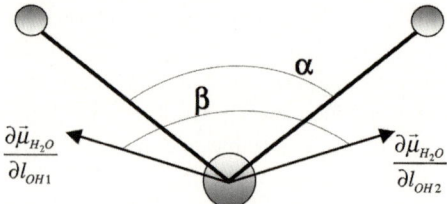

Fig. 5.1 Geometrical polarization of water molecule. Change in molecular dipole with change in bond length can not be described in the effective atom paradigm, $\alpha \approx 104°$, $\beta \approx 150°$

model (like QMPFF2 [65]), the electronic structure of both atoms can be approximated by spherically symmetric clouds which feel exchange and electrostatic repulsion from the interacting cloud and electrostatic attraction with the nucleus of the interacting atom. As a result of the interactions the cloud can shift from the nuclear position at the expense of induction energy quadratically dependent on the shift. Figure 5.2 shows qualitatively the radial dependence for electrostatic, exchange-repulsion and total interaction energy in the QMPFF model. The penalty caused by induction and unfavorable dipole-dipole electrostatic interaction is lower than the gain from exchange-repulsion attenuation due to the clouds drifting apart, overall resulting in the slower rise of the repulsion barrier at small distances than that without the polarization and the negative quadrupole moment of the atom pair (Fig. 5.2). Thus, as expected the electronic relaxation provides negative contribution to the total energy, while the induced electrostatic interaction in this case is unfavorable which is rather counterintuitive.

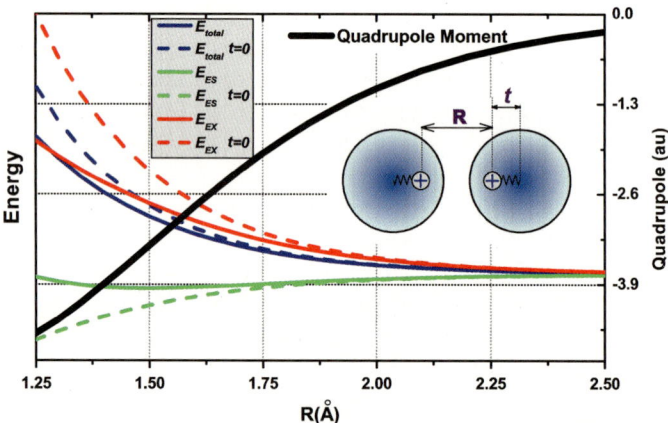

Fig. 5.2 Interaction energies of two rare gas atoms obtained with the QMPFF2 force field [65]. The dependencies of the total energy E_{tot}, its exchange-repulsion E_{ex} and electrostatic E_{es} component on the interatomic distance R are shown in *blue*, *red* and *green* color, respectively. In QMPFF2 model the atoms are described as nuclei with attached by spring spherical exponential electronic clouds. *Solid lines* represent the system with polarized electronic clouds (i.e. with optimal cloud shift t = t_{opt}) while *dashed lines* stand for the system without polarization (clouds are fixed at nuclei positions, t = 0). Inset shows the polarization shifts of cloud positions

In a similar way the mechanical polarization (i.e. exchange-repulsion between electrons of different molecules) can weaken the strength of ordinary electronic polarization in response to an external electric field. It is believed to be the main factor that causes a decrease of molecular polarizability in dense systems as compared to the gas phase [66, 67].

Figure 5.2 also illustrates another electrostatic effect—penetration, which to some degree imitates polarization. Like polarization, penetration acts even between neutral interaction centers and provides a gain in the interaction energy. It arises because of shielding of the electrostatic interaction of overlapped electronic clouds with a more steep weakening electrostatic repulsion between clouds than opposite twofold electrostatic attraction of clouds to nuclei. Penetration, for example, makes a significant contribution to the energetics of hydrogen bonding [68]. The question of what is the principal stabilization factor for H-bonding network in water—attractive electrostatic interaction, polarization or penetration effect—is debated to this day.

Different polarization mechanisms are not independent. E.g. interaction of a water molecule with surrounding molecules will change its dipole moment via electronic polarization thus changing also the orientational response. As a result, the averaging in Eq. (5.1) includes not only the simple orientational one, but also an effective averaging over configurations of the environment around a molecule, and these two statistics are not independent. This simple consideration shows why it is more difficult to build a polarizable model than a non-polarizable one.

In the latter case the two averages are believed to be independent and the molecular dipole moment in the condensed phase is taken as an "effective" model parameter. In parallel, parameters of the intermolecular potentials are chosen to reproduce some "effective" structure of the medium, which, after averaging with a chosen "effective" dipole moment, provides accurate macroscopic properties. To be successful the model should not necessarily reproduce the real structure as well as the true average dipole moment. "Observables" can be matched by tuning an arbitrary molecular dipole moment to reach an agreement. It is also significant that the model is additive. Thus, though parameterized on properties of small molecules, it reasonably reproduces the properties of more complex systems as well.

In contrast, the task of constructing a polarizable model is much more ambitious. Ideally, a model parameterized just on the basis of ab initio data for monomer properties and intermolecular interactions in dimers and small clusters should be able to reproduce real statistics of molecular structure configurations in the condensed phase, accurate response of each molecule to a particular environment, and accurate response of such obtained molecular ensemble to the external electric field (which is equivalent to accurate macroscopic fluctuations of the ensemble dipole moment and implies accurate thermodynamic behavior).

Moreover, a polarizable FF is non-additive. Thus having been parameterized on small systems it is not guaranteed to work well on larger ones. To achieve such transferability the model should not depend on effective parameters and should represent all of the underlying physics. We shall see below that available experience in development of polarizable FFs clearly supports this idea.

The aforementioned coupling of different polarization mechanisms also produces difficulties for polarizable FFs with the well-established empirical parameterization approaches. These approaches either cause the polarizable model to be over-parameterized and parameterization becomes unstable or the resulting parameters are appropriate just for the narrow range of conditions close to those used for the parameterization.

5.3 Basic Approaches to Include Electronic Polarization in Classical FFs

In a practical realization of the model one needs to make several choices, and the first one is of the relative weight between levels of physical correctness and computational efficiency. Up to now, in most cases, computational efficiency has been considered as the only critical issue, as conventional wisdom states that low numerical efficiency is the main obstacle for use of polarizable models. However, in our opinion, this position is currently unwarranted. First of all, as was shown above, polarizable models are much more sensitive to the overall physical accuracy of the model. Second, computational power of modern computers is growing so fast and is currently at a high enough level that it is no longer a predominant bottleneck for many of the MM simulations. Rather the accuracy of the model has become of principal importance. Third, the computational cost incurred by more physically reasonable models is small in comparison to the inclusion of the polarization itself. Fourth, there are many ways to improve numerical efficiency without a loss of the physical soundness of the model.

In conclusion, while the choices pointed out below are well known [48], the prevailing ongoing emphasis on speed, in our opinion, is becoming misguided. Repeated preference for numerical efficiency has prevented some physically sound avenues from being explored. As a result, improvements in the physical correctness of the models still await further investigation.

There are three basic approaches to incorporate electronic polarization in classical FF paradigm—point inducible dipoles, Drude (or shell, or charge-on-spring) model, and fluctuating charges. For all these approaches, the technical aspects and the history of their development and applications have been perfectly reviewed by Rick and Stuart [42]. Here we shall focus on just a few details that, in our opinion, require more caution.

5.3.1 Inducible Dipoles

Inducible dipoles are the most popular and most developed method of introducing molecular polarizability into the FF. This approach is the method of choice in polarizable versions of AMBER [69, 70] and OPLS [67], as well in PIPF-CHARMM [71], NEMO [72], SIBFA [73], EFP [54], AMOEBA [74], and

QMPFF3 [75]. Thus by this example we can follow all the characteristic details and difficulties that arise in polarizable models. In this approach the polarizable centers (which are most frequently atoms) are represented as inducible dipoles with some center-specific polarizability:

$$\vec{\mu}_c = \alpha_c \vec{E}. \tag{5.4}$$

For such a system the potential energy becomes:

$$U_{pot} = U_0 - \sum_a \vec{\mu}_a \vec{E}_a^{per} - \frac{1}{2}\sum_a \sum_{b \neq a} \vec{\mu}_a \mathbf{T}_{ab} \vec{\mu}_b + \frac{1}{2}\sum_a \frac{\mu_a^2}{\alpha_a}, \tag{5.5}$$

where the first term is the potential energy without polarization, the second and third terms describe interaction of induced dipoles with permanent electric moments (producing permanent electric field $\vec{E}_a^{per}$) and with each other, and the last term is a penalty for perturbing the basic electronic state of the molecule (induction energy in the narrow sense) and is always positive. The dipole field tensor $\mathbf{T}_{ab}$ is a function of inter-center vector $\vec{R}_{ab}$ and is defined so that in the center a the electric field from induced dipoles in all other centers is

$$\vec{E}_a^{ind} = \sum_{b \neq a} \mathbf{T}_{ab} \vec{\mu}_b \tag{5.6}$$

For point dipoles

$$T_{ab,ij} = R_{ab}^{-3}(3n_{ab,i}n_{ab,j} - \delta_{ij}), \tag{5.7}$$

$$\vec{n}_{ab} = \vec{R}_{ab}/R_{ab}. \tag{5.8}$$

In accord with the Born-Oppenheimer approximation the electronic system is assumed to be totally relaxed in any nuclear configuration and its state corresponds to a minimum of potential energy. This is provided by values of induced dipoles

$$\vec{\mu}_a = \alpha_a \left(\vec{E}_a^{per} + \sum_{b \neq a} \mathbf{T}_{ab} \vec{\mu}_b \right). \tag{5.9}$$

Under this condition the potential energy can be simplified

$$U_{pot} = U_0 - \frac{1}{2}\sum_a \vec{\mu}_a \vec{E}_a^{per}. \tag{5.10}$$

5.3.1.1 Choice of Polarization Centers

When introducing molecular polarizability into the MM model one should decide whether to use a one-center (unimolecular) or a multi-center (distributed) description of polarizability. From a physical point of view it is clear that since some changes occur in all points of molecular electronic cloud in response to an external electric field, the distributed description is more natural [76]. However, the observable values, e.g. change in dipole moment, characterize the whole molecule only and just like there is no unique physically based way to distribute the electric charge between the atoms in a molecule, there is no unique way to distribute polarizability [77]. All possible ways of redistributing the polarizable response to the field over different centers will have some errors close to the molecule and will be approximately equivalent in the far region. The distributed models are more capable of providing a larger region of convergence and smaller corresponding errors.

Description of polarization by induced dipoles introduces additional computational expenditures compared to those with a point charge electrostatic model. Consequently, since it has long been accepted that the electrostatic potential (ESP) of small molecules can be reproduced reasonably well by one-center multipoles, computationally effective models use a restricted set of polarizable centers, placing the centers on heavy atoms only or on group centers of small quasi-neutral groups [78, 79].

This choice has obvious disadvantages in the description of polarization for close atom contacts, particularly for H-bonding. In many respects the situation resembles the treatment of hydrogen atoms in classical FFs [43]. For numerical efficiency the first generation of FFs did not include hydrogen atoms at all, parameterizing all heavy atoms as effective "united" atoms. Subsequently it was recognized that at least polar hydrogen atoms should be treated explicitly. Finally, current FFs describe all hydrogen atoms explicitly and sometimes add additional centers for lone pairs or bonds.

The use of an all-atom model of induced dipole polarization not only provides a more accurate polarization field, but also permits the improvement of permanent electrostatics. In this model a permanent dipole can be added at each atom without any additional computational expense [67]. Nevertheless, for some unknown reason, this possibility for complimentary permanent dipoles is not widely used.

5.3.1.2 How to Determine Values of Induced Dipoles

The exact calculation of induced dipoles from Eq. (5.9) requires an $N \times N$ matrix inversion where N is the number of polarizable centers. For large N this becomes intolerably inefficient (scales as N^3). Consequently the values of induced dipoles are typically determined by either iterative methods or by an extended Lagrangian formalism [42, 80]. In the latter approach the induced dipoles are considered as additional degrees of freedom whose dynamics are described by Newton-like equations with forces proportional to the difference between left and right hand sides of Eq. (5.9). Effective inertia ("mass") is chosen to provide the dynamics, which should be faster than any nuclear dynamics but not so fast as to interfere with a reasonable time step for integration [80].

Another group of methods is the so-called non-iterative methods. They were the first approaches used when induced dipoles were employed to describe polarization [16, 81, 82]. Considering polarization formally as a small parameter one can construct a solution of Eq. (5.9) as a formal expansion in degrees of polarization:

$$\vec{\mu}_a^{(0)} = \alpha_a \vec{E}_a^{per}, \tag{5.11}$$

$$\vec{\mu}_a^{(1)} = \alpha_a \vec{E}_a^{per} + \alpha_a^2 \sum_{b \neq a} \mathbf{T}_{ab} \vec{E}_b^{per}, \tag{5.12}$$

and so on. Substitution of relations (5.11) and (5.12) in Eqs. (5.5) or (5.10) provides the corresponding series for the interaction energy. These series will differ in the largest order terms as Eqs. (5.5) and (5.10) are equivalent only with the exact solution (5.9) for induced multipoles. It is interesting that substitution of the lowest approximation (5.11) into Eq. (5.5) gives the same expression for the energy as a substitution of the next approximation (5.12) into Eq. (5.10). This had resulted in some confusion with the terminology. The same expression for the energy is called zero-approximation in Ref. [83] and the second order interaction model in Refs. [84–87].

Recent works show that such non-iterative models can efficiently account for the polarization effects and simultaneously be rather accurate if parameterized appropriately [41, 84–88]. Consequently they can be considered as a method of choice for modern large-scale simulations that treat molecular polarization explicitly.

5.3.1.3 Overpolarization Problem

A long known unpleasant feature of the point induced dipole model is the existence of a polarization catastrophe for close polarization centers [89, 90]. If two centers are near each other and lie along the direction of an external electric field their mutual interaction will act to increase the induced dipoles with overall axial polarizability

$$\alpha_{||} \propto 1 / \left(1 - 4\alpha_a\alpha_b / R_{ab}^6\right) \tag{5.13}$$

and solution of Eq. (5.9) is going to infinity for

$$R_{ab} = (4\alpha_a\alpha_b)^{1/6}. \tag{5.14}$$

For inter-center distances larger than the critical value in Eq. (5.14) the solution of Eq. (5.9) is finite, but still, the obtained values of dipoles may be artificially amplified and therefore be unphysical. Such unphysical behavior is a consequence of several simplifications in the model: the neglect of diffuse character of the electronic clouds, nonlinear polarization, etc.

Thus, to be more reliable and stable, a polarization model based on induced dipoles should be complemented by some mechanism to dampen interaction for short inter-center distances. The most popular model used currently in the majority of polarizable FFs [49, 52, 91–98] is one proposed by Thole [90].

Thole begins the development of his model with a clear physical argument that the interaction of diffuse electronic clouds at small distances is weaker than that of point dipoles. As a result the tensor $\mathbf{T}_{ab}$ in Eq. (5.6) is modified by a damping factor in comparison with that of point dipoles in Eq. (5.7). Despite this initial argument, the Thole model is purely mathematical. First, he substitutes the interaction of two diffuse dipole clouds with the interaction of one diffuse cloud with a point dipole. Then, more significantly, Thole does not use a physically reasonable cloud width. Instead, he solves the inverse problem and determines widths that exclude the possibility of the catastrophe for any inter-center distance. Such an approach results in characteristic widths of electronic clouds exceeding 1 Å that are much larger than the typical size of molecular orbitals (about 0.25 Å).

Elking has tried to develop a more physically reasonable model by representing induced dipoles as Gaussian dipole clouds [99]. However he also parameterizes the cloud widths by requiring the suspension of the catastrophe for all distances and obtains unphysically large widths greater than 1 Å.

Figure 5.3 compares the dependence of the axial polarization and its denominator [see Eq. (5.13) for example] on the interatomic distance in several models of diffuse clouds. As an example of a physically realistic model we present the QMPFF3 [75] model where induced dipoles are represented as exponential dipole clouds with widths parameterized on the base of real electronic density distribution (through reproduction of spatial dependence of dimer electrostatic interaction energies including small intermolecular distances). The realistically distributed dipole clouds dampen the dipole interaction very similarly to Thole's and Elking's

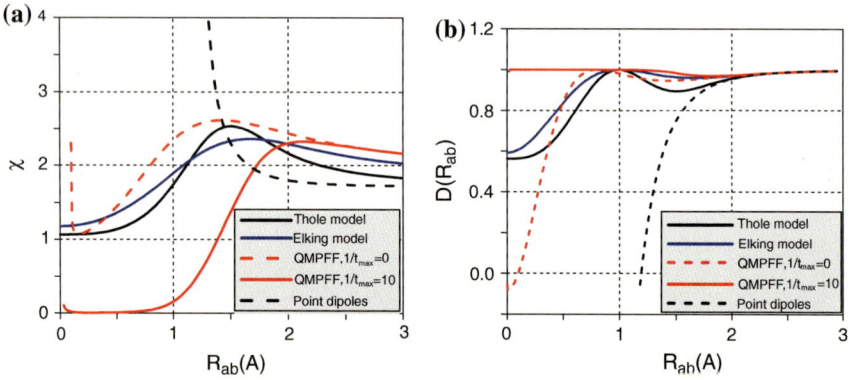

Fig. 5.3 a Dependence of the axial polarizability, $\alpha_{\parallel}^{ab}$, on the interatomic distance of two nearby rare-gas atoms a and b in point dipoles Thole, Elking, and QMPFF3 (including linear, $1/t_{max} = 0$, and nonlinear polarization, $t_{max} = 0.1$ Å) models. To be able to compare different screening models, polarizabilities of these atoms are arbitrarily chosen to be equal to C and N polarizabilities in the correspondent model. Parameters for the first three are from Ref. [99] for the last two from Ref. [75]. **b** The same dependence of the denominator from the formula for polarizability indicating the possibility of a polarization catastrophe

models, thus preventing the polarization catastrophe for all physically meaningful intermolecular distances. Yet the avoidance of the catastrophe does not automatically guarantee a solution of the overpolarization problem (see Sect. 5.4).

Unrealistic cloud width is not the only physical deficiency of Thole's model. As was pointed out in the previous section, the diffuse character of electronic clouds changes the atom interaction at small distances not only quantitatively but also qualitatively resulting, additionally, in mechanical polarization and penetration effects. As a result, the polarization state of close atoms can differ significantly from that predicted if one accounts for the external electric field only (Fig. 5.4).

In conclusion we point out that Thole's model should be considered as one of many possible mathematical approaches to restricting overpolarization. Other examples are application of soft-core potentials [48] or the truncated non-iterative description of polarization [83, 84], which is close to the exact solution for small values of induced dipoles and systematically underestimates the self-consistent solution of Eq. (5.9) in the high polarization limit. A more physically rigorous treatment of diffuse character of electronic clouds goes much further than Thole's model, requiring inclusion of clouds with realistic scales, and of all types of interactions caused by nonzero values of these scales [75].

Treating the electronic clouds as diffuse entities cannot by itself solve the overpolarization problem. This conclusion is supported by the consideration of the non-additive energy in water clusters [100] and by problems with description of the polarization state in large structurally organized molecules such as proteins [49]. One of the popular methods of solving this problem is to bring in nonlinear polarization, so that a linear relationship like in Eq. (5.4) is just the first term valid for only small values of the induced dipole and/or applied electric field [49].

Typically nonlinear polarization is obtained rather formally by introducing additional nonlinear terms that saturate Eq. (5.4) at large field amplitudes or by additional terms that lead to faster growth of the penalty induction term in Eq. (5.5) for larger induced dipoles [49, 101].

Fig. 5.4 Dependence of the angle between the dipole moment of one of the closely situated rare-gas atoms and interatomic radius vector on the interatomic distance in an external field of 0.01 a.u. perpendicular to this vector with and without account for diffuse cloud interactions

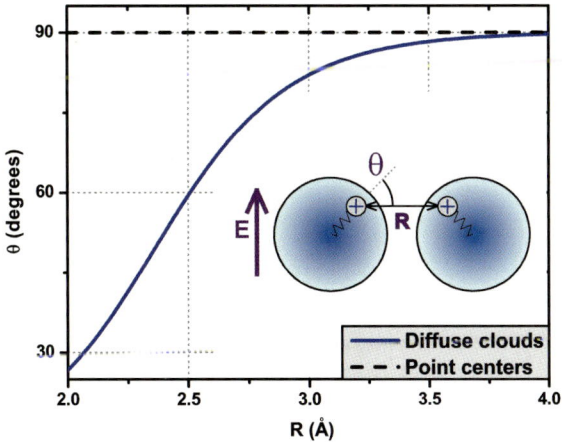

A more natural way to introduce the nonlinearity was used in the QMPFF force field [65, 75, 100, 102], which uses simply the properties of the wave function with a dipole component.

The simplest normalized anisotropic wave function with nonlinear dipole moment can be written as:

$$\Psi_a = \sqrt{\frac{1}{\pi \left(1+p_a^2\right)w_a^3}} \left[1 + (\vec{p}_a \cdot \vec{n})\right] e^{-r/w_a}. \tag{5.15}$$

Here $\vec{r}$ is radius vector in the frame with the origin in the cloud center, $\vec{n} = \vec{r}/r$, w_a is characteristic width of the cloud and $\vec{p}_a$ is the weight of the p-state component providing the nonzero dipole moment.

The energy of the cloud depends on the weight of the p-state:

$$E_a = \frac{E_s + p_a^2 E_p}{1 + p_a^2} = E_s + \frac{p_a^2}{1 + p_a^2} E^*, \tag{5.16}$$

where E_s and E_p are respectively s- and p-state energies whereas $E^* = E_p - E_s > 0$ is the excitation energy.

The dipole moment of the cloud with respect to the center (induced dipole) is

$$\vec{\mu}_a = \int \vec{r}\rho_a(\vec{r})dV = Q_a \vec{t}_a, \tag{5.17}$$

where Q_a is the total charge of the cloud and $\vec{t}_a$ its "shift" related with the weight of p-state by:

$$\vec{t}_a = \frac{2w_a \vec{p}_a}{1 + p_a^2}. \tag{5.18}$$

From this equation it is evident that absolute value of the shift of the cloud (and induced dipole as well) is always smaller than

$$t_{\max} = w_a. \tag{5.19}$$

The induction (penalty) term U^{IN} is associated with the increase of intra-atomic energy given by Eq. (5.16):

$$U_a^{IN} = E_a - E_s = \frac{\mu_a^2}{\alpha_a} \frac{1}{1 + \sqrt{1 - (t_a/t_{\max})^2}}. \tag{5.20}$$

where the polarizability parameter α_a is

$$\alpha_a = \frac{Q_a^2 w_a^2}{2E_a^*}. \tag{5.21}$$

For small induced dipoles (shifts) the induction energy coincides with that in Eq. (5.5) and describes the linear polarization, while for $t_a \rightarrow t_{max}$ $\partial U_a^{IN}/\partial t_a \rightarrow \infty$ corresponding thus to zero polarizability.

Modification of the denominator in the expression for axial polarization like Eq. (5.13) with account for finite t_{max} with a value from Ref. [75] is also shown in Fig. 5.3. It is seen that the problem of polarization catastrophe is totally avoided.

5.3.2 Drude Models

An alternative approach to introduce electronic polarization into a FF is based on the classical Drude oscillator model. This approach is the method of choice in polarizable versions of GROMOS FF [103], as well in CHARMM Drude-2013 [49], and QMPFF2 [65] FFs.

The origin of the Drude model can be traced back to the work of Paul Drude who introduced the method in 1902 to describe the dispersive properties of materials [14]. In this model (called also shell or charge-on-spring model) polarizability of a given atom with a partial point charge q_a is represented by introducing an "auxiliary" mobile Drude particle with a point charge q_{aD} attached to the atomic particle by a harmonic spring with a force constant k_{aD}. While rationally the Drude particle represents an electronic cloud which can move off-center in an external electric field, its charge is usually considered to be a model parameter and can be even positive in some models. To preserve the net charge of the atom–Drude pair the charge of the atom is replaced by $\tilde{q}_a = q_a - q_{aD}$.

The electrostatic interaction of two atoms in a non-polarizable FF is substituted by the four site Coulomb interaction with addition of induction energy of the spring:

$$U_{el} = \frac{1}{2}\sum_a \sum_{b \neq a} \left(\frac{\tilde{q}_a \tilde{q}_b}{r_{ab}} + \frac{\tilde{q}_a q_{bD}}{\left|\vec{r}_{ab} + \vec{d}_{bD}\right|} + \frac{q_{aD}\tilde{q}_b}{\left|\vec{r}_{ab} - \vec{d}_{aD}\right|} + \frac{q_{aD}q_{bD}}{\left|\vec{r}_{ab} - \vec{d}_{aD} + \vec{d}_{bD}\right|} \right) + \frac{1}{2}\sum_a k_{aD}d_{aD}^2.$$
$$(5.22)$$

Relative positions $\vec{d}_{aD} = \vec{r}_{aD} - \vec{r}_a$ of all Drude particles are then adjusted self-consistently to minimize energy for any given atomic configuration of the system.

An important practical aspect in favor of the Drude model is that this method deals only with point charges and as such its implementation does not require significant modification of computer codes developed for standard non-polarizable models. In an external electric field E the Drude particle is shifted, producing the induced dipole

$$\vec{\mu}_{aD} = q_{aD}\vec{d}_{aD} = \frac{q_{aD}^2}{k_{aD}}\vec{E}.$$
$$(5.23)$$

This results in the expression for the isotropic atomic polarizability as follows

$$\alpha_a = q_{aD}^2/k_{aD}. \tag{5.24}$$

Anisotropic polarizability can be introduced by requiring the anisotropy of the spring.

For a given α_a, the force constant k_{aD} can be chosen so that the displacement of the Drude particle remains much smaller than any interatomic distance, and that the resulting induced dipole $\vec{\mu}_{aD}$ is almost equivalent to a point dipole. Under these conditions the Drude model is very close in all properties to the induced point dipole model with the same atomic polarizability.

For the sake of computational efficiency the Drude particle positions can be treated via the extended Lagrangian technique as additional dynamical degrees of freedom [48, 49]. In this realization a small mass, m_{aD} taken from the mass of the parent atom m_a, is attributed to all Drude particles. The value of m_{aD} is chosen to provide a balance between the maximum reachable time step for converged MD integration and sufficient decoupling of the Drude particle motions from atomic motions [104].

In an analogy with the induced dipole model, the permanent shift of the cloud can also be used to include permanent dipoles in the electrostatic model [65, 102].

The point charge Drude model also suffers from the polarization catastrophe and the overpolarization problem for close atom contacts. Purely mathematical solution of this problem can be achieved by damping the nearby interactions in the spirit of the Thole model or using alternative forms of damping functions. Realistic account for the diffuse character of the electronic cloud is already sufficient to avoid the polarization catastrophe [65, 102] (Fig. 5.5). Additional stabilization appears with account for nonlinear induction in accord with a relation like Eq. (5.20).

Several water potentials with diffuse Drude particles are presented in the literature [105, 106]. However, typically the charge of the Drude cloud is erroneously chosen to be simply equal the excessive charge of the atom. With such a choice

Fig. 5.5 Cloud shift of the electron clouds described as a diffusive Drude particle in QMPFF2 model for different values of t_{max} (see Eq. 5.20) in the neon dimer with different interatomic distances

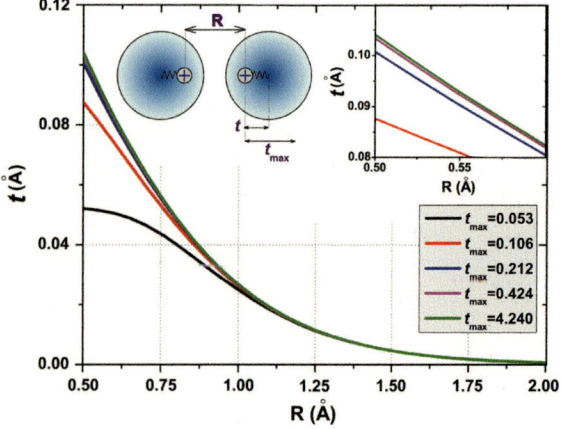

of parameters for the spring force constant and the cloud characteristic width it is possible to obtain the target polarizability and also to avoid the polarization catastrophe. However, the interaction of such an atom with others at small distances will be unrealistic and, specifically, the description of the penetration effect will be inaccurate. To obtain a realistic model the cloud charge should be of the order of the total charge of the valence electronic shell with a realistically shaped distribution. The promise of such an approach was clearly demonstrated by the QMPFF2 model [65], which is currently the only FF model of water able to reproduce fluid properties in wide temperature range having been parameterized only on the quantum data on the dimer and small clusters properties without any use of macroscopic properties of water.

5.3.3 Fluctuating Charge Models

The third popular category of methods that simulate molecular electronic polarizability are the fluctuating charge approaches (FQ). They are the method of choice in CHARMM-FQ [47, 107, 108] and ABEEM [55] FFs. In these models the charges on the chosen molecular centers are not permanent but are allowed to vary in response to variations in the environment. Depending on the model the variable charges obey charge conservation for the atom pair, for some molecular subgroup (e.g. protein residue), for each molecule, or for the whole system.

Because of the relatively small cost of describing polarizability the FQ approach is attractive from computational point of view—the polarizable model has the same number and the same type (point charges) of interaction centers as a non-polarizable prototype. If the charge variation degrees of freedom are simulated by the extended Lagrangian method, corresponding FQ models were only approximately 10 % slower than "classical" FFs in case of water [109], and slightly more than that in the case of proteins and condensed systems due to the necessity of using a smaller time step [107, 108].

In some cases, intra-molecular charge transfer is a rather natural way of representing electronic polarization, particularly along the plane of conjugated systems [110, 111]. On the other hand charge transfer can provide polarization just along bonds. This restricts the polarization of planar molecules to the plane of the molecule, and this, in turn, contradicts to experimental data—for example molecular polarizability of water is approximately isotropic. The problem can be resolved by the introduction of additional off-atom centers, e.g. in lone pair and pi-orbital locations [55].

A variety of methods to describe the charge fluctuations have been developed so far (see [112] and references therein), the most popular of which is based on the expansion of the self-energy of a charged center to the second order in charge [113]:

$$U_a^{self}(q_a) = E_{a0} + \chi_a q_a + \frac{1}{2} J_{aa} q_a^2. \tag{5.25}$$

Parameters of this expansion—electronegativity χ_a and hardness J_{aa}, can be related to the ionization potential and electron affinity of the atom [114], but more frequently they are considered as parameters of the particular atom type and are chosen in the course of the model parameterization [115].

At this approximation the whole electrostatic energy of the system is

$$U_{el} = \sum_a \left(E_{a0} + \chi_a q_a + \frac{1}{2}J_{aa}q_a^2 \right) + \frac{1}{2}\sum_a\sum_{b \neq a} J_{ab}(r_{ab})q_a q_b. \quad (5.26)$$

The second-order coefficient, J_{ab} depends on the distance r_{ab} between the centers a and b, and at large distances it should be equal to r_{ab}^{-1}. At shorter distances, there may occur a screening of the interactions, as discussed for the dipole–dipole and Drude particle-charge interactions in the earlier sections. Some authors (e.g. for general force fields [113] or for water models [109]) introduce the screening in accord with the interaction of Slater-orbitals yet with partial charges of the interacting atoms, other authors introduce more simple formulas with r_{ab}^{-1} asymptotics at large distances and a constant limited value at zero distance (see e.g. the works of Patel et al. [107, 108]).

For any particular geometry of the system, charge values are assumed to minimize the energy subject to a constraint that the total charge is conserved. That yields a set of conditions

$$\partial U_{el}/\partial q_a = \chi_a + J_{aa}q_a + \sum_{b \neq a} J_{ab}(r_{ab})q_b = \chi = const. \quad (5.27)$$

Here χ is a Lagrangian multiplier accounting for a conditional extreme value and simultaneously it is equal to the constant electronegativity (negative of chemical potential [114]) of the system of charges. For this reason the method is also called electronegativity or chemical potential equalization method.

The electronegativity equalization approach suffers from two problems related with incorrect behavior at large and small distances. The large-scale problem can be easily demonstrated for a system of two distant atoms. If the charge conservation is applied to the whole atom pair the solution for the atoms charges is

$$q_a = -q_b = \frac{\chi_b - \chi_a}{J_{aa} + J_{bb} - 2J_{ab}(r_{ab})} \xrightarrow[r_{ab} \to \infty]{} const. \quad (5.28)$$

Thus the model predicts a finite charge transfer for infinitely distant atoms with different electronegativities. The same difficulty will appear for a long molecule if the charge conservation is applied to the whole molecule. Specifically the problem was demonstrated for long alkanes [116] and proteins [108], where superlinear polarizability scaling was observed with the chain elongation. To solve this problem formulations allowing just atom-bonded atom charge transfer (or bond increment) have been proposed [115, 117]. Another solution is a restriction of charge conservation to some small chemically related group of atoms (e.g. a residue in the case of proteins) [116, 118].

More seriously overpolarization occurs in condensed phases and is sometimes explained by the lower molecular polarizability in a dense environment. To avoid this effect some authors of FQ models increase hardness by a factor of 1.15 [108] or introduce a special function which makes the hardness "cross-dependent" on charge values (which is e.g. lower in the "gas" phase) [47].

A more physically relevant approach is an introduction in Eq. (5.26) of a distance dependent functions penalizing long-range charge transfer [119–121]. Unfortunately, available variants of such an approach give only qualitatively and not quantitatively accurate results [119].

At small distances interpenetration of diffuse electronic clouds changes the interaction of the nearby atoms. One aspect of this effect, as mentioned above, is the damping of the interaction coefficients J_{ab}. But the penetration effect renormalizes the electronegativity coefficient as well. As was shown by Itskowitz and Berkowitz in the frame of the derivation the electronegativity equilibration formalism from density functional theory the electronegativity coefficient has a form [122]:

$$\chi_a = \chi_a^\infty + \sum_{b \neq a} \int \left[v_b(\vec{r}_1) + \int \frac{\rho_b(\vec{r}_2)}{r_{12}} d\vec{r}_2 \right] f_a(\vec{r}_1) d\vec{r}_1, \qquad (5.29)$$

where $\vec{r}_1 = \vec{r} - \vec{r}_a$ and $\vec{r}_2 = \vec{r} - \vec{r}_b$ are radius vectors in frames with the center on atoms a and b respectively, $v_b(\vec{r}_1)$ is the electrical potential produced by the nucleus of atom b and the normalized function $f_a(\vec{r}_1)$ determines the structure of the electron cloud of atom a. For a neutral atom b at large interatomic distances the expression in brackets converges exponentially to zero. Yet, due to the penetration effect, for close contacts the second term in Eq. (5.29) can be comparable with the first one.

The authors of Ref. [122] demonstrated that simply accounting for the contribution of the second term allows them to obtain partial charges correct in sign and value in methane. Note that this contribution may be significant also for close nonbonded atom–atom contacts.

5.3.4 Alternative Approaches to Describe Polarization

Conformationally Dependent Charges. One of the clearest manifestations of polarization is the change in the electrostatic state of a molecule induced by a change of its conformation. The change contains contribution of two effects. First, for large enough molecules, the change of the conformation changes the mutual positions of distant atoms resulting in ordinary electronic polarization. This part can be described by one of the approaches considered above. The second contribution is related with geometric polarization and emerges when the relative positions of bonded atoms change. This effect has a complex many-body character and is

omitted by the approaches described earlier. Nevertheless it could be rather important, especially if simulated molecules are not taken to be rigid.

Without an accurate classical physical model of an intrinsically quantum effect,
the only way to include this is to construct a model where the parameters depend
directly on the molecule geometry. Several such models have been published by
Dinur and Hagler [61] and by Krimm's group [123, 124]. Universally, the atomic
charges are allowed to vary linearly with deformations of internal coordinates.
Charge transfer between any pair of bonded atoms depends on the length of the
bond as well as on the magnitude of the valence coordinates that surround that
bond:

$$dq_b = dq_b^0 + j_b(b - b_0) + \sum_{b'} j_{b'}(b' - b'_0) + \sum_{\theta} j_\theta(\theta - \theta_0) + \sum_{\tau} j_\tau(1 \pm \cos n\tau),$$

(5.30)

where b is the bond under discussion, θ is a valence angle that contain the bond, b'
is any bond connected to b, and τ is a torsion angle that contains b. b_0 and θ_0 are
reference values, and the sign in front of the cosine term depends on the periodicity n. dq_b^0, j_b, j_θ and j_τ are parameters that characterize the pair of atoms [61].

It was demonstrated [61] that for a wide set of alkanes, aldehydes, ketones, and
amides all terms besides the torsion term are significant for correct description
of the conformational change in the molecular dipole moment both in absolute
value and in direction. The torsion term was found to be sizable for amides only.
Accounting for dipole changes induced by the intra-molecular dynamics is principal for an accurate reproduction of infrared spectra of the molecules [123].

A model that can describe geometric polarization is also able to correctly
describe atomic intermolecular forces for non-rigid molecules. If the charge on
an atom depends on geometry, the atomic force should include an additional term
depending on the derivatives of the charge on the coordinates. It was shown [61]
that for this reason the quantum mechanically calculated atomic forces can be several times (up to six) bigger in amplitude in comparison with the forces predicted
by a model with constant charges, while the model that accounts for the geometric
polarization provides an accurate evaluation of the forces [61].

Available investigations with models accounting for the geometric polarization
imply that it can help to resolve at least two longstanding contradictions between
observations and predictions of FFs. Particularly a well known problem is that
all standard water models either with or without electronic polarization predict a
decrease of the water bending angle in water clusters and liquid phase, in direct
disagreement with experimental data [125] and the results of quantum calculations
[64, 126, 127]. It is notable that models that account for geometric polarization all
uniformly predict an increase of the bending angle and its temperature dependency
in perfect agreement with experiment [64, 124].

Proper parameterization of the φ (CNC$^\alpha$C) and ψ (NC$^\alpha$CN) torsion potentials
is critical to representing the correct conformations and flexibility of the proteins
properly. However standard FFs are not able to describe these torsions correctly
without ad hoc map-like corrections [128]. It was found [124] that inclusion of

conformationally dependent charges resolves this problem with a single 3-fold torsion term with a barrier about a few tenth of kcal per mole.

An interesting alternative to the explicit description of the geometric polarization, such as that given by Eq. (5.30), was proposed recently in Ref. [129]. It was shown that geometry dependent charges, dipoles and quadrupoles as well as the related part of the intramolecular electrostatic energy can be successfully predicted by appropriately chosen and trained neural networks.

More complex interplay of geometrical and usual electronic polarization occurs if the molecule can occupy different resonant structures [130, 131]. Such structures possess both different geometry and charge distribution while the contribution of each structure to the averaged state of the molecule depends on their relative energies. The latter in turn will depend on the surroundings of the molecule. Particularly, in water solution a more polarized structure would be more preferable while in the gas phase the more neutral structure would be favoured. As a result, the geometry and charge distribution of the molecule, e.g. an amide containing molecule, turn out to be dependent on the environment [130]. In this case the effects of the usual electronic polarization induced by an external electric field and of the geometric one, induced by conformational changes, should be considered self-consistently.

Mean-Field Polarizable Model. The success of modern non-polarizable FFs in many biophysical applications indicates that the effective mean-field description of the polarization effect could be rather accurate if the parameters of the electrostatic model are chosen properly. Yet, these fixed parameters are unable to reflect changes in a polar environment. This is important in non-uniform systems, e.g. proteins. A new mean-field paradigm developed recently [132] allowed to extend the approach to describe these changes.

The new model is based on a very basic physical observation that despite the inherent non-additivity of polarizable interactions, in the condensed phase, the effect, in fact, includes a significant additive component related to the electronic dielectric screening of pairwise Coulomb's interactions. The dielectric constant ε_∞ corresponding to the screening effect is experimentally measurable as the high-frequency or optic dielectric constant $\varepsilon_\infty = n^2$ (n is the refractive index) of a matter with a typical value of about 2 across different materials. Thus, in this model all charges are considered not to be in vacuum as in standard non-polarizable models but in a polarizable dielectric of ε_∞.

In this framework the effects of polarization are naturally partitioned [132, 133] into two components: those related to the generic electronic dielectric screening of electrostatic interactions and those related to additional adjustment of the molecular charge distribution to local polar environment. In the developed approach [132] the first component is treated implicitly by using *effective*, i.e. scaled by the electronic dielectric factor atomic charges [39, 133] $q_i^{eff} = q_i / \sqrt{\varepsilon_\infty}$. The remaining part of the polarization is modeled explicitly by adjusting the molecule *effective* dipole (or equivalent point charges q_i^{eff}) not to the instantaneous field, as it is usually done in standard polarizable models, but to the *effective* mean-field acting on the molecule.

The adaptive approach was exemplified in the Mean-Field Polarizable (MFP) water model [132]. The parameters were optimized using non-polarizable TIP3P potential as a reference for liquid water environment. It was shown that MFP water model is identical to its non-polarizable prototype in a polar environment, such as water, while it is able to readjust the effective parameters to match environments with different polarity: those inside a protein, at interfaces, or in any other environment different from the bulk water conditions [132]. The main advantage of the MFP model is that it is essentially as computationally efficient as the standard non-polarizable models. This makes it attractive for large scale biological simulations.

5.3.5 Physical (Mathematical) Truncation: Higher Rank Polarizabilities, Higher Order Polarizabilities, Isotropic Versus Anisotropic Polarizabilities, Point or Distributed Dipoles, Concerted Mechanisms of Polarization

Up to now we have discussed only dipole polarization and small range charge transfer that are very similar to dipoles in the far field. Both of these effects are just the first terms of the general decomposition of the molecular response to the external small amplitude electric field [7, 76, 134]. All discussed models and thus practically all modern FFs use some truncation of this decomposition. Is this choice physically grounded or just convinient mathematical approximation?

There are two very similar multipole decompositions of the structure of the molecular charge density. One is the decomposition of the unperturbed density and this is the usual multipole decomposition. The second one is the decomposition of the density perturbation induced by some external impact. In both cases each component of the decompositions has a specific structure (symmetry) type characterized by a specific angular dependency and a specific radial behavior at large distances from the center of the charge distribution. The higher is the rank of the multipole the faster potential produced by the corresponding component decreases with distance. The only overhead of induced density decomposition is a necessity to carry two sets of indices—the first one characterizing the structure of the perturbation moment and the second one characterizing the structure of the potential moment, which induce the decomposing density perturbation (or vise versa) [76, 135].

Change in the uniform potential does not cause change of the molecular structure. Thus the first nonzero term in the perturbation decomposition is the response to the first derivative of the potential (electric field). The total charge of an isolated molecule is constant, so the first term in the decomposition of the perturbation of the molecular charge density is of dipole character—this is the so called dipole-dipole polarizability. If a distributed (multi-center) polarizability is applied, the first non-vanishing terms will be dipole-charge and dipole–dipole ones, which describe the change in partial charges and dipoles of the centers in response to the external electric field. As they are the leading terms in the decomposition, they dominate in the far region. However this does not guarantee that other terms of the decomposition are negligible at distances characteristic for condensed phase applications of the FFs [76, 77, 135],

Unfortunately the question of significance of higher rank polarizabilities under conditions of realistic molecular environment has no definite answer up to now. It was shown that molecular polarizabilities up to rank 3 (octupole-octupole) are necessary to estimate electrostatic interaction energies with an accuracy higher than several tenths of kilocalories per mole even in the far region (the exchange-repulsion energy is less than 0.01 kcal/mol, at 3.5–5.0 Å closest atom-atom separation) [135]. However this does not prohibit a possibility of restricting the model to leading terms only when the multi-center polarizability is applied.

Several models which include higher rank multi-center polarizabilities have been presented, but they are not numerous and restrict the applications mainly to small molecule potentials [77, 136–138]. On the other hand, available general purpose polarizable FFs demonstrate reasonable accuracy in reproduction of a perturbed electrostatic potential when a potential was used for their parameterization [85, 87, 139, 140].

The question of the role of higher rank distributed multipoles of an unperturbed molecular charge density has been investigated much more thoroughly. It has been demonstrated over and over that including higher rank multipoles is crucial for the accurate reproduction of the molecular electrostatic potential and the corresponding component of the intermolecular interactions [9, 56, 57, 141–143]. In some respects, this inclusion has not only a simply quantitative but even a qualitative effect. Because of the difference in the angular behavior of different multipole components the models using only low rank multipoles (e.g. partial charges) are, in principle, unable to reproduce the geometry of some molecular clusters correctly. For example, only models with a partial quadrupole placed on the oxygen can correctly reproduce the structure of the optimal water dimer [40].

In analogy to restriction to dipole polarization only, the restriction of the model to linear polarization is just an arbitrary truncation of the corresponding series expansion of the molecular energy in an external electric field [7, 144]. Higher order polarizabilities (hyper-polarizabilities) are experimentally measurable values [145] and have been used in some water models to restrict molecular polarizability in the condensed phase [101]. However, such values are introduced through the formal (purely mathematical) decomposition (like the multipole decomposition discussed above) and lack any clear physical sense. This makes them difficult to parameterize and thus limits their transferability.

Dipole molecular polarizabilities are generally significantly anisotropic. However, it was shown in pioneering works of Applequist and Thole that anisotropy, for the most part, is a consequence of the molecular geometry and can be well reproduced by a model with isotropic partial atomic polarizabilities [89, 90]. Subsequent works addressed the cases, which are intrinsically aniso tropic, particularly aromatic [110, 111], halide [99] and hydrogen [146] molecules. To describe molecular polarizabilities in these cases one needs to take into account other mechanisms: charge transfer in the plane of aromatic molecules

[110, 111], anisotropic atomic polarizabilities [75, 100] or additional polarizable centers [49, 139].

Currently some FFs use one of the above mechanisms to account for the anisotropy of polarizability [49, 75, 100] but this is still not a well-established element of the FFs. Again, the main obstacle for modeling this effect well is the absence of clear physical models, inspired by the electronic state or electronic structure of an atom, that dictate when and how anisotropy should be included.

A sister topic is the anisotropy of exchange-repulsion [147]. This effect is related to the anisotropic shape of the atomic electronic cloud in a molecule and thus will vary with the variation of the electronic structure caused by polarization. An inclusion of coupling of exchange-repulsion and polarization is possible by a model directly that directly relates the repulsive force with the state of the electronic degrees of freedom [65, 75, 100, 102, 136, 147].

We have already mentioned several effects crucially affecting intermolecular interactions at small scales. Most significant among them are exchange-repulsion, penetration effect and screening of electronic interactions. The significance of these effects for adequate modeling of the condensed phase is well established and adopted by developers of modern FFs. However, for the most part, these effects are not accounted for in a self-consistent manner but rather by an introduction of different corrections or independent terms in the potential. For example, penetration is accounted for by re-parameterization of the exchange-repulsion [9], while screening is incorporated by damping functions, which are generally different and not consistent for different types of interactions [9, 48, 148]. More direct and consistent way to account for all these effects is explicit incorporation of diffuse electronic clouds into the FF.

While this idea is not new [149], up to now it has not became an intrinsic feature of modern FFs. For the most part, available experience with explicit accounting for the diffuse nature of electronic clouds jumps between oversimplifications, e.g. Gaussian clouds with charge about the partial charge of the atom [68, 106] and over-complications, e.g., description of the electrons based on density fitting using basis sets with many thousands of basis function *a la* their quantum mechanical (density functional theory) representation [143, 150]. According to published results exponential Slater orbitals of different rank representing symmetric S-core, dipole and quadrupole components of the electronic cloud seem a promising compromise between computational efficiency and physical reliability [65, 75, 143].

The different approaches to incorporate explicit electronic polarizability into a FF formalism discussed in the beginning of this section are not just different mathematical procedures; they also reflect different physical mechanisms of the polarization. It is reasonable to ask whether there are any physical arguments in favor of one of them or whether they are all equivalent. From a purely theoretical point of view, these mechanisms are not mutually exclusive but rather complimentary to each other [76, 77, 151]. Which specific polarization mechanisms exist and dominate depends on the particular system under consideration. In certain cases, like aromatic molecules, according to phenomenological arguments and results of the correspondent parameterization different mechanisms can act in concert [110, 111]. Promising results have been obtained for water by combining geometric polarization and induced dipole polarization, which are clearly complimentary [64]. On the other hand, attempts at inclusion of several polarization mechanisms in general purpose FFs did not result in clear arguments in favor of such an approach [152, 153].

Intermolecular charge transfer gives another example of a long debatable mechanism which complements an ordinary electronic polarization [7]. While some FFs include this effect as a component from the very beginning [54, 73] there are no clear arguments for its universal importance, nor is there a widely accepted method for its description. One of the possible reasons is that it is computationally difficult to estimate this effect accurately outside the quantum consideration of the system [54]. As a result, calculation of the corresponding FF terms becomes the most expensive part of calculation while the accuracy of the obtained estimate is questionable [54].

5.3.6 Parameterization Aspects

Because any FF, even one based on physically grounded energy component terms, is an approximation, adequate parameterization of the FF is of principle importance. Several recently developed thorough and systematic approaches to the parameterization have demonstrated that parameterization is at least as significant for the resultant FF quality as the functional form [88, 98, 154–159].

Formally speaking explicit inclusion of the polarization in a FF adds an additional problem to the usual task—that of choosing the parameters of the polarization model. Earliest developments of the polarizable FFs put the parameterization methodology exactly in this paradigm [107, 115, 160]. Most of the parameters were simply transferred from the non-polarizable prototype FF while the electrostatic part of the model was reparameterized with simultaneous parameterization of the polarization.

Subsequently it was recognized that the electrostatic part of a non-polarizable FF is not independent because of the inevitable partial compensation of errors between different energy components. Thus introduction of polarizability requires the reparameterization of the whole FF [67, 92, 159, 161, 162].

However, up to now it is not well recognized that the addition of the explicit electronic polarization opens promising novel possibilities for FF parameterization. Ideally the inclusion of explicit polarizability makes the FF transferable to any environment—for example to both gas and condensed phase conditions. This potential transferability makes it possible to parameterize such an FF exclusively on the basis of the results of high-level quantum mechanics calculations [9]. That provides not only an extensive and physically well-grounded foundation for design of the FF, but also an effective avenue for further FF improvement with advances in QM methodologies and increase of computer speed [75].

Moreover, available procedures of molecular interaction energy decomposition [10, 11, 102, 163–166] open a possibility to parameterize different FF components independently providing an additional base for the FF transferability. Up to now, the only general purpose FF following this concept of parameterization is the ab initio QM polarizable FF QMPFF [65, 75, 100, 102, 167], which was fitted exclusively to high-quality QM data and, in contrast to other modern general purpose polarizable FFs, neither uses experimental data nor applies artificial tuning of the parameters (as in Ref. [107]) or modification of the functional form (as in Ref. [80]) when transferring from the gas to the condensed phase.

As for the parameterization of the polarization model, the basic difficulty, as has been mentioned earlier, is the absence of a unique a priori valid way to distribute molecular polarizability between different centers [76, 77]. As a result, corresponding parameters are derived on the basis of some observable properties: molecular polarizability, change in the molecular dipole, change in the molecular electrostatic potential, induction component of the interaction energy, etc., as a result of solution of the corresponding inverse problem. Like the determination of partial atomic charges from the molecular electrostatic potential such inverse problems can be ill-conditioned without a unique well-defined solution. Particularly, it was shown that partial polarizabilities derived from molecular polarizability data and data on the perturbed electrostatic potential for the same set of molecules are different, while both describe molecular polarizabilities equally well [99].

Another similarity between atomic polarizabilities and atomic partial charges is the possible dependency on the molecular conformation. It has been shown that parameterization choices are critical in this respect. Like partial charges derived from the electric potential data in far zone, atomic polarizabilities derived from the data on molecular polarizabilities turn out to be dependent on conformation [99]. More comprehensive approaches are able to provide parameters valid for any conformation [67, 85, 87, 91, 99, 140].

Ideally, a polarizable FF should predict the energetics of molecular interactions both in gas and condensed phases as well as basic molecular properties like dipole and higher moments, polarizability etc. The majority of current polarizable FFs do not meet this requirement. It was found that the parameterization of a polarizable model to gas phase polarizability is able to reproduce related properties in that phase, but also results in over-polarization in the condensed one [47, 49, 67, 108]. To solve this problem, the polarizability in the condensed phase is typically scaled by an empirical factor ranging down to 0.6. The physical nature of such a scaling is yet to be justified.

For example, the value of the polarizability parameter in the SWM4-NDP water model [168] is about 30 % lower than the experimental gas phase value [169] ($\alpha_{exp} = 1.47$ Å^3), while most of the recent QM studies of bulk water polarizability [145, 170, 171] do not support such significant damping. Moreover, the experimentally known value of the high-frequency dielectric constant of liquid water, $\varepsilon_\infty = 1.78$, accurately reproduces the gas phase value via the Clausius-Massotti relation. The use of the SWM4-NDP polarizability results in a reduction of ε_∞ by about 20 % as compared to experiment.

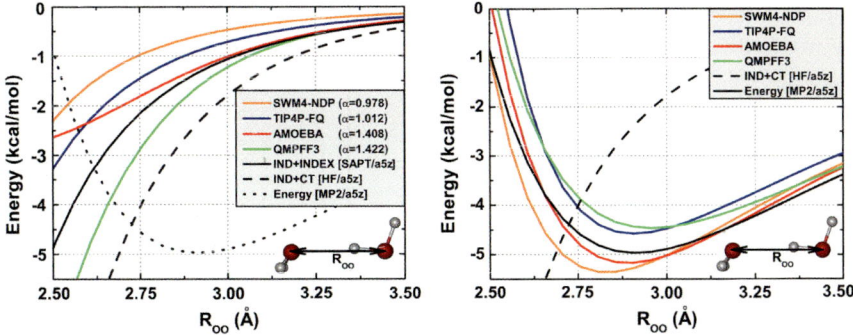

Fig. 5.6 Induction (polarization, *left*) and total interaction (*right*) energy in the water dimer at different distances in several polarizable water models: SWM4-NDP [168]—Drude oscillator, TIP4P-FQ [109]—Fluctuating charges, AMOEBA [141]—Induced Dipole with Thole damping, QMPFF3 [75]—Diffuse induced dipoles. On the legend the average water polarizabilities for each of the model are presented (MP2/a5Z value is 1.445 A^3). There is a clear correlation between the induction component of the energy and the corresponding polarizabilities and the use of damping. For reference we also present a pure SAPT calculation of induction (IND + INDEX components) at the HF/a5Z level and part of the HF/a5z energy which is responsible for induction and charge-transfer effects (obtained by full HF energy minus electrostatic and exchange-repulsion components). Despite a huge difference in representation of induction energies components all of these water models describe the condensed phase of water relatively well. This is due to a compensation of errors in different energy components, as can be seen from general agreement of all total energy dependencies. Total MP2/a5z energy is presented to get a proper scale

Moreover, it was shown that the scaling does not solve overpolarization problems satisfactorily in some cases [172–174].

On the other hand, use of arbitrary scaling makes purely QM parameterization insufficient and requires FF parameterization using some condensed phase properties [47, 49, 67, 108]. An important drawback of this approach is that the force field is no longer transferable and incorporates the error compensation (Fig. 5.6). This strongly contradicts the idea that the inclusion of polarization is required to obtain transferable force fields.

5.4 Examples of Polarizable Protein Force Fields

In this section, we give several examples of polarizable FFs, which were used or can be used for protein simulations. Again, we are not trying to describe all currently available FFs as most of them are well described in the literature including several reviews. Instead by these examples we would like to demonstrate weak and strong facets of modern polarizable FFs and their specific choices of functional form and parameterization methodology. The field is in permanent development and new FFs appear constantly. Some of them provide interesting ideas with respect to functional form and/or parameterization but are only parameterized for a small set of molecules [98, 156, 158]. It should be revealed by future investigations if such approaches result in a new generation of general purpose FFs.

5.4.1 OPLS/PFF

Since the late 1990s Friesner, Berne and co-workers have worked on creating a comprehensive force field for proteins [109, 175, 176]. The basic idea was to construct a FF primarily based on parameterization on results obtained by ab initio quantum chemistry. The first version used the fluctuating charge approach to incorporate electronic polarizability [115], and after trying fluctuating charges, induced dipoles and combined schemes, the authors concentrated exclusively on the induced dipole model [67, 80, 152, 153, 177].

The electrostatics of a system of molecules is represented by a collection of interacting bond-charge increments q_{ab} and dipoles $\vec{\mu}_a$:

$$U_{el} = \sum_a \left(\vec{\chi}_a \vec{\mu}_a + \frac{\mu_a^2}{2\alpha_a} \right) + \frac{1}{2} \sum_{ab \neq cd} q_{ab} J_{ab,cd} q_{cd} + \sum_{ab,c} q_{ab} \vec{S}_{ab,c} \vec{\mu}_a + \frac{1}{2} \sum_{a \neq b} \vec{\mu}_a \mathbf{T}_{ab} \vec{\mu}_b.$$
(5.31)

Here $J_{ab,cd}$, $\vec{S}_{ab,c}$, and $\mathbf{T}_{ab}$ are correspondent coupling factors. The bond-charge increments are FF parameters while dipoles are determined by minimizing the electrostatic energy in Eq. (5.31). Introduction of the linear term with the parameter $\vec{\chi}_a$ is a way to introduce a "permanent" nonzero dipole moment in an isolated molecule [67, 177].

One specific feature of the model is the introduction of massless virtual sites representing lone pairs on oxygen atoms with account to atom hybridization. Bond-charge increments are placed on all sites, but permanent dipoles are only induced on polarizable atoms.

Polarizability parameterization is based on a set of changes in the molecular electrostatic potential induced by a dipole probe with a dipole moment of 2.17 D placed at various locations and computed using the density-functional theory (DFT) with the B3LYP functional parameterization. The polarizabilities α_a are assumed to be isotropic and are chosen to minimize the mean-square deviation between the change in the ESP as given by the model and by the DFT calculations. This approach provides stable results insensitive to the exact form of the perturbations, i.e. the magnitude or position of the probe charges [67].

As in the most modern polarizable FFs authors of OPLS/PFF decreased the gas phase polarizability to avoid overpolarization in the condensed phase. Yet the method is unique and nontrivial. The authors assumed that the electronic density described by diffuse functions does not contribute to the molecular polarization in condensed phase because Pauli repulsion from neighboring molecules raises the energies of diffuse functions and so diminishes their contribution to the polarization. To account for this effect the polarization parameterization is performed on DFT results obtained with the cc-pVTZ (-f) basis set [67, 177].

Parameters of the permanent electrostatic model, $\vec{\chi}_a$ and q_{ab}, are fitted to best reproduce results of DFT calculations of the electrostatic potential of the charge distribution of the unperturbed target molecule. The vectors $\vec{\chi}_a$ are expressed as a sum of vector parameters pointing along bonds connecting adjacent atoms, and as such, will change during the course of a simulation as a flexible molecule changes conformation.

To avoid instabilities in solution of the inverse problems related to parameterization special techniques have been developed [115].

Short-range van der Waals parameters are determined by fitting the binding energies of gas phase molecular dimers, enabling the development of these parameters for a wide range of atom types. To enable independent variation of attractive dispersion and repulsive exchange terms an exponential term was added to the standard 6–12 function:

$$U_{nb,a} = \sum_{b \neq a} \left[A_{ab}/r_{ab}^{12} - B_{ab}/r_{ab}^{6} + C_{ab}\exp(r_{ab}/\alpha_{ab}) \right]. \tag{5.32}$$

Here the first term provides positive values for the interaction energy for small distances, while the exponential term is parameterized using quantum mechanical interaction energies and for the most part it is responsible for the repulsive part at optimal interatomic distances [177].

The dispersion term is parameterized to reproduce by MD simulation the thermodynamic properties of liquids and is considered as an atomic rather than FF atom type characteristic. Finally, stretching and bending terms are retained from the OPLS-AA fixed charge force field [25–27], whereas torsional parameters are fitted to reproduce high-level quantum chemical data for conformational energy differences [27]. To avoid problems with short-range interactions 1–2 and 1–3 interactions are dismissed and 1–4 interactions are scaled.

Thorough parameterization procedure and an elaborate functional form resulted in promising results of the FF assessment and testing. OPLS/PFF was successfully applied to calculations of dimer energies in vacuum [178], peptide minimization [177], MD simulation of protein (bovine pancreatic trypsin inhibitor) dynamics in water [179], calculation of pKa values of a number of solutes [180] and residues in turkey ovomucoid third domain protein [181], calculation of hydration free energies for a set of small molecules. Polarizable FF has demonstrated significantly better ability in pKa and hydration energy predictions. For example while the non-polarizable FF yields errors of about 5 units in the absolute pKa values for phenols and methanol, the polarizable FF produces the acidity constant values within a 0.8 unit accuracy [180]. The polarizable FF average error in absolute free energies of hydration calculations was only about 0.13 kcal/mol [85]. Crucial role for explicit account for the polarization has been demonstrated with OPLS/PFF in simulations of monovalent Cu(I) interaction energies and hydration [182, 183]. It was shown that while the fixed-charge model leads to a relatively small error in the gas-phase interaction between Cu(I) and water, the solvation free energy in bulk water is underestimated by about 22 %. At the same time, the performance of the polarizable model was uniformly adequate [182]. The same polarizable model without any additional fitting was found to be successful in reproduction of the Cu(I) binding energy to copper chaperone from Bacillus subtilis [183].

The recent development of the FF is devoted to improvement of numerical efficiency by employing a non-iterative approach of calculating induced dipoles. This resulted in the development of POSSIM FF [84–87]. It was found that simulations of gas-phase dimers, quantum mechanical electrostatic three-body energies, pure liquids, solutions and peptides have given no indication that the used second-order approximation leads to any deficient physical results, and authors have always been able to produce fitting to quantum mechanical and experimental data which was as good as for the full-scale polarization [86]. Recently POSSIM FF has been extended to include parameters for alanine peptides and protein backbones [182] and a set of small molecules serving as models for peptide and protein side-chains [85]. One notable accomplishment is the confirmed transferability of the electrostatics between the POSSIM NMA model and the protein backbone, with accurate reproduction of the interaction energy and geometry of the alanine dipeptide complex with a water molecule, as well as the alanine di- and tetra-peptide conformational energies [182].

5.4.2 CHARMM Drude-2013

All three aforementioned approaches to electronic polarizability have been implemented as extensions of the FF CHARMM and reviewed in Ref. [51]. An implementation of the induced dipole method in CHARMM based on the polarizable intermolecular potential functions (PIPF) model of Gao et al. [184] has been reported in Ref. [71]. The polarizable fluctuating charge model in CHARMM results from the work of Patel, Brooks and co-workers [108, 185]. The water model is based on the TIP4PFQ model of Rick, Stuart and Berne [109]. The polarizable Drude model in CHARMM results from the work of MacKerell, Roux and co-workers [168, 186] and it was geared at developing polarizable force fields for biological macromolecules [51]. We shall describe this version of the polarizable CHARMM FF in more detail.

Development of the Drude polarizable force field in CHARMM has been ongoing since the early 2000s [162, 187, 188]. Current available parameters include series of alkanes [189], alcohols [190], ethers [191, 192], aromatics [193, 194], amides [195], sulfur-containing compounds [196] and ions [197, 198]. The collection of model compounds covers functional groups in proteins, nucleic acids [199], lipids [200, 201], and carbohydrates [202]. A significant effort is now being made in extending the polarizable force field from small molecules to biologically relevant macromolecular systems. Most recently, a refined version of the force field for peptides and proteins, called Drude-2013, has been completed [188].

Optimized parameter sets for DPPC lipids [203] and acyclic polyalcohols [202] have also been released.

In general the polarizable FF uses the functional form of the prototype CHARMM FF with addition of Drude particles on non-hydrogen atoms to allow explicit polarizability. Standard exclusion of 1–2 and 1–3 interactions is applied while nearby induced dipole-dipole interactions are allowed with Thole-like screening with atomically based screening parameters.

It was found that the introduction of explicit polarization into an empirical force field requires a complete re-optimization of all parameters of both bonded and non-bonded interactions to attain a fully consistent polarizable force field in which the different energy components are properly balanced [162, 187, 188]. In the parameterization procedure, parameters from previously developed CHARMM FF versions were used as initial trials and modified when necessary. The parameterization strategy, however, essentially was kept the same as previously used for the non-polarizable FF. In this respect the approach to deal with the conformational dependency of the electrostatic model should be pointed out. In CHARMM different conformations are fitted independently and the resulted model is obtained by averaging the charges obtained for each conformation.

Generalization of the FF from small molecules to macromolecules encountered transferability problems. While parameters obtained for heterocycles were well transferable to nucleic acid bases [199], parameters for N-methylacetamide (NMA) were found not to be transferable to the protein backbone [49]. NMA parameters resulted in the overpolarization of the backbone, overestimation of the peptide dipole moments and incorrect relative population of different protein conformations. Hence, an additional parameter optimization using larger model compounds and a large body of target data was necessary. The development of the refined force field focused on the parameter optimization for the polypeptide backbone and the connectivity between the backbone and side chains.

Five conformations of the alanine dipeptide were selected covering different relative orientations of the peptide bonds (C7eq, C5, PPII, αR, and αL) and subjected to RESP fitting yielding five electrostatic models, which were then averaged. For each conformation, the partial atomic charges, atomic polarizabilities, and atom-based Thole factors were fitted to QM ESP potential. To achieve correct conformational energies in aqueous solution and balance different electrostatic contributions the parameters were also optimized for reproduction of gas phase conformational energies of acetyl-(Ala)$_5$-amide and its interactions with water. The QM polarizabilities were scaled by 0.85 in accord with the scaling used previously in the development of the CHARMM Drude parameters for small compounds [194, 204]. Then, a parameter adjustment was made to reproduce in MD simulations the experimental NMR J-coupling data for the (Ala)$_5$ polypeptide in solution. The final model tuning involved an additional adjustment to the relative potential energies of different regions of the φ, ψ diagram for crambin, lysozyme, and the GB1 hairpin peptide.

Besides the empirical re-parameterization, rescaling of polarizabilities and Drude screening, additional means to restrict polarization of the protein backbone were introduced in the model. The first involves a "hyperpolarization" term where higher order anharmonic terms are added to the bond between the atomic core and the Drude particle [49].

$$E_{hyp} = \sum K_{hyp}(R - R_0)^n, \qquad (5.33)$$

where R is the distance between the nucleus and the Drude particle, n is the order of the term, typically 4 or 6, K_{hyp} is the force constant, and R_0 defines the distance at which the term starts to impact the Drude particle, typically 0.2 Å, such that the normal trajectory of the Drude is not impacted by the higher order term. More recently, a Drude reflective "hard wall" term has been added to more rigorously avoid a polarization catastrophe [49].

The Drude-2013 force field has been used in simulations of several peptides and full proteins, and shown to maintain the stable folded state of the studied systems on the 100 ns time scale in explicit solvent MD simulations. However, the FF results in larger deviation of the predicted protein structures [188] in comparison with the non-polarizable CHARMM36 model [32]. Thus the authors concluded on the necessity of further improvements in the model.

5.4.3 QMPFF

Several versions of general purpose polarizable FF have been developed in Algodign, LLC. The principle basic idea of the FF development was to make a physically based highly transferable FF, which will be able to reproduce properties of all phases and all environments while being parameterized exclusively on high quality quantum mechanical data. This fact was reflected in the FF name— QMPFF—ab initio quantum mechanics polarizable FF.

The first version, QMPFF1 was a proof of the concept and described properties and interactions of small molecules considered as rigid [102]. The basic FF structure consisted of four non-bonded interaction components: electrostatic (ES), exchange-repulsion (EX), induction (IN), and dispersion (DS). The functional form of the components imitates that of their QM counterparts. Specifically, electrostatics represents the classical Coulomb interaction of point charges and exponential "electronic clouds" by allowing the latter to shift off-center to represent permanent and induced dipoles. Exchange-repulsion is a result of cloud–cloud interactions that decay almost exponentially with distance. Induction is simulated by a "spring" attaching each mobile electron cloud to a reference position. The restraint potential provided by the spring is close to harmonic at small distances, but the stiffness becomes infinite as the distance approaches a limiting value, thus preventing the "polarization catastrophe" (Eqs. 5.15–5.21). Dispersion is represented by a term decaying as r^{-6} at large distances with Tang-Toennies damping at small distances [102]. (See the original papers for details of the functional form.)

The training set for the FF parameterization includes molecular properties (components of the polarizability tensor, dipole, and quadrupole moments) for all representative molecules as well as components of intermolecular energy of representative dimers with proper energy decomposition (allowing to parameterize different components of the potential independently and thus to avoid the error compensation).

Parameterization of QMPFF1 achieved for the separate components in the energy decomposition scheme the rms errors 0.21, 0.17, and 0.27 kcal/mol for ES, EX, and DS, respectively. The correlation coefficients between the QMPFF and QMdata were 0.97, 0.98, and 0.87 for ES, EX, and DS, respectively. Good transferability of the model was demonstrated on large test set of dimer energies not included in the training set, particularly on the reproduction of the variation of the dimer energy with distance between the monomers. Additionally, the model provided a perfect estimate of non-additive effects in different multimers, dimerization Gibbs energies and radial distribution functions of some liquids [102].

In the second version—QMPFF2—the bonded interactions are conventionally subdivided into stretching, bending, and torsion terms. The first two terms use a quadratic function, whereas the torsion term uses a threefold cosine function. Also, more atom types are used in QMPFF2 to allow better resolution of related, but electronically distinct, atom types, and to cover other biologically important elements not treated by QMPFF1. The functional form of non-bonded terms was slightly modified from QMPFF1 by including the precise form of exchange-repulsion between electronic clouds; more accurate description of the dispersion term by addition of a r^{-8} component; and allowing the induction to be anisotropic [65].

QMPFF2 parameters were fitted to QM data on properties of 144 molecules and 79 molecular dimers, the total number of dimer conformations being 2,916. The relative rms deviations (RMSD) for absolute values of dipole moments and quadrupole and polarizability tensors were 6.2, 23 and 3.7 %, respectively. QMPFF2 fits the QM energies of all of the dimers in the training set well with a RMSD of 0.38 kcal/mol. For the 165 training dimers that have been optimized to have minimum energy, the RMSD of the energy and geometry calculated by QMPFF2 relative to ab initio QM were 0.50 kcal/mol and 0.09 Å. QMPFF2 also reasonably predicts dimerization Gibbs energies: rms deviation between calculated and experimental values was 0.35 kcal/mol for 27 different homodimers and heterodimers, consistent with the overall QMPFF2 accuracy [65].

A high transferability of QMPFF provided by its physically grounded functional form and parameterization technique was demonstrated by reproduction of non-additive energies of water multimers and thermodynamic properties of bulk water. Agreement of QM and QMPFF2 energies of water multimers including even correct ranking of the hexamer energies, which is a challenging test because of the role of many-body effects, was excellent considering that no systems larger than the dimer were used in the parameterization of QMPFF2.

QMPFF2 with classical molecular dynamics reproduced the density of water over the entire investigated range −25–100 °C to within 0.6 % (better than provided by the TIP5P model specifically parameterized to reproduce this dependency) including reproduction of the maximum at the temperature curve (8 °C) close to the experimental value (4 °C). Liquid binding energy, diffusion coefficient and radial distribution functions were also well reproduced especially after accounting for quantum effects by means of path integral molecular dynamics [65].

The quality of QMPFF2 predictions of water properties is overall at least as good as the best specially designed water models. The most exciting feature of this result is that the model is based strictly on ab initio-calculated QM data, unlike empirical water potentials, which are fitted to experimental data and therefore cannot be said to be *predicting* the properties of water. This conclusion indicates that careful, physically based simulation of intermolecular interaction in vacuum, which takes into account its main features while avoiding oversimplification, does accurately reproduce the properties of the condensed phase, and even subtle properties of water such as the anomalous density-temperature dependence.

In the final version of the QMPFF3 torsional parameterization was incorporated into the overall scheme of non-bonded parameterization providing much better accuracy for the torsion potential. That allowed this version of the FF to perform successful simulations of proteins including calculations of relative binding affinity directed to pro-drug optimization [100].

Specific features of QMPFF3 also include change in the description of permanent electrostatics and polarization—permanent and induced dipoles now are described by the direct inclusion of a p-type component in the electronic density. This modification introduces the one-center atom model instead of the earlier two-center model, resulting in a more economical numerical scheme in multi-particle applications (e.g. molecular dynamics). Additionally atom based permanent diffuse quadrupole components were introduced for better representation of molecular quadrupole moments and the electrostatic potential [75]. Bonded potentials in QMPFF3 have been modified to account for anharmonicity of internal degrees of freedom similarly to the MMFF94 force field [205].

Refinement in the parameterization procedure of QMPFF3 included the use of QM data on representative multimers in their stationary conformations. The correspondent training set contains the total intermolecular energy and its non-additive component calculated as a difference between the total oligomer energy and the sum of energies of all the dimers taken in the same geometry as in the oligomer. Mainly, these data increase the reliability of the parametrization of the IN component, especially its anharmonicity. Also the decomposition scheme was modified slightly to account better the difference between MP2 and HF monomer densities [100].

Finally, MP2 results traditionally used in QMPFF parameterization cannot be considered as reliable data for the dispersion component of intermolecular interactions of aromatic compounds. For this reason, QMPFF3 parameterization utilized a two-step procedure in this special case. At the first (preliminary) step, the standard QMPFF parametrization is performed using the MP2-based training set to find a preliminary parameter set. Then, in the second step, the parameters of the dispersion component are refined to fit the most accurate, CCSD(T), QM data found in the current literature on the total energy of benzene dimer in stationary conformations. It was found that a reasonable agreement with CCSD(T) calculations can be attained by varying only one force constant of the r^{-6} term for carbon atom type, whereas all other parameters were kept fixed [100]. According to the general QMPFF philosophy, all parameterization was based exclusively on high level QM data without any empirical fitting.

The total number of QM values used for parameterization of non-covalent interactions for the atom/bond types was more than 20,000 and for valence potentials about 80,000. These numbers demonstrate the abilities of modern QM methods in the construction of accurate, diverse and representative training sets.

QMPFF3 deviations for molecular polarizability, dipole and quadrupole moments from the experimental data are mainly due to the imperfection of QM data used for the parameterization [75]. QMPFF3 provides a rather accurate fit of QM data for dimers, generally within a few percent. This agreement corresponds to absolute RMSD values of 0.1 Å for the optimal distance, 0.39 and 2.8 kcal/mol for dimerization energies of neutral and charged dimers. The latter value is considered as a good accuracy considering that the charged dimers are bonded by an order of magnitude more strongly than the neutral ones. As for multimers, the absolute RMSD turned out to be about 1.2 kcal/mol for the total energy and 0.52 kcal/mol for the non-additive energy. As for the quality of parameterization of bonded interactions, it is characterized by RMSD values of 0.006 Å for bond lengths, 1.28° for bond angles, and 0.35 kcal/mol for torsion angles [75].

To validate the modifications made to the DS component for aromatics, benzene second virial coefficient, benzene and naphthalene thermodynamic properties of the liquid phase, and cohesion energies and specific volumes for 15 PAH crystals have been successfully evaluated [100]. The overall QMPFF concept was also confirmed by successful simulations of graphite and fullerene crystals and their interactions with PAH compounds and graphite [206]. QMPFF3 also provided an accurate description of bulk graphite and solid C_{60} properties. In all the studied systems the electrostatics due to the penetration effect was found to be important and comparable in magnitude with the total interaction energy [206]. QMPFF3 predicted the graphite exfoliation energy of 55 meV/atom in agreement with the relatively large experimental value of 52 ± 5 meV/atom recently suggested by Zacharia et al. [207].

The general QMPFF3 model was thoroughly validated in gas, liquid, and solid phases [75]. QMPFF3 transferability in the gas phase was illustrated by an accurate reproduction of the dimerization Gibbs energies (actually second virial coefficients) for pure vapors and simple gas-phase optimization of protein geometry. For liquid phase validation MD simulations have been performed for almost 60 test liquids, in which the densities and enthalpies of vaporization were evaluated. Validation in the solid-phase was performed by comparing QMPFF3 predictions with experimental cohesion energies and geometry characteristics of the unit cell such as volume and lengths and orientations of the lattice vectors. Characteristics of 78 molecular crystals have been calculated.

In all cases very good transferability was demonstrated. In particular, this was manifested by the fact that QMPFF3 provided an accuracy in liquid and solid phases comparable with that for the FF explicitly fitted to data on these phases.

Thus, for the first time, it was demonstrated that a physically well-based general purpose FF fitted exclusively to a comprehensive set of high level vacuum quantum mechanical data, when applied without any tunning to simulation of condensed phase, provided high transferability for a wide range of chemical compounds [75].

Finally transferability and quality of QMPFF3 was exemplified by the application to calculations of relative binding affinities of ligands to proteins. Five ligands, differing by replacement of an atom or functional group, in complexes with three serine proteases—trypsin, thrombin, and urokinase-type plasminogen activator—with available experimental binding data were used as test systems. The calculated results were found in excellent quantitative (rmsd $= 1.0$ kcal/mol) and qualitative ($R^2 = 0.90$) agreement with experimental data (15 experimental relative affinities). The potential of the methodology to explain the observed differences in the ligand affinities was also demonstrated [167].

5.4.4 AMOEBA 2013

The development of AMOEBA (Atomic Multipole Optimized Energetics for Biomolecular Applications) FF commenced in the beginning of this century (by Ren and Ponder [91, 141]) with the goal of substantially improving the electrostatic model. First, AMOEBA uses multipoles up to quadrupole, rather than point charges only, to provide accurate description of permanent electrostatics. Second, an induced dipole polarization scheme was incorporated with a balanced description of both intra- and intermolecular polarizability, damped by Thole scaling to avoid the polarization catastrophe. These two features allowed AMOEBA's electrostatic model to accurately reproduce the conformational dependence of the molecular electrostatic potential in different environments [91, 140].

Other specific features of the AMOEBA potential are: nonzero van der Waals parameters and reduction factors on hydrogen atoms which produce an improved molecular "shape", buffered 14–7 potential to model pairwise additive vdW interactions, anharmonic functional forms for bond stretching and angle bending from the MM3 force field, Wilson-Decius-Cross function for restraining the out-of-plane bending, and Bell torsion for dihedral angles involving two joined trigonal centers. All induced dipoles interact with each other, while permanent multipoles only polarize atoms outside their polarization group [74, 140].

A parameterization methodology that has been evolving throughout the whole period of the AMOEBA development finally includes the following critical aspects. First, a trial set of partial charges, dipoles and quadrupoles is obtained from distributed multipole analysis (DMA) by Stone in its initial form using ab initio calculations at the MP2/6-311G** level. Then partial charges are fixed, while dipoles and quadrupoles are optimized against MP2/aug-cc-pVTZ electrostatic potential values computed on a grid of points around each model compound at different conformations. The ESP for all conformations is fitted simultaneously with the aim of deriving a parameter set conformationally independent as much as possible. Some additional conformations not used in the training set are applied as test cases to check the applicability of fitted parameters [91, 140].

The AMOEBA uses atomic polarizabilities derived by Thole for all atoms except for aromatic carbon and hydrogen atoms, which have been systematically refined using a series of aromatic systems, including a small carbon nanotube [140]. Reducing the damping factor from Thole's original value of 0.572 to AMOEBA's 0.39 was found to be critical for correctly reproducing water cluster energetics [91, 92]. On the other hand, AMOEBA's stronger damping leads to a slight systematic underestimation of molecular polarizabilities [93].

The vdW parameters are optimized to gas phase cluster structures and energetics as well as to condensed phase properties. A critical strategy in deriving the vdW parameters, due to their empirical nature, is to ensure chemical consistency among different elements. This is achieved by simultaneously parametrizing multiple compounds sharing the same vdW "classes" to improve transferability [93].

The valence parameters (bond stretching, angle bending, bond angle cross terms, out of plane bending) are transferred from organic small molecules, where they were derived by matching the QM geometries and vibrational frequencies [140].

In the course of torsion parameterization the scaling factors for the intramolecular electrostatic and vdW interactions are chosen to minimize the contribution of the explicit torsional terms and ensure maximal transferability of parameters between dipeptides and tetrapeptides. The alanine dipeptide is used to parametrize backbone torsional terms for all amino acids, with the exception of glycine and proline. The difference between AMOEBA and MP2/CBS energy is taken as the fitting target for the torsional parameters, using a standard three-term Fourier expansion. The backbone torsion parameters are further improved by comparing the AMOEBA PMF in solution to a statistically derived alanine backbone potential of mean force derived from the PDB database. For the side-chain torsions for all other residues, the parameters are obtained by fitting to the MP2/CBS conformational energy of the corresponding dipeptides [140].

Earlier versions of the AMOEBA protein force field, like many other protein FFs, included a torsion-torsion coupling term implemented via a cMAP style two dimensional bicubic spline [74]. This term is essentially a grid-based correction to match the force field $\phi - \psi$ conformational energies to those of a target QM based potential surface. In the last AMOEBA 2013 version, due to more thoroughly parameterized electrostatic model, the "traditional" three term Fourier expansion function is used for all torsion angles, except for the backbone of glycine [140].

The AMOEBA FF has been successfully applied to modeling water [208], mono- and divalent ion solvation [209, 210], organic molecules [93] and peptides [92, 140, 211, 212], small-molecule hydration free energies [213], trypsin–ligand binding prediction [214], and computational X-ray crystallography [215] with promising results. The results obtained suggest the AMOEBA force field performs well across different environments and phases.

As a validation of predicted properties of small organic molecules, the hydrogen bonding energies and structures of gas phase heterodimers with water have been evaluated. For 32 homo- and heterodimers, the association energy agrees with ab initio results to within 0.4 kcal/mol. The RMS deviation of hydrogen bond distance from QM optimized geometry was less than 0.06 Å. In addition, liquid self-diffusion and static dielectric constants computed from molecular dynamics simulation were consistent with experimental values. The FF was also used to compute the solvation free energy of 27 compounds not included in the parameterization process, with a RMS error of 0.69 kcal/mol [92].

Molecular dynamics simulations with AMOEBA FF were performed with double decoupling of benzamidine from both water and the trypsin binding site with free energy perturbation. The computed absolute binding free energy was well within experimental accuracy. It was found critical to treat polarization explicitly to achieve chemical accuracy in predicting the binding affinity of charged systems. On mutation from benzamidine to diazamidine, the binding weakens by 1.21 kcal/mol, consistent with 1.59 kcal/mol obtained by experiment [214].

The protein version of AMOEBA FF was validated in reproduction of dipole moment components and the electrostatic potential of dipeptide model compound for each amino acid. The dipole moment components of all the dipeptides are accurately reproduced, regardless of the conformation and the residue type, resulting in a correlation coefficient of 0.998. The average RMSE between the ab initio and AMOEBA electrostatic potential is 0.45 kcal/mol per unit charge on a grid surrounding the neutral amino acids, and only slightly higher (0.64) for charged dipeptides, with the absolute value of the potential for the latter being orders of magnitude higher. Thus, thanks to the intramolecular polarization model, the transferability of backbone and side chain electrostatic multipoles of AMOEBA is quite satisfactory [140].

The solvation of the unblocked and protonated $(Ala)_5$ peptide has been examined. The chi square (χ^2) difference between simulation and experiment spin–spin coupling (J coupling) constants, computed using the experimental uncertainties, was about 0.994, while the overall RMS difference is 0.33 [140].

Ten well studied proteins were chosen as the validation set to evaluate the AMOEBA 2013 parameters. The stability of each protein was characterized by its backbone RMSD value relative to the PDB structure over 10 ns MD simulation. The overall average RMSD of the ten simulated protein structures is 1.33 Å, and seven of them are close to 1.0 Å [140]. Impressive results for AMOEBA FF have been obtained in reproduction of experimentally observed structures of four β-hairpin peptides depending on degree of methylation of structure stabilizing lysine group [57, 216].

5.5 Conclusions

There are several compelling reasons for inclusion of polarization into molecular models. Clearly, a more detailed model can potentially be more precise in predicting the desired results and behaviors: liquid properties, free energy of binding, and many others. More importantly, polarization is the main physical mechanism describing the difference in atomic and molecular behavior in transition between the gas and condensed phase. Therefore a proper polarizable FF may be parameterized in the gas phase—namely by ab initio QM methods—and successfully transferred into use in the condensed phase, where it is most useful.

Though this sentiment is widely accepted in the field, more than 10 years of active development have not produced a polarizable FF that is clearly superior to non-polarizable ones. In our opinion future progress depends on three major points.

First, when parameterizing a polarizable FF, one needs to consider the specifics of the model and tailor the fitted data, conformations and properties accordingly. Second, one needs to use QM data for parametrization, and to trust it to a greater extent than is currently done. When a desired prediction is incorrect, adjusting an ensemble of parameters to get a better value will more often than not ruin transferability and cause a multitude of other answers to be qualitatively wrong. Finally, and most importantly, the model itself should represent the underlying physics as much as is computationally feasible. This last point will bring about good transferability, good precision, and wide coverage of phase and chemical space. The exact description of 'physically correct' will be the main undertaking of molecular modeling in the next decade.

References

1. Levitt M (2001) Nat Struct Mol Biol 8:392
2. Nygaard R, Zou Y, Dror RO, Mildorf TJ, Arlow DH, Manglik A, Pan AC, Liu CW, Fung JJ, Bokoch MP, Thian FS, Kobilka TS, Shaw DE, Mueller L, Prosser RS, Kobilka BK (2013) Cell 152:532
3. Kruse AC, Hu J, Pan AC, Arlow DH, Rosenbaum DM, Rosemond E, Green HF, Liu T, Chae PS, Dror RO, Shaw DE, Weis WI, Wess J, Kobilka BK (2012) Nature 482:552
4. Stouch TR (2012) J Comput Aided Mol Des 26:1
5. Karplus M, McCammon JA (2002) Nat Struct Biol 9:646
6. Friesner RA (2005) Proc Natl Acad Sci USA 102:6648
7. Stone A (2013) The theory of intermolecular forces, 2nd edn. Oxford University Press, Oxford
8. Stone AJ (1990) Studies in physical and theoretical chemistry. In: Rivail J-L (ed) Modeling of molecular structures and properties, vol 71. Elsevier Science, p 27
9. Stone AJ, Misquitta AJ (2007) Int Rev Phys Chem 26:193
10. Kitaura K, Morokuma K (1976) Int J Quant Chem 10:325
11. Jeziorski B, Moszynski R, Szalewicz K (1994) Chem Rev 94:1887
12. Lorentz HA (1880) Ann Phys 9:641
13. Lorenz LV (1880) Ann Phys 11:70
14. Drude P, Mann CR, Millikan RA (1902) The theory of optics. Longmans, Green and Co., New York
15. Jacucci G, McDonald IR, Singer K (1974) Phys Lett A 50:141
16. Warshel A, Levitt M (1976) J Mol Biol 103:227
17. Vesely FJ (1977) J Comput Phys 24:361
18. Warshel A, Russell ST, Churg AK (1984) Proc Natl Acad Sci USA 81:4785
19. Stillinger FH, David CW (1978) J Chem Phys 69:1473
20. Barnes P, Finney JL, Nicholas JD, Quinn JE (1979) Nature 282:459
21. Guillot B (2002) J Mol Liq 101:219
22. Cornell W, Cieplak P, Bayly C, Gould I, Merz K, Ferguson D, Spellmeyer D, Fox T, Caldwell J, Kollman P (1995) J Am Chem Soc 117:5179

23. Salomon-Ferrer R, Case DA, Walker RC (2013) Wiley Interdiscip Rev: Comput Mol Sci 3:198
24. Wang J, Wolf RM, Caldwell JW, Kollman PA, Case DA (2004) J Comput Chem 25:1157
25. Jorgensen WL, Tirado-Rives J (1988) J Am Chem Soc 110:1657
26. Jorgensen WL, Maxwell DS, Tirado-Rives J (1996) J Am Chem Soc 118:11225
27. Kaminski GA, Friesner RA, Tirado-Rives J, Jorgensen WL (2001) J Phys Chem B 105:6474
28. van Gunsteren WF, Billeter SR, Eising AA, Hunenberger PH, Kruger P, Mark AE, Scott WRP, Tironi IG (1996) Biomolecular simulation: the GROMOS96 manual and user guide. Vdf Hochschulverlag AG an der ETH Zurich, Groningen
29. Oostenbrink C, Villa A, Mark AE, Van Gunsteren WF (2004) J Comput Chem 25:1656
30. MacKerell AD, Bashford D, Bellott M, Dunbrack R, Evanseck J, Field M, Fischer S, Gao J, Guo H, Ha S, Joseph-McCarthy D, Kuchnir L, Kuczera K, Lau FTK, Mattos C, Michnick S, Ngo T, Nguyen DT, Prodhom B, Reiher WE III, Roux B, Schlenkrich M, Smith JC, Stote R, Straub J, Watanabe M, Wiorkiewicz-Kuczera J, Yin D, Karplus M (1998) J Phys Chem B 102:3586
31. MacKerell AD (2004) J Comput Chem 25:1584
32. Best RB, Zhu X, Shim J, Lopes PEM, Mittal J, Feig M, MacKerell AD Jr (2012) J Chem Theory Comput 8:3257
33. Piana S, Lindorff-Larsen K, Shaw DE (2011) Biophys J 100:L47
34. Jensen MO, Jogini V, Eastwood MP, Shaw DE (2013) J Gen Physiol 141:619
35. Jorgensen WL (2007) J Chem Theory Comput 3:1877
36. Jungwirth P, Tobias DJ (2006) Chem Rev 106:1259
37. Wu Y, Yang Z-Z (2004) J Phys Chem A 108:7563
38. Lynden-Bell RM (2010) Phys Chem Chem Phys 12:1733
39. Leontyev IV, Stuchebrukhov AA (2011) Phys Chem Chem Phys 13:2613
40. Halgren TA, Damm W (2001) Curr Opin Struct Biol 11:236
41. Demerdash O, Yap EH, Head-Gordon T (2014) Annu Rev Phys Chem 65:149
42. Rick SW, Stuart SJ (2002). Reviews in computational chemistry. In: Lipkowitz KB, Boyd DB (eds). Wiley, New York, p 89
43. Ponder JW, Case DA (2003) Adv Protein Chem 66:27
44. Warshel A, Sharma PK, Kato M, Parson WW (2006) Biochim Biophys Acta 1764:1647
45. Warshel A, Kato M, Pisliakov AV (2007) J Chem Theory Comput 3:2034
46. Patel S, Brooks CL (2006) Mol Simul 32:231
47. Bauer B, Patel S (2012) Theory Chem Acc 131:1
48. Yu HB, van Gunsteren WF (2005) Comput Phys Commun 172:69
49. Lopes PEM, Huang J, Shim J, Luo Y, Li H, Roux B, MacKerell AD (2013) J Chem Theory Comput 9:5430
50. Cisneros GA, Darden TA, Gresh N, Pilmé J, Reinhardt P, Parisel O, Piquemal JP (2009). York D, Lee T-S (eds) Multi-scale quantum models for biocatalysis, vol 7. Springer, Netherlands, p 137
51. Lopes PM, Harder E, Roux B, MacKerell A Jr (2009). In: York D, Lee T-S (eds) Multi-scale quantum models for biocatalysis, vol 7. Springer, Netherlands, p 219
52. Cieplak P, Dupradeau FY, Duan Y, Wang J (2009) J Phys Condens Matter : Inst Phys J 21:333102
53. Gordon MS, Fedorov DG, Pruitt SR, Slipchenko LV (2011) Chem Rev 112:632
54. Gordon MS, Smith QA, Xu P, Slipchenko LV (2013) Annu Rev Phys Chem 64:553
55. Gong L (2012) Sci. China Chem. 55:2471
56. Cisneros GA, Karttunen M, Ren P, Sagui C (2013) Chem Rev 114:779
57. Marshall GR (2013) J Comput Aided Mol Des 27:107
58. Landau LD, Pitaevskii LP, Lifshitz EM (1984) Electrodynamics of continuous media, 2nd edn, vol 8. Butterworth-Heinemann, Oxford
59. Buchner R, Barthel J, Stauber J (1999) Chem Phys Lett 306:57
60. Yu H, van Gunsteren WF (2005) Comput Phys Commun 172:69
61. Dinur U, Hagler AT (1995) J Comput Chem 16:154
62. Rothman LS, Rinsland CP, Goldman A, Massie ST, Edwards DP, Flaud JM, Perrin A, Camy-Peyret C, Dana V, Mandin JY, Schroeder J, McCann A, Gamache RR, Wattson RB,

Yoshino K, Chance KV, Jucks KW, Brown LR, Nemtchinov V, Varanasi P (1998) J Quant Spectrosc Radiat Transfer 60:665
63. Partridge H, Schwenke DW (1997) J Chem Phys 106:4618
64. Burnham CJ, Xantheas SS (2002) J Chem Phys 116:5115
65. Donchev AG, Galkin NG, Illarionov AA, Khoruzhii OV, Olevanov MA, Ozrin VD, Subbotin MV, Tarasov VI (2006) Proc Natl Acad Sci USA 103:8613
66. Morita A, Kato S (1999) J Chem Phys 110:11987
67. Kaminski GA, Stern HA, Berne BJ, Friesner RA (2004) J Phys Chem A 108:621
68. Guillot B, Guissani Y (2001) J Chem Phys 114:6720
69. Wang Z-X, Zhang W, Wu C, Lei H, Cieplak P, Duan Y (2006) J Comput Chem 27:781
70. Wang J, Cieplak P, Li J, Wang J, Cai Q, Hsieh M, Lei H, Luo R, Duan Y (2011) J Phys Chem B 115:3100
71. Xie W, Pu J, MacKerell AD, Gao J (2007) J Chem Theory Comput 3:1878
72. Engkvist O, Åstrand P-O, Karlström G (2000) Chem Rev 100:4087
73. Gresh N, Cisneros GA, Darden TA, Piquemal J-P (2007) J Chem Theory Comput 3:1960
74. Ponder JW, Wu C, Ren P, Pande VS, Chodera JD, Schnieders MJ, Haque I, Mobley DL, Lambrecht DS, DiStasio RA, Head-Gordon M, Clark GNI, Johnson ME, Head-Gordon T (2010) J Phys Chem B 114:2549
75. Donchev AG, Galkin NG, Illarionov AA, Khoruzhii OV, Olevanov MA, Ozrin VD, Pereyaslavets LB, Tarasov VI (2008) J Comput Chem 29:1242
76. Stone AJ (1985) Mol Phys 1065:56
77. Lillestolen TC, Wheatley RJ (2007) J Phys Chem A 111:11141
78. Claverie P (1988). In: Maruani J (ed) Molecules in physics, chemistry, and biology, vol 2. Springer, Netherlands, p 393
79. Dykstra CE (1993) Chem Rev 93:2339
80. Harder E, Kim B, Friesner RA, Berne BJ (2005) J Chem Theory Comput 1:169
81. Straatsma TP, McCammon JA (1990) Mol Simul 5:181
82. Straatsma TP, McCammon JA (1991) Chem Phys Lett 177:433
83. Palmo K, Krimm S (2004) Chem Phys Lett 395:133
84. Kaminski GA, Friesner RA, Zhou R (2003) J Comput Chem 24:267
85. Kaminski GA, Ponomarev SY, Liu AB (2009) J Chem Theory Comput 5:2935
86. Ponomarev SY, Kaminski GA (2011) J Chem Theory Comput 7:1415
87. Li X, Ponomarev SY, Sa Q, Sigalovsky DL, Kaminski GA (2013) J Comput Chem 34:1241
88. Wang LP, Head-Gordon T, Ponder JW, Ren P, Chodera JD, Eastman PK, Martinez TJ, Pande VS (2013) J Phys Chem B 117:9956
89. Applequist J (1977) Acc Chem Res 10:79
90. Thole BT (1981) Chem Phys 59:341
91. Ren PY, Ponder JW (2002) J Comput Chem 23:1497
92. Ren P, Wu C, Ponder JW (2011) J Chem Theory Comput 7:3143
93. Wang J, Cieplak P, Li J, Hou T, Luo R, Duan Y (2011) J Phys Chem B 115:3091
94. Hermida-Ramón JM, Brdarski S, Karlström G, Berg U (2003) J Comput Chem 24:161
95. Xie W, Pu J, MacKerell AD, Gao J (2007) J Chem Theory Comput 3:1878
96. Borodin O (2009) J Phys Chem B 113:11463
97. Yan TY, Burnham CJ, Del Popolo MG, Voth GA (2004) J Phys Chem B 108:11877
98. McDaniel JG, Schmidt JR (2013) J Phys Chem A 117:2053
99. Elking D, Darden T, Woods RJ (2007) J Comput Chem 28:1261
100. Donchev AG, Galkin NG, Pereyaslavets LB, Tarasov VI (2006) J Chem Phys 125:244107
101. Baranyai A, Kiss PT (2011) J Chem Phys 135:234110
102. Donchev AG, Ozrin VD, Subbotin MV, Tarasov OV, Tarasov VI (2005) Proc Natl Acad Sci USA 102:7829
103. Kunz AP, Allison JR, Geerke DP, Horta BA, Hunenberger PH, Riniker S, Schmid N, van Gunsteren WF (2012) J Comput Chem 33:340
104. Lamoureux G, Roux B (2003) J. Chem. Phys. 119:3025
105. Paricaud P, Predota M, Chialvo AA, Cummings PT (2005) J Chem Phys 122:244511

106. Baranyai A, Kiss PT (2010) J Chem Phys 133:144109
107. Patel S, Brooks CL (2004) J Comput Chem 25:1
108. Patel S, MacKerell AD Jr, Brooks CL 3rd (2004) J Comput Chem 25:1504
109. Rick SW, Stuart SJ, Berne BJ (1994) J Chem Phys 101:6141
110. Olson ML, Sundberg KR (1978) J Chem Phys 69:5400
111. Applequist J (1993) J Phys Chem 97:6016
112. Cho K-H, Kang YK, No KT, Scheraga HA (2001) J Phys Chem B 105:3624
113. Rappe AK, Goddard WA III (1991) J Phys Chem 95:3358
114. Parr RG, Donnelly RA, Levy M, Palke WE (1978) J Chem Phys 68:3801
115. Banks JL, Kaminski GA, Zhou R, Mainz DT, Berne BJ, Friesner RA (1999) J Chem Phys 110:741
116. Warren GL, Davis JE, Patel S (2008). J Chem Phys 128
117. Chelli R, Procacci P, Righini R, Califano S (1999) J Chem Phys 111:8569
118. Davis JE, Warren GL, Patel S (2008) J Phys Chem B 112:8298
119. Chen J, Martínez TJ (2007) Chem Phys Lett 438:315
120. Chen J, Hundertmark D, Martinez TJ (2008) J Chem Phys 129:214113
121. Chen J, Martínez TJ (2009). J Chem Phys 131
122. Itskowitz P, Berkowitz ML (1997) J Phys Chem A 101:5687
123. Palmo K, Krimm S (1998) J Comput Chem 19:754
124. Palmo K, Mannfors B, Mirkin NG, Krimm S (2006) Chem Phys Lett 429:628
125. Ichikawa K, Kameda Y, Yamaguchi T, Wakita H, Misawa M (1991) Mol Phys 73:79
126. Nymand TM, Åstrand P-O (1997) J Phys Chem A 101:10039
127. Silvestrelli PL, Parrinello M (1999) J Chem Phys 111:3572
128. MacKerell AD, Feig M, Brooks CL (2003) J Am Chem Soc 126:698
129. Darley MG, Handley CM, Popelier PLA (2008) J Chem Theory Comput 4:1435
130. Ingrosso F, Monard GR, Hamdi Farag M, Bastida A, Ruiz-López MF (2011). J Chem Theory Comput 7:1840
131. Mannfors BE, Mirkin NG, Palmo K, Krimm S (2003) J Phys Chem A 107:1825
132. Leontyev IV, Stuchebrukhov AA (2012) J Chem Theory Comput 8:3207
133. Leontyev IV, Stuchebrukhov AA (2014) J Chem Phys 141:014103
134. Buckingham AD (1967) Advances in chemical physics. Wiley, New York, p 107
135. Elking DM, Perera L, Duke R, Darden T, Pedersen LG (2011) J Comput Chem 32:3283
136. Wheatley RJ, Price SL (1990) Mol Phys 69:507
137. Holt A, Karlström G (2008) J Comput Chem 29:2033
138. Holt A, Boström J, Karlström G, Lindh R (2010) J Comput Chem 31:1583
139. Kaminski GA, Stern HA, Berne BJ, Friesner RA (2004) J Phys Chem A 108:621
140. Shi Y, Xia Z, Zhang J, Best R, Wu C, Ponder JW, Ren P (2013) J Chem Theory Comput 9:4046
141. Ren P, Ponder JW (2003) J Phys Chem B 107:5933
142. Williams DE (1988) J Comput Chem 9:745
143. Elking DM, Cisneros GA, Piquemal JP, Darden TA, Pedersen LG (2010) J Chem Theory Comput 6:190
144. Buckingham AD, Orr BJ (1967) Q Rev, Chem Soc 21:195
145. Gubskaya AV, Kusalik PG (2001) Mol Phys 99:1107
146. Donchev AG (2007) J Chem Phys 126:124706
147. Stone AJ, Tong CS (1994) J Comput Chem 15:1377
148. Slipchenko† LV, Gordon MS (2009) Mol Phys 107:999
149. Wheatley RJ (1993) Mol Phys 79:597
150. Piquemal JP, Cisneros GA, Reinhardt P, Gresh N, Darden TA (2006) J Chem Phys 124:104101
151. Wheatley R, Lillestolen T (2008) Mol Phys 106:1545
152. Stern HA, Kaminski GA, Banks JL, Zhou R, Berne BJ, Friesner RA (1999) J Phys Chem B 103:4730
153. Stern HA, Rittner F, Berne BJ, Friesner RA (2001) J Chem Phys 115:2237

154. Wang L-P, Chen J, Van Voorhis T (2013) J Chem Theory Comput 9:452
155. McDaniel JG, Schmidt JR (2012) J Phys Chem C 116:14031
156. Huang L, Roux B (2013) J Chem Theory Comput 9:3543
157. Ansorg K, Tafipolsky M, Engels B (2013) J Phys Chem B 117:10093
158. Cacelli I, Cimoli A, Livotto PR, Prampolini G (2012) J Comput Chem 33:1055
159. Dehez F, Ángyán JG, Gutiérrez IS, Luque FJ, Schulten K, Chipot C (2007) J Chem Theory Comput 3:1914
160. Cieplak P, Caldwell J, Kollman P (2001) J Comput Chem 22:1048
161. Wang J, Cieplak P, Li J, Cai Q, Hsieh M-J, Luo R, Duan Y (2012) J Phys Chem B 116:7088
162. Zhu X, Lopes PEM, MacKerell AD Jr (2012) Wiley Interdiscip Rev: Comput Mol Sci 2:167
163. Misquitta AJ, Szalewicz K (2002) Chem Phys Lett 357:301
164. Misquitta AJ, Jeziorski B, Szalewicz K (2003) Phys Rev Lett 91:033201
165. Heßelmann A, Jansen G (2002) Chem Phys Lett 357:464
166. Heßelmann A, Jansen G (2002) Chem Phys Lett 362:319
167. Khoruzhii O, Donchev AG, Galkin N, Illarionov A, Olevanov M, Ozrin V, Queen C, Tarasov V (2008) Proc Natl Acad Sci USA 105:10378
168. Lamoureux G, Harder E, Vorobyov IV, Roux B, MacKerell AD (2006) Chem Phys Lett 418:245
169. Murphy WF (1977) J Chem Phys 67:5877
170. Morita A, Kato S (1999) J Chem Phys 110:11987
171. Tu YQ, Laaksonen A (2000) Chem Phys Lett 329:283
172. Masia M, Probst M, Rey R (2005) J Chem Phys 123:164505
173. Masia M, Probst M, Rey R (2006) Chem Phys Lett 420:267
174. Masia M (2008) J Chem Phys 128:184107
175. Rick SW, Stuart SJ, Bader JS, Berne BJ (1995) J Mol Liq 65/66:31
176. Rick SW, Berne BJ (1996) J Am Chem Soc 118:672
177. Kaminski GA, Stern HA, Berne BJ, Friesner RA, Cao YX, Murphy RB, Zhou R, Halgren TA (2002) J Comput Chem 23:1515
178. Maple JR, Cao Y, Damm W, Halgren TA, Kaminski GA, Zhang LY, Friesner RA (2005) J Chem Theory Comput 1:694
179. Kim B, Young T, Harder E, Friesner RA, Berne BJ (2005) J Phys Chem B 109:16529
180. Kaminski GA (2005) J Phys Chem B 109:5884
181. Click TH, Kaminski GA (2009) J Phys Chem B 113:7844
182. Ponomarev SY, Click TH, Kaminski GA (2011) J Phys Chem B 115:10079
183. Click TH, Ponomarev SY, Kaminski GA (2012) J Comput Chem 33:1142
184. Gao J, Habibollazadeh D, Shao L (1995) J Phys Chem 99:16460
185. Patel S, MacKerell AD Jr, Brooks CL III (2004) J Comput Chem 25:1504
186. Lamoureux G, MacKerell AD, Roux B (2003) J Chem Phys 119:5185
187. Lamoureux G, MacKerell AD, Roux B (2003) J Chem Phys 119:5185
188. Lopes PEM, Huang J, Shim J, Luo Y, Li H, Roux B, MacKerell AD Jr (2013) J Chem Theory Comput 9:5430
189. Vorobyov IV, Anisimov VM, MacKerell AD (2005) J Phys Chem B 109:18988
190. Anisimov VM, Vorobyov IV, Roux B, MacKerell AD Jr (2007) J Chem Theory Comput 3:1927
191. Vorobyov I, Anisimov VM, Greene S, Venable RM, Moser A, Pastor RW, MacKerell AD (2007) J Chem Theory Comput 3:1120
192. Baker CM, MacKerell AD Jr (2010) J Mol Model 16:567
193. Lopes PEM, Lamoureux G, Roux B, MacKerell AD (2007) J Phys Chem B 111:2873
194. Lopes PEM, Lamoureux G, Mackerell AD Jr (2009) J Comput Chem 30:1821
195. Harder E, Anisimov VM, Whitfield TW, MacKerell AD Jr, Roux B (2008) J Phys Chem B 112:3509
196. Zhu X, Mackerell AD Jr (2010) J Comput Chem 31:2330
197. Luo Y, Jiang W, Yu H, MacKerell AD Jr, Roux B (2013) Faraday Discuss 160:135
198. Yu H, Whitfield TW, Harder E, Lamoureux G, Vorobyov I, Anisimov VM, MacKerell AD Jr, Roux B (2010) J Chem Theory Comput 6:774

199. Baker CM, Anisimov VM, MacKerell AD Jr (2011) J Phys Chem B 115:580
200. Anisimov VM, Lamoureux G, Vorobyov IV, Huang N, Roux B, MacKerell AD (2005) J Chem Theory Comput 1:153
201. Harder E, MacKerell AD Jr, Roux B (2009) J Am Chem Soc 131:2760
202. He X, Lopes PEM, MacKerell AD Jr (2013) Biopolymers 99:724
203. Chowdhary J, Harder E, Lopes PEM, Huang L, MacKerell AD Jr, Roux B (2013) J Phys Chem B 117:9142
204. Lopes PE, Roux B, MacKerell AD Jr (2009) Theory Chem Acc 124:11
205. Halgren TA (1999) J Comput Chem 20:730
206. Donchev AG (2006) Phys Rev B 74:235401
207. Zacharia R, Ulbricht H, Hertel T (2004) Phys Rev B 69:155406
208. Chipman DM (2013) J Phys Chem B 117:5148
209. Jiao D, King C, Grossfield A, Darden TA, Ren P (2006) J Phys Chem B 110:18553
210. Piquemal J-P, Perera L, Cisneros GA, Ren P, Pedersen LG, Darden TA (2006) J Chem Phys 125
211. Liang T, Walsh TR (2006) Phys Chem Chem Phys 8:4410
212. Liang T, Walsh TR (2007) Mol Simul 33:337
213. Shi Y, Wu C, Ponder JW, Ren P (2011) J Comput Chem 32:967
214. Jiao D, Golubkov PA, Darden TA, Ren P (2008) Proc Natl Acad Sci USA 105:6290
215. Schnieders MJ, Fenn TD, Pande VS, Brunger AT (2009) Acta Crystallogr Sect D, Biol Crystallogr 65:952
216. Zheng X, Wu C, Ponder JW, Marshall GR (2012) J Am Chem Soc 134:15970

Chapter 6
Continuum Electrostatic Analysis of Proteins

G. Matthias Ullmann and Elisa Bombarda

6.1 Introduction

Electrostatic interactions play an important role in many biochemical systems especially because of their long range. Electrostatic interactions often guide the association of binding partners, but they also tune catalytic properties of the active site of enzymes. For instance, the protonation and redox behavior of residues and prostetic groups in protein is heavily influenced by the electrostatics of the surrounding. Moreover, regulation of biochemical processes is often mediated by electrostatic modifications of proteins such as for instance by phosphorylation of serine, threonine or tyrosine residues or by acetylations of lysine residues. However, the electrostatics of a protein is not only influenced by the distribution of charged and polar aminoacid residues in the protein but also by the surrounding solvent and the ions that are dissolved therein. The solvent may screen charge-charge interactions and stabilizes the structure of protein by solvating charged aminoacids. Another example of such effect are membrane potentials, that can influence the conformation of proteins.

There are different ways to describe solvent effects theoretically [1]. In molecular dynamics simulations, the solvent is described explicitly in form of individual solvent molecules, normally water molecules [2]. In such simulations, ions are

G.M. Ullmann (✉)
Structural Biology/Bioinformatics, University of Bayreuth, Universitätsstr. 30,
NW I, 95447 Bayreuth, Germany
e-mail: ullmann@uni-bayreuth.de

E. Bombarda
Experimental Physics IV, University of Bayreuth, Universitätsstr. 30, 95447 Bayreuth, Germany

© Springer International Publishing Switzerland 2014   135
G. Náray-Szabó (ed.), *Protein Modelling*, DOI 10.1007/978-3-319-09976-7_6

represented explicitly as well. These simulations require a considerable amount of computing time mainly to simulate the water and the ions solvating the protein. For this reason, molecular mechanics cannot be applied when information about longer time scales (more than micro-seconds) or many different states of a molecule are required. In such cases, continuum electrostatics, which relies on Maxwell's equations, is the approach of choice [3]. In continuum electrostatics, the protein and the surrounding solvent are described as dielectric continua. Since electrostatic interactions play a major role in biomolecular systems, continuum electrostatics has a broad range of applications in biomolecular modeling. Moreover in combination with a master equation approach, continuum electrostatics can even be used to describe the reaction kinetics of complex systems.

In this article, we describe some methods that are based on continuum electrostatic calculations. In the beginning, we will introduce the electrostatic model based on the Poisson-Boltzmann equation. In this part, we try to make the underlying theory understandable in order to give a feeling for its strength but also for its limitations. Moreover, we explain how the linearized Poisson-Boltzmann equation can be solved numerically. Afterwards, we discuss how the continuum electrostatic model was used to analyze the association of biomolecules and the thermodynamics of biochemical reactions. The major purpose of this review is to give a general overview of methods that rely on continuum electrostatics and discuss their physical basis.

6.2 The Continuum Electrostatic Model Based on the Poisson-Boltzmann Equation

6.2.1 The Physical Basis of the Poisson-Boltzmann Equation

The conceptual idea of modeling proteins using continuum electrostatics is relatively simple (Fig. 6.1). The protein is assumed to have a fixed structure defining a region of low polarizability which is embedded in a region with high polarizability representing the solvent. The polarizability is related to the relativedielectric constant of the medium, also called relative permittivity, the higher the dielectric constant the higher the polarizability. This model is mathematically represented by the Poisson equation with a spatially varying dielectric constant or actually better said dielectric permittivity (Eq. 6.1).

$$\nabla[\varepsilon(\mathbf{r})\nabla\phi(\mathbf{r})] = -4\pi\rho(\mathbf{r}) \tag{6.1}$$

where $\varepsilon(\mathbf{r})$ is the permittivity of the medium which varies spatially (inside and outside of the protein), ∇ is the differential operator, $\phi(\mathbf{r})$ is the electrostatic potential, $\rho(\mathbf{r})$ is the charge distribution within the protein and the solvent. The low dielectric region of the protein is delimited by assigning atomic radii to each atom and determining the solvent excluded volume by rolling a sphere over the protein [4]. Physically, the polarizability of a medium depends mainly on the mobility and the polarity of the molecules or molecular groups of the medium. Thus, solvents

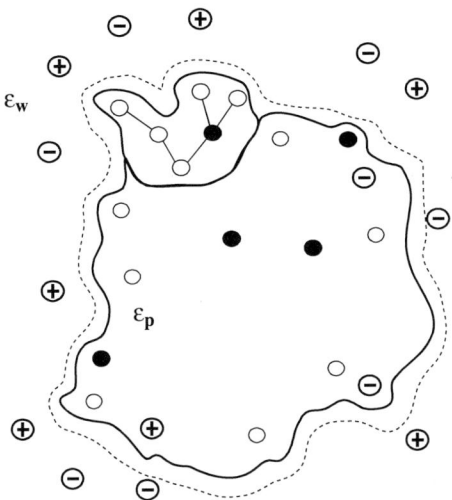

Fig. 6.1 Conceptual model of the continuum electrostatic approach. The protein is modeled as a dielectric continuum of low permittivity ε_p with fixed point charges. The protein is embedded in an environment with a high permittivity ε_w representing the solvent. In the continuum with a high permittivity, a charge density represents the ions dissolved in the aqueous solution. The *dotted line* marks the so-called Stern layer or ion-exclusion layer. Mobile ions are not allowed inside the protein volume and the Stern layer

with freely movable molecules and large dipoles have a high dielectric constant. A charged solute in a solvent induces a so-called reaction field (see Fig. 6.2). This reaction field is caused by the orientation of solvent molecules towards the charge, the so-called orientational polarization, and by the electronic polarization of the solvent molecules, i.e. by the deformation of the electron clouds. Apolar solvents such as octane have a relative dielectric constant of about 2, since the shielding due to orientational polarization is negligible while polar solvents such as water possess a large molecular dipole and consequently a high dielectric constant.

There are two types of charges in this model, spatially fixed charges representing the charge distribution within the protein and mobile charges representing the ions in the solvent. The ions dissolved in the solvent are excluded from the volume of the protein. The charges of the protein are usually represented by point charges at the position of the nucleus of the atoms. These point charges allow to represent charged aminoacids in the protein such as for instance aspartate and glutamate residues but also dipoles like for instance in the protein backbone or in the side chains of uncharged aminoacids. Mathematically, the charge distribution is represented as

$$\rho(\mathbf{r}) = \rho_f(\mathbf{r}) + \rho_{ion}(\mathbf{r}) \tag{6.2}$$

where ρ_f represents the charge distribution due to the point charges in the solute, i.e. the protein, and ρ_{ion} represents the charge distribution of the ions dissolved

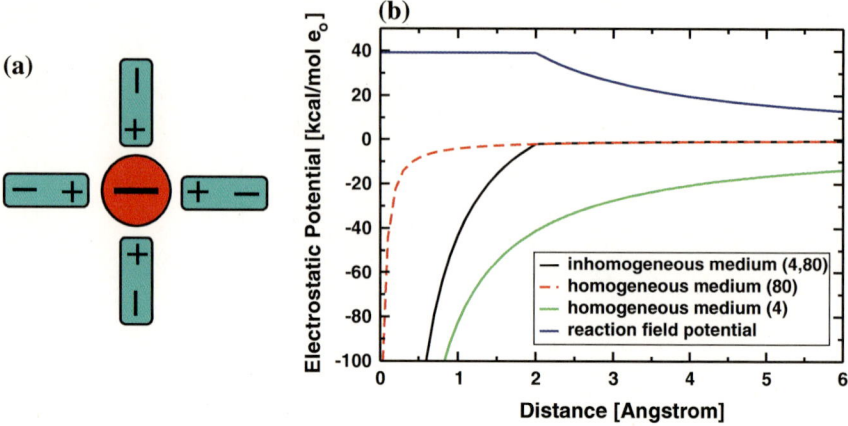

Fig. 6.2 Electrostatic potential of an anion in solution. **a** The solvent (*blue*) generates a solvation shell around the solvated anion (*red*). This solvation shell gives rise to the reaction field which counter-acts the electrostatic potential of the anion. **b** Electrostatic potentials of an anion with a radius of 2 Å in solution calculated with software APBS [5]. The *black line* shows the electrostatic potential for an inhomogeneous dielectric medium with a dielectric constant of 4 inside the ion and a dielectric constant of 80 in the solvent. The *green line* describes the electrostatic potential in a homogeneous medium with a dielectric constant of 4. The reaction field potential (*blue line*) is obtained as the difference between the *green line* and the *black line*. The *red dashed line* describes the electrostatic potential in a homogeneous medium with a dielectric constant of 80

in the solvent. These ions are represented by a charge density that adopts a Boltzmann distribution. This distribution can be approximated by

$$\rho_{ion}(\mathbf{r}) = \sum_{i=1}^{K} c_i^{\text{bulk}} Z_i e_0 \exp\left(\frac{-Z_i e_0 \phi(\mathbf{r})}{RT}\right) \tag{6.3}$$

assuming that there is no correlation between the ions in the solution. In Eq. 6.3, Z_i is the charge number of the ion of type i, K is the number of the different ion types in the solution, e_0 is the elementary charge, i.e. the charge of a proton, and c_i^{bulk} is the bulk concentration of the ion, i.e., the concentration where the protein electrostatic potential vanishes. Substituting Eqs. 6.2 and 6.3 in 6.1, the Poisson-Boltzmann equation for a medium with a spatially varying permittivity assumes the following form

$$\nabla[\varepsilon(\mathbf{r})\nabla\phi(\mathbf{r})] = -4\pi\left(\rho_f(\mathbf{r}) + \sum_{i=1}^{K} c_i^{\text{bulk}} Z_i e_0 \exp\left(\frac{-Z_i e_0 \phi(\mathbf{r})}{RT}\right)\right) \tag{6.4}$$

This equation is a non-linear partial differential equation since the potential $\phi(r)$ occurs not only on the left side of the equation but also in a non-linear term, namely in the exponential, on the right side of the equation, which describes the ion distribution around the protein. Generally, non-linear partial differential equations are difficult to solve even numerically. However, by approximating the exponential as

$$\sum_{i=1}^{K} c_i^{bulk} Z_i e_o \exp\left(\frac{-Z_i e_o \phi(\mathbf{r})}{RT}\right) \approx \sum_{i=1}^{K} c_i^{bulk} Z_i e_o - \sum_{i=1}^{K} c_i^{bulk} Z_i^2 e_o^2 \frac{\phi(\mathbf{r})}{RT} \quad (6.5)$$

and realizing that the first term on the right side is zero because of charge balance, one obtains the linearized Poisson-Boltzmann equation, i.e. the potential occurs on the right side of the equation only in a linear term.

$$\nabla[\varepsilon(\mathbf{r})\nabla\phi(\mathbf{r})] = -4\pi \left(\rho_{prot}(\mathbf{r}) - \sum_{i=1}^{K} c_i^{bulk} Z_i^2 e_o^2 \frac{\phi(\mathbf{r})}{RT} \right) \quad (6.6)$$

With the common definitions of the ionic strength $I = \frac{1}{2}\sum_{i=1}^{K} c_i^{bulk} Z_i^2$ and a modified inverse Debye length $\bar{\kappa} = \sqrt{\frac{8\pi N_A e_o^2 I}{k_B T}}$ the linearized Poisson-Boltzmann equation assumes the form that is found in some biophysics text books (Eq. 6.7).

$$\nabla[\varepsilon(\mathbf{r})\nabla\phi(\mathbf{r})] = -4\pi\rho_{prot}(\mathbf{r}) + \bar{\kappa}^2(\mathbf{r})\phi(\mathbf{r}) \quad (6.7)$$

As can be seen from Eq. 6.4, the Poisson-Boltzmann equation depends explicitly on temperature. However, this temperature dependence describes only the temperature dependence of the ion distribution. The temperature dependence of the dielectric constant is not explicitly included. Therefore, varying only the temperature in the Poisson-Boltzmann equation is not physically meaningful. Normally, room temperature is assumed in these kind of calculations and when another temperature is chosen, the dielectric constants should be adapted.

The linearity of Eq. 6.7 implies that the potentials of two charge distributions $\rho_1(r)$ and $\rho_2(r)$ are additive as long as the spatial distribution of the dielectric permittivity does not change, i.e., for the charge distribution $\rho(r) = \rho_1(r) + \rho_2(r)$ one can obtain the total potential as sum of the partial potentials $\phi(r) = \phi_1(r) + \phi_2(r)$ as long as the spatial distribution of the dielectric permittivity $\varepsilon(r)$ stays the same. This property has important consequences for the various applications. For instance the calculation of relative binding constants, which is a typical application of continuum electrostatic calculations, relies on this property. Namely, the calculation of pH titration curves is only possible because of this linearity as will be explained in a later section.

6.2.2 Solving the Linearized Poisson-Boltzmann Equation Numerically

For a few simple geometries analytical solutions of the linearized Poisson-Boltzmann equation exist [6, 7]. For irregular geometries, this equation can be solved by numerical methods. The most popular methods to solve Poisson-Boltzmann equation rely on regular finite difference methods [3, 5, 8–10], but also adaptive-grid methods [11], multi-grid-level based methods [12–15], boundary element methods [16], or finite element methods [17] can be used.

The principle idea of finite difference methods is to replace the differential $\nabla f(x)$ by a quotient of finite differences $\frac{f(x+h)-f(x)}{h}$, where h is a discretization coarse graining. The approximation approaches the exact result, when h goes to zero. In finite differences methods, every linear differential equation becomes a system of linear equations, which can be solved by numeric algorithms. The space in which the potential should be determined is discretized and the potential is calculated for each volume element. In order to obtain a numerical approach to solve the linearized Poisson-Boltzmann equation, we rearrange it

$$\nabla[\varepsilon(\mathbf{r})\nabla\phi(\mathbf{r})] - \bar{\kappa}^2(\mathbf{r})\phi(\mathbf{r}) + 4\pi\rho_{prot}(\mathbf{r}) = 0 \tag{6.8}$$

and integrate it over the descritezed volume elements

$$\int \nabla[\varepsilon(\mathbf{r})\nabla\phi(\mathbf{r})]\,dr - \int \bar{\kappa}^2(\mathbf{r})\phi(\mathbf{r})dr + 4\pi \int \rho_{prot}(\mathbf{r})\,dr = 0 \tag{6.9}$$

One the basis of Gauss's theorem, the first integral can be transformed into a surface integral that can be approximated with a finite difference expression.

$$\int \nabla[\varepsilon(\mathbf{r})\nabla\phi(\mathbf{r})]dr = \int [\varepsilon(\mathbf{r})\nabla\phi(\mathbf{r})]dA \tag{6.10}$$

$$= \sum_{i=1}^{6} \frac{h^2\varepsilon_i(\phi_i - \phi_0)}{h} \tag{6.11}$$

$$= \sum_{i=1}^{6} h\varepsilon_i(\phi_i - \phi_0) \tag{6.12}$$

The volume integrals on Eq. 6.9 can be written as

$$\sum_{i=1}^{6} h\varepsilon_i(\phi_i - \phi_0) - h^3\bar{\kappa}_0^2\phi_0 + 4\pi q_0 = 0$$

where $\bar{\kappa}_0^2$ is the modified inverse Debye length (related to the ionic strength) that is associated with this grid point. Rearranging this equation gives

$$\phi_0 = \frac{\left(\sum_{i=1}^{6} h\varepsilon_i\phi_i\right) + 4\pi q_0}{\left(\sum_{i=1}^{6} h\varepsilon_i\right) + h^3\bar{\kappa}_0^2} \tag{6.13}$$

Equation 6.13 is the numeric solution of the linearized Poisson-Boltzmann equation. It says that the potential ϕ_0 in the grid cell depends on the electrostatic potential ϕ_i of the six surrounding grid cells, the dielectric constant ε_i between the present and neighboring grid cells, the charge q_o and the ionic strength parameter $\bar{\kappa}_o^2$ assigned to the grid cell. Such an equation exists for almost all grid cells in the lattice, except for those at the boundary of the box. For the grid points at the boundary of the box, a good initial value needs to be determined for instance from an analytical approximation or a from numerical solution that was obtained with a coarser grid resolution. The set of equations like Eq. 6.13 form a system of linear equations, which can be solved. A common way is to obtain the potential iteratively. First a value of the potential is assigned to each grid cell, for instance from an analytical approximation. Then the potential is iteratively calculated, i.e., the potential of the present iteration is calculated from the potential of the previous iteration. The iteration is continued until the potential is sufficiently accurate. In practice, the iteration is stopped when the difference between the electrostatic potentials that were determined in two subsequent iteration steps is sufficiently small.

To solve the Poisson-Boltzmann equation for molecular systems practically, a flowchart of the type represented in Fig. 6.3 is followed. Different implementations of Poisson-Boltzmann solvers may vary in details. Here, the description focuses on the standard finite difference methods.

As a first step, the parameters of the molecule are read in. In particular, the parameters are the coordinates, the radii and the partial charges of the atoms. Also the dielectric constant of the solvent and of the solute needs to be defined as well as the ionic strength and the probe sphere radii for defining the solvent accessible surface and the ion exclusion layer. The temperature, which influences only the ionic distribution in the solution (see the discussion above), can also be defined. Moreover, the parameters for the numeric solvers needs to be read such as the number of grid points, the position of the grid, and the grid spacing.

In the next step, the boundaries of the dielectric regions are calculated. On the basis of the coordinates and radii of atoms, molecular surfaces are calculated for each dielectric regions. Usually surface is defined by rolling balls over the atoms of the molecule, which are represented as spheres of a defined coordinates and radii [18] (Fig. 6.4). The rolling ball represents a solvent molecule. For water, a radius of 1.4 Å is generally assumed as radius of the molecule. The step of defining the boundaries between the high and low dielectric regions is crucial, since it defines at which positions the reaction field is formed. A too small solvent radius may lead to unrealistically small cavities inside the protein that are filled with high dielectric continuum. A too large solvent radius may cause that internal cavities and surface clefts of the protein, that are actually filled with water, are not filled with a high dielectric continuum and thus no reaction field can arise from these cavities.

To solve the Poisson-Boltzmann equation numerically, all physical properties of the system (charge, electrostatic potential, electrical permittivity and ion accessibility) have to be mapped onto a grid. An easy way to map charges to the grid is a linear interpolation scheme. In this approach, a charge q_p at a given position $\mathbf{r}_p$ is fractioned to the eight surrounding grid points at the positions $\mathbf{r}_a (a = 1 \cdots 8)$ as follows: $q_p = q_a \left(1 - \frac{r_{ax} - r_{px}}{h}\right)\left(1 - \frac{r_{ay} - r_{py}}{h}\right)\left(1 - \frac{r_{az} - r_{pz}}{h}\right)$, where r_{ax} is the x

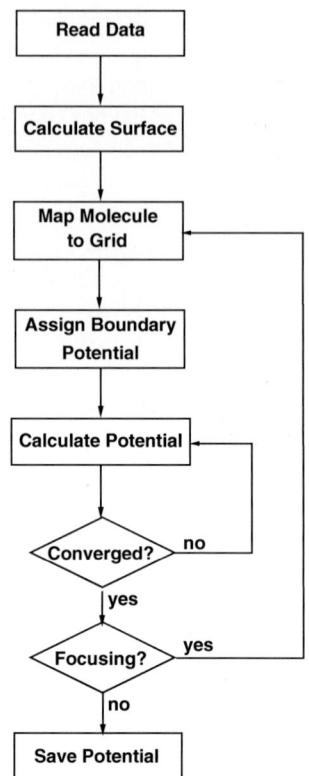

Fig. 6.3 Solving the Poisson-Boltzmann equation by a finite difference methods. First, all parameters for the calculation are read (coordinates, charges, radii, dielectric constants and grid definitions). Then the surface of the molecule is calculated as dielectric boundary. According to the boundaries, dielectric constants are mapped to the grid. The atomic charges are distributed over the surrounding grid points. Electrostatic potentials are assigned to the grid points based on an initial guess. It is important, that the initial potentials at the boundaries of the grid are very good approximations, because they remain constant during the calculation. The finite difference formulation of the Poisson-Boltzmann equation is solved by an iterative scheme, until the electrostatic potential does not change significantly anymore between two subsequent iterations. The computations require on one hand a large initial grid to minimize the error due to the approximated boundary potential and on the other hand a fine final grid to minimize the error due to the finite difference approximation. Both requirements can usually not be fulfilled directly due to a limited amount of memory. Therefore, consecutively smaller and finer grids are calculated using the previous grid to define the potentials at the boundaries. This method is called focusing. Finally, electrostatic energies are calculated as product of charge and electrostatic potential

-component of the vector a and h the grid spacing (mesh size). Analogously, the other neighboring grid points get the remaining fraction of the charge assigned according to their distance. The spatial dependent dielectric permittivity is defined on a grid, which is shifted by half a grid unit compared to the charge grid (ε_1 to ε_6 in Fig. 6.5). The surface of the molecule is used to assign the dielectric constant of the region, if the point is inside the surface. In a similar way, also the ionic

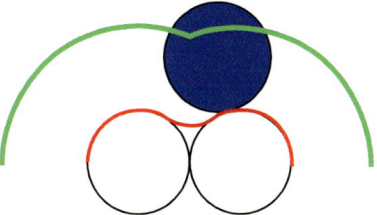

Fig. 6.4 Calculation of the solvent accessible surface. The solvent accessible surface (*red line*) of the two atoms (*white circle*) is calculated by a 'rolling ball' (*blue circle*). The surface of the Stern layer is shown by the *green line*

Fig. 6.5 Representation of one grid cell for solving the linearized Poisson-Boltzmann equation by a finite difference method. To the point in the center of the box, the electrostatic potential ϕ_0, the inverse Debye length κ, and the charge q_0 are assigned. The *filled points* represent the centers of the neighboring grid cells to which the potentials ϕ_i are assigned. The dielectric constants ε_i are assigned to the lines connecting two neighboring grid points

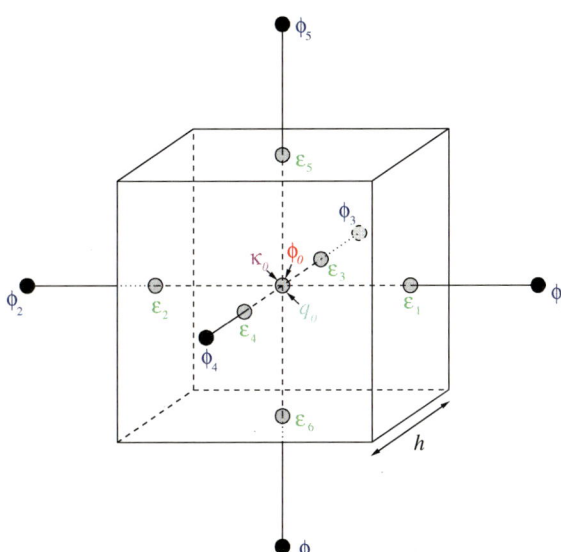

accessibility is defined on a grid with the only difference that a larger probe sphere radius (usually of 2 Å) is used to define the surface.

In order to actually calculate the potential numerically, the electrostatic potential grid needs to be initialized with an initial guess. A reasonable starting point is the Debye-Hückel expression. It is important that the initial guess of the potential at the outer boundaries of the grid is given accurately enough, since these potential values will not change during the calculation. For this reason, it is important that the distance between the protein and the outer boundary of the grid is large enough (at least 10 Å). In order to obtain a sufficient precision and numerical stability of the calculated potentials, the grids used in the calculation should have a resolution of at least 0.25 Å. Even for relatively small proteins, the number of grid points needed at this resolution to cover the protein and an adequate part of the solvent would require a huge amount of memory. Therefore, the grid is refined in several steps. One starts with a grid that is large enough to hold the whole protein and has a distance of at

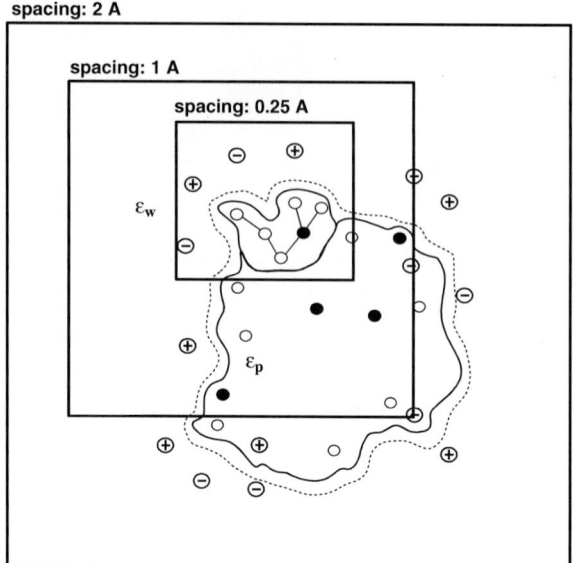

Fig. 6.6 Focusing of the grid for the calculation of the electrostatic potential using the Poisson-Boltzmann Equation. In order to calculate the electrostatic potential numerically, the protein needs to be mapped on a grid. To get a reasonable solution, the grid needs to be large enough to initialize the outer boundary of the grid with a reasonable analytic approximation. However, to get the good electrostatic potential, the grid needs to have a fine resolution. This fine resolution is obtained by first solving the Poisson-Boltzmann equation on a coarse grid. This solution is then used to initialize a smaller grid with a better resolution and so on until the desired resolution is reached. The outer grid is usually centered on the geometric center of the protein. Instead, the finest grid is usually centered on the geometric center of the group at which the exact electrostatic potential is desired

least 10 Å between the protein and the outer boundary of the grid. The center of the grid is usually chosen as the geometric center of the protein. This grid will be rather coarse (for instance 2.0 Å). Once the potential of this grid is converged, it can be used to initialize a finer grid (for instance 1.0 Å), which is embedded in the coarser grid. This procedure, called focusing [19], is repeated until a sufficiently fine grid can be used (Fig. 6.6). The finer grids are centered on the center of interest, for instance the set of atoms that form the site. Once the calculation is converged, the potential is stored and can be used for subsequent visual analysis using molecular visualization software or for calculations of electrostatic energies.

6.2.3 Electrostatic Potentials and Electrostatic Energies

The solution of the Poisson-Boltzmann equation is the electrostatic potential $\phi(\mathbf{r})$ which can be expressed as a potential that is composed of two parts.

$$\phi(\mathbf{r}) = \sum_{i=1}^{M} \frac{q_i}{4\pi \varepsilon_p |\mathbf{r} - \mathbf{r}_i'|} + \phi_{\mathrm{rf}}(\mathbf{r}) \tag{6.14}$$

The first term describes the Coulomb electrostatic potential at the position $\mathbf{r}$ caused by M point charges q_i at positions $\mathbf{r}_i'$ in a medium with a permittivity ε_p, the term $\phi_{rf}(\mathbf{r})$ describes the reaction fieldpotential originating from the charge distribution and the dielectric boundary between the protein and the solvent as well as from the distribution of ions in the solution. The reaction field is always oriented opposite to the field of the solute and therefore shields the field of the solute. This reaction field is of great importance for understanding structural and functional properties of proteins. For instance, in aqueous solution the dipole of a peptide α-helixis counteracted by the reaction field, which drastically reduces the strength of the helix dipole compared to its value in vacuum [20]. Moreover, reaction field effects can explain the orientation of helices in membrane proteins [21].

The electrostatic potential that is obtained by solving the Poisson-Boltzmann equation has already a great value by itself. Visualization of this potential can give first insights into the interaction between molecules as shown for instance in Fig. 6.7 where different representations of the electrostatic potential of Cytochrome c Peroxidase are shown. By convention, red shows negative electrostatic potentials and blue shows positive electrostatic potentials.

Probably more important than visualizing the electrostatic potential is using it to calculate electrostatic energies. Such calculations can give quantitative insights into biochemical mechanisms. Two different kind of electrostatic energies can be distinguished: interaction energies and reaction field energies.

The interaction energy G_{inter} energy is obtained by charging charge set ρ_2 in the presence of the electrostatic potential caused by charge ρ_1. Assuming that the charge set ρ_2 consists of a single charge q_f, the interaction energy becomes

$$G_{inter} = \int_0^{qf} \phi(\rho_1, \mathbf{r}_q)dq = \phi(\rho_1, \mathbf{r}_q)q_f \tag{6.15}$$

Since the potential $\phi(\rho_1, \mathbf{r}_q)$ at the position $\mathbf{r}_q$ of the charge q_f is totally independent of the charge q_f itself, the integration in Eq. 6.15 reduces to a simple multiplication. Equation 6.15 can be generalized to the interaction between two disjunct sets of charges $\{q\}$ and $\{p\}$, which is given by

$$G_{inter} = \sum_{i=1}^{N_q} q_i \phi(\{p\}, \mathbf{r}_{q_i}) = \sum_{i=1}^{N_p} p_i \phi(\{q\}, \mathbf{r}_{p_i}) \tag{6.16}$$

where N_q and N_p are the number of charges in the charge sets $\{q\}$ and $\{p\}$, respectively, $\phi(\{p\}, \mathbf{r}_{q_i})$ is the potential caused by the charge set $\{p\}$ at the position of the charge q_i and $\phi(\{p\}, \mathbf{r}_{q_i})$ is the potential caused by the charge set $\{q\}$ at the position of the charge p_i. As can be seen from Eq. 6.16, this interaction energy is symmetric.

The reaction field energy in continuum electrostatics, which is also sometimes called self-energy, is the interaction energy of the charge set $\{q\}$ with its own reaction field potential ϕ_{rf}. To obtain this energy, one imagines the charging of a

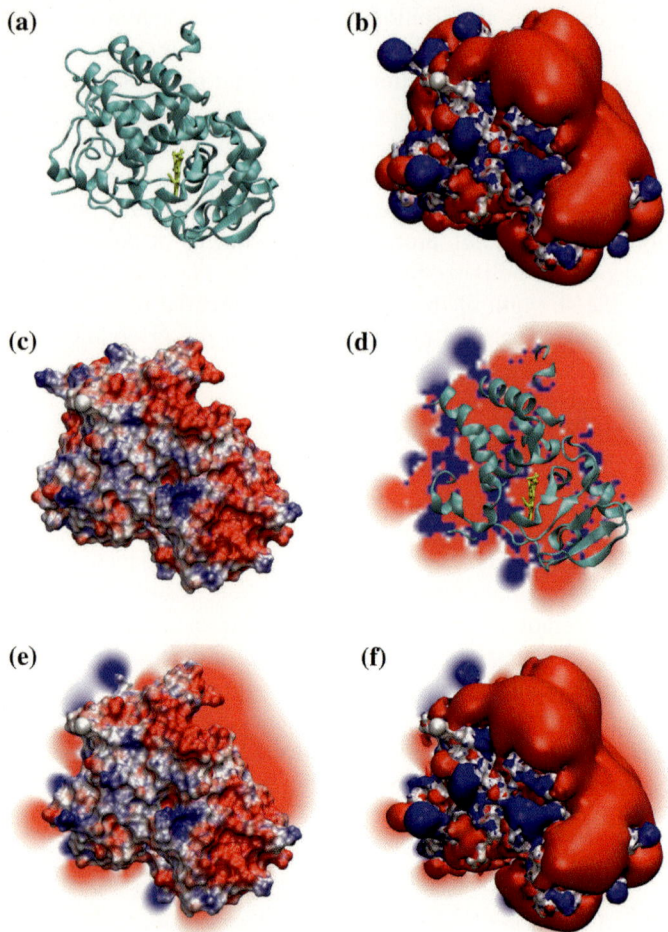

Fig. 6.7 Different visualizations of the electrostatic potential of cytochrome c peroxidase (CcP). The protein is shown in the same orientation in all pictures. **a** Cartoon representation of CcP showing the orientation of the protein. The heme is shown in a stick representation. **b** Isosurfaces of the electrostatic potential. *Blue* represent positive potentials, *red* negative potentials. The *red* and *blue surfaces* show where the potential has the value of -1 k_BT/e_o and 1 k_BT/e_o, respectively. **c** Electrostatic potential mapped to the molecular surface of the protein. The potential on the surface shows values between -3 k_BT/e_o (*red*) and 3 k_BT/e_o (*blue*). **d** Slice thought the electrostatic potential. The potential on the slice is scaled between -1 k_BT/e_o (*red*) and 1 k_BT/e_o (*blue*). **e** A combination of the representation shown in **c** and **d** gives an impression how the electrostatic potential fill the space but also allows to see more details on the molecular surface. **f** A combination of the representation shown in **c** and **b** gives a better impression how the electrostatic potential fill the space but lacks details on the molecular surface. The potentials were calculated using APBS [5] and visualized with VMD [22]

particle in a dielectric medium, and asks what is the energy of this charging process. Analogous to Eq. 6.15, we can write

$$G_{\mathrm{rf}} = \int_0^{q_f} \phi_{rf}\left(q, \mathbf{r}_q\right) \mathrm{d}q \tag{6.17}$$

in contrast to before, the reaction field potential ϕ_{rf} depends on the charge of the particle. For simplicity, one assumes a linear response, i.e. $\phi_{rf} = Cq_f$. Thus, from Eq. 6.17, we obtain

$$G_{\mathrm{rf}} = \int_0^{q_f} Cq \, \mathrm{d}q = \frac{1}{2}Cq_f^2 = \frac{1}{2}\phi_{rf}(q_f, \mathbf{r}_q)q_f \tag{6.18}$$

The last term $\frac{1}{2}\phi_{rf}\left(q_f, \mathbf{r}_q\right)q_f$ is obtained by using $\phi_{rf} = Cq_f$. Equation 6.18 can be generalized to obtain the reaction field energy of a charge set $\{q\}$

$$G_{\mathrm{rf}} = \frac{1}{2}\sum_{i=1}^{N_q} q_i \phi_{\mathrm{rf}}\left(\{q\}, \mathbf{r}_{q_i}\right) \tag{6.19}$$

As shown above, the factor $\frac{1}{2}$ in this equation is a consequence of the linear response ansatz.

Although simple at a first sight, the continuum electrostatic model is surprisingly successful in describing the properties and processes that are connected to the electrostatics of biomolecules. Such properties are for instance the association of proteins or redox and pH titration behavior. Its success relies probably on the fact that solvent degrees of freedom, which are normally difficult to sample in molecular dynamics simulations, are averaged from the beginning by assuming that the medium can be described by a dielectric continuum. To use the full power of continuum electrostatics, it needs to be combined with other techniques from statistical thermodynamics such as free energy calculations and Monte Carlo techniques as will be detailed below.

6.3   Electrostatic Association of Proteins

Since electrostatic force is long range, it plays a particular important role in the interaction of proteins. Especially in the case of electron transfer proteins, electrostatics plays a major role in the association process. In order to ensure a fast turnover, electron transfer proteins often associate transiently. This feature implies that electron transfer complexes are often relatively loose and dynamic [23]. In these complexes, electrostatics helps to get the right balance between specificity and flexibility. The structural flexibility has been observed for electron transfer complexes and studied extensively especially for the complex of cytochrome c peroxidase and cytochrome c or the complex of plastocyanin and cytochrome f [24–28].

6.3.1 Electrostatic Docking of Proteins

The docking of proteins can be simulated by Metropolis Monte Carlo [24]. For this purpose, the electrostatic potential of one protein, usually the larger protein, is mapped to a three dimensional grid surrounding the protein. The second molecule is placed in a random orientation at a certain distance from molecule one, i.e., randomly on the surface of a sphere which surrounds molecule one. The sphere should be large enough that the electrostatic potential of protein one on the surface of this sphere is zero or at least equipotential, i.e., everywhere the same on the surface. Then the second protein is randomly rotated and translated in the electrostatic potential of the first protein. The new configurations are accepted according to the Metropolis criterion [29]. That means that configurations with energies that are lower than that of the previous configuration are always accepted and configurations with a higher energy are accepted with a probability that is proportional to $\exp(-\Delta E/RT)$, where ΔE is the interaction energy difference between the new and the old configuration, R is the gas constant and T is the absolute temperature. The interaction energy is calculated by multiplying the charge distribution of the second molecule with the potential of the first molecule according to Eq. 6.16. If the second molecule moves too far away from the first molecule, the simulation is restarted on the sphere mentioned above. The Metropolis Monte Carlo procedure generates a Boltzmann ensemble which describes the equilibrium distribution of molecule two around molecule one. Such an ensemble is experimentally accessible by paramagnetic relaxation enhancement (PRE) NMR spectroscopy using spin labels attached at defined positions of the protein [23].

We recently applied this Monte Carlo docking procedure to the complex of Cytochrome c and Cytochrome c Peroxidase [30, 31]. As can be seen in Fig. 6.8a, b, the electrostatics of the docking surface of the two proteins is complementary. While Cytochrome c Peroxidase shows a clear negative potential (Fig. 6.8a), Cytochrome c shows a negative potential (Fig. 6.8b). Figure 6.8c, d shows the density of the center of mass of Cytochrome c generated in the Monte Carlo simulation. The large extension of this density shows that the complex of Cytochrome c and Cytochrome c Peroxidase allows a larger flexibility. This flexibility is often observed in the case of electron transfer proteins, since it allows a fast turnover because of a larger dissociation rate [32].

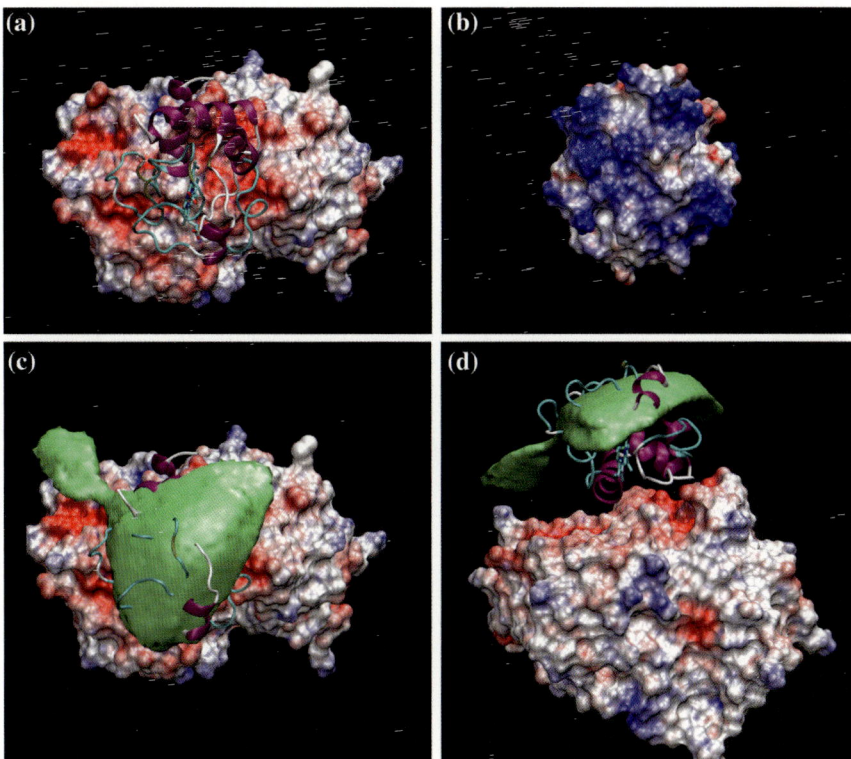

Fig. 6.8 Complex of cytochrome *c* (Cc) and Cytochrome *c* Peroxidase (CcP). The interaction of these two proteins is governed by electrostatic interactions. **a** Complex of Cc and CcP. Cc is shown as a cartoon. CcP is shown in a representation in which the electrostatic potential is mapped to its surface. **b** Cc with its electrostatic potential mapped to its surface. Cc is rotated by 180° around the y-axis (axis from *left* to *right* within the plain) compared to **a**. It can be seen that the electrostatic potential of Cc is positive and thus complementary to the electrostatic potential of CcP. **c** Simulated docking of Cc and CcP. The green density shows the region in which Cc can be found with a high probability. The density represents the frequency with which the center of mass of Cc is found in this certain volume area. The extension of the *green volume* indicates that the complex is not well defined but shows a high flexibility. This behavior was also found for other electron transfer complexes and is in agreement with experimental findings. **d** Simulated docking of Cc and CcP as in **c** but rotated by 90° around the x-axis

Brownian dynamics simulations are in some respect similar to the Monte Carlo approach [33–37]. In Brownian dynamics simulations, the association of two proteins is simulated using Newton's equations of motion combined with additional random and friction terms and the interaction is calculated based on electrostatic potentials obtained from the Poisson-Boltzmann equation. Brownian dynamics simulations enable to determine relative association rate constants and thus to study for instance the influence of mutations or of ionic strength on association rates. These simulations have been also applied to a variety of electron transfer complexes and allowed to interpret experimental findings on the association between electron transfer partners [38–41].

6.3.2 Similarity of Electrostatic Potentials of Proteins

Proteins with similar function show often similar structures. A special case, which at the first sight seems to be an exception to this rule, can be found for the electron transfer between cytochrome b_6f and photosystem I in photosynthesis. This transfer is usually mediated by the blue copper protein plastocyanin. Under copper deficiency however, plastocyanin is replaced by the heme protein cytochrome c_6. Although plastocyanin and cytochrome c_6 differ considerably in composition and structure, they perform the same function in the photosynthetic electron-transport chain.

The functional equivalence of the two proteins can be understood on the basis of similarity of their electrostatic potentials [42, 43]. The similarity of the electrostatic potentials is defined on the basis of the integral-based Hodgkin index H_{ab}^{elec} [44].

$$H_{ab}^{elec} = \frac{2 \int \phi_a \phi_b \, dV}{\int \phi_a^2 \, dV + \int \phi_b^2 \, dV} \tag{6.20}$$

The potentials ϕ of the structurally different molecules a and b are integrated over the whole volume V. The numerator quantifies the spatial overlap of the electrostatic potentials ϕ, while the denominator normalizes this value such that the resulting similarity index H_{ab}^{elec} falls in the interval between -1 and $+1$. The value $+1$ corresponds to molecules with identical potentials, whereas -1 corresponds to electrostatic complementarity, i.e., potentials of the same magnitude but opposite sign. In order to optimize the superposition of the two molecules, the Coulomb potentials are approximated by Gaussian potentials and Eq. 6.20 is minimized with respect to the relative orientation of the two molecules [43]. The structural superposition, which was obtained by minimizing Eq. 6.20, was used to identify functionally equivalent residues in plastocyanin and cytochrome c_6 [42]. Interestingly, it can be seen that functional analogous aminoacids enable the specific recognition for the two isofunctional proteins. Figure 6.9 shows the electrostatic potential and the cartoon representations of plastocyanin and cytochrome c_6 of *Chlamydomonas rheinhardii* in the orientation that corresponds to the superposition of the two proteins.

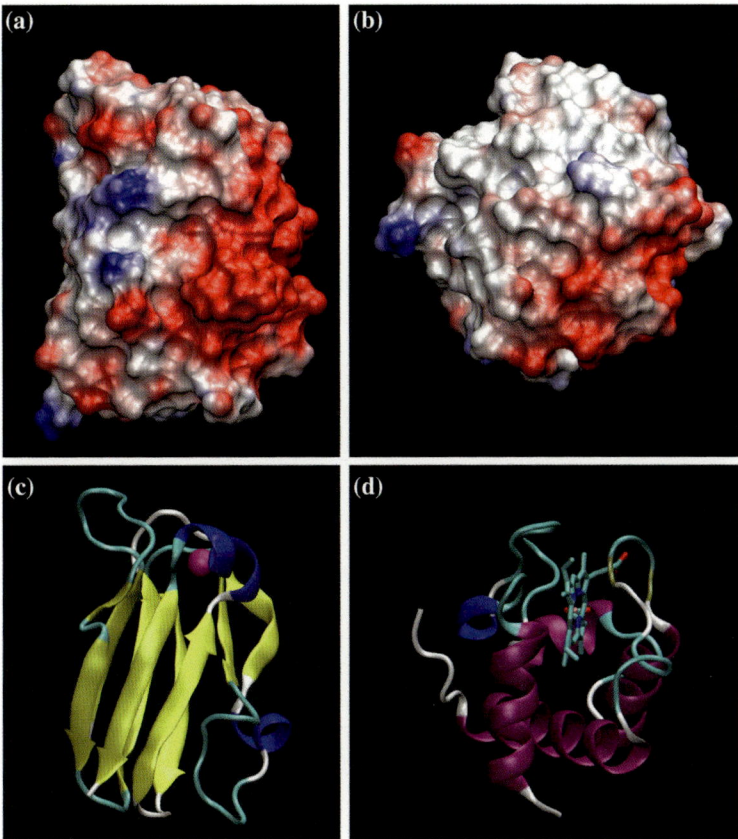

Fig. 6.9 Comparison of the electrostatic potentials of the isofunctional proteins plastocyanin and cytochrome c_6. **a** Electrostatic potential of plastocyanin mapped to its surface. **b** Electrostatic potential of cytochrome c_6 mapped to its surface. **c** Cartoon representation of plastocyanin. The copper ion is shown in a space-filled model. **d** Cartoon representation of cytochrome c_6. The heme is shown in a stick model

6.4 Titration Behavior of Proteins

The electrostatics of a protein is determined by the charge states of its protonatable and redox-active groups. However, often the description of proton binding and redox equilibria in proteins is considerably more complicated than that of small molecules because of the mutual interaction of the many protonatable and redox-active groups in one protein. Here we outline the methods how to describe such equilibria.

6.4.1 Shifted Titration Curves in Proteins

The titration curve of aminoacids in proteins can be shifted compared to the titration curve in aqueous solution. This shift has two main causes. First, the charges and dipoles of the protein may stabilize or destabilize a charge state in the protein. Second, the desolvation when a charged group is brought from the highly polar aqueous solution into the usually apolar interior of the protein destabilizes charged states. The actual direction of the shift depends on the balance of the different effects. Generally, it can be said that buried charges in the protein are unfavorable and thus the aminoacids tend to be uncharged under such circumstances. However, this rule of thumb may not apply if ion pairs are buried.

One way to obtain the protonation energy of a site within a protein is to calculate the difference between the pK_a value of an appropriate model compound in aqueous solution and the pK_a value of the protonatable group in the protein. The pK_a value associated with this protonation energy is called intrinsic pK_a value. In proteins with many interacting titratable residues, the intrinsic pK_a value is the pK_a value the titratable group would have if all other titratable groups in the protein are in a defined reference protonation state. To compute intrinsic pK_a values, pK_a values of a model compounds are required, such as for instance an aminoacid with blocked terminal amino- and carboxyl group. These values can be obtained either from experiment or from quantum chemical calculations. Sometimes only quantum chemical calculations are able to obtain the pK_a values for model compounds of prosthetic groups in proteins, because appropriate model compounds cannot be synthesized.

If only electrostatic contributions cause the difference between the protonation energies of a titratable group in a protein and in aqueous solution, the Poisson-Boltzmann equation provides a reasonable approximation of this energy difference. Transferring the protonatable group i with a given pK_a value $pK_{a,i}^{\text{model}}$ from aqueous solution into a protein causes an energy shift. This energy shift can be separated into two contributions. The first energy contribution $\Delta\Delta G_{\text{Born}}$ is a Born-energy-like term (Eq. 6.21), which arises from the interaction of the charges of the protonatable group with its reaction field.

$$
\Delta\Delta G_{\text{Born},i}^{\text{prot}} = \frac{1}{2} \sum_{a=1}^{N_{Q,i}} Q_{a,i}^{h} [\phi_p(\mathbf{r}_a; Q_i^h) - \phi_m(\mathbf{r}_a; Q_i^h)]
$$
$$
- \frac{1}{2} \sum_{a=1}^{N_{Q,i}} Q_{a,i}^{d} [\phi_p(\mathbf{r}_a; Q_i^d) - \phi_m(\mathbf{r}_a; Q_i^d)]
$$

(6.21)

lThe second energy contribution $\Delta\Delta G_{\text{back}}$ arises from the interaction of the charges of the protonatable group with non-titrating background charges (Eq. 6.22).

$$
\Delta\Delta G_{\text{back},i}^{\text{prot}} = \sum_{a=1}^{N_p} q_a [\phi_p(\mathbf{r}_a; Q_i^h) - \phi_p(\mathbf{r}_a; Q_i^d)] - \sum_{a=1}^{N_m} q_a [\phi_m(\mathbf{r}_a; Q_i^h) - \phi_m(\mathbf{r}_a; Q_i^d)]
$$

(6.22)

The summations in Eq. 6.21 run over the $N_{Q,i}$ atoms of group μ that have different charges in the protonated (h) $(Q_{a,i}^h)$ and in the deprotonated (d) $(Q_{a,i}^d)$ form. The first summation in Eq. 6.22 runs over the N_p charges of the protein that belong to atoms in non-titratable groups or to atoms of titratable groups (not i) in their uncharged protonation form. The second summation in Eq. 6.22 runs over the N_m charges of atoms of the model compound that do not have different charges in the different protonation forms. The terms $\phi_m(\mathbf{r}_a, Q_i^h)$, $\phi_m(\mathbf{r}_a, Q_i^d)$, $\phi_p(\mathbf{r}_a, Q_i^h)$, and $\phi_p(\mathbf{r}_a, Q_i^d)$ denote the values of the electrostatic potential at the position $\mathbf{r}$ of the atom a. The electrostatic potential was obtained by solving the Poisson-Boltzmann equation numerically using the shape of either the protein (subscript p) or the model compound (subscript m) as dielectric boundary and assigning the charges of the titratable group μ in either the protonated (Q_i^h) or the deprotonated (Q_i^d) form to the respective atoms. These two energy contributions and the pK_a value of the model compound $pK_{a,\mu}^{model}$ are combined to obtain the so-called intrinsic pK_a value pK_μ^{intr} (Eq. 6.23) of the residue.

$$pK_i^{intr} = pK_{a,i}^{model} - \frac{1}{RT\ln 10}\left(\Delta\Delta G_{Born,i}^{prot} + \Delta\Delta G_{back,i}^{prot}\right) \tag{6.23}$$

The intrinsic pK_a value is the pK_a value that this group would have, if all other protonatable groups are in their reference protonation form.

6.4.2 Microstate Model

An additional complication in proteins is that proteins usually contain more than one titratable group. Also the interaction W_{ij} between the two groups i and j in their charged form can be calculated using the electrostatic potential obtained from the PBE (Eq. 6.24).

$$W_{ij} = \sum_{a=1}^{N_{Q,i}}\left[Q_{i,a}^h - Q_{i,a}^d\right]\left[\phi_p\left(\mathbf{r}_a, Q_j^h\right) - \phi_p\left(\mathbf{r}_a, Q_j^d\right)\right] \tag{6.24}$$

The interaction between the titratable groups can lead to titration curves that do not show a standard sigmoidal shape that can be fitted with the Henderson-Hasselbalch equation [45]. The difficulties can be resolved if the problem is formulated in terms of well-defined microstates of the protein which have a certain probability, instead of considering the protein as a system of groups with a certain protonation probability, as will be outlined now. Since the formalism can be easily extended to treat not only protonation equilibria but also redox equilibria [43], we explain it here in a more generalized form.

Let us consider a system that possesses N protonatable sites and K redox-active sites. Such a system can adopt $M = 2^{N+K}$ states assuming that each sites can exist in two forms. The interaction between them can be modeled purely electrostatically, i.e. the electronic coupling is negligible. Each state of the system can be written as an $N + M$-dimensional vector $\vec{x} = (x_1, \ldots, x_{N+K})$, where x_i is 0 or 1 if site i is deprotonated (reduced) or protonated (oxidized), respectively. Each state of the system has a well-defined energy which depends on the energetics of the individual sites and the interaction between sites. The energy of a state $\vec{x}_v$ is given by [46–50]:

$$G(\vec{x}_v) = \sum_{i=1}^{N} (x_{v,i} - x_i^\circ) RT \ln 10 \left(\text{pH} - pK_{a,i}^{\text{intr}} \right)$$

$$- \sum_{i=1}^{K} (x_{v,i} - x_i^\circ) F \left(E - E_i^{\text{intr}} \right) \tag{6.25}$$

$$+ \frac{1}{2} \sum_{i=1}^{N+K} \sum_{j=1}^{N+K} \left(x_{v,i} - x_i^\circ \right) \left(x_{v,j} - x_j^\circ \right) W_{ij}$$

where R is the gas constant; T is the absolute temperature; F is the Faraday constant; $x_{v,i}$ denotes the protonation or redox form of the site i in state $\vec{x}_v$, x_i° is the reference form of site i; $pK_{a,i}^{\text{intr}}$ and E_i^{intr} are the pK_a value and redox potential, respectively, that site i would have if all other sites are in their reference form (intrinsic pK_a value and intrinsic redox potential); E is the reduction potential of the solution; pH is the pH value of the solution; W_{ij} represents the interaction of site i with site j.

Equilibrium properties of a physical system are completely determined by the energies of its states. To keep the notation concise, states will be numbered by Greek indices, i.e., for state energies we write G_v instead of $G(\vec{x}_v)$. For site indices, the roman letters i and j will be used. The equilibrium probability of a single state $\vec{x}_v$ is given by

$$P_v^{eq} = \frac{e^{-\beta G_v}}{Z} \tag{6.26}$$

with $\beta = 1/RT$ and Z being the partition function of the system.

$$Z = \sum_{v=1}^{M} e^{-\beta G_v} \tag{6.27}$$

The sum runs over all M possible states. Properties of single sites can be obtained from Eq. 6.26 by summing up the individual contributions of all states. For example, the probability of site i being oxidized is given by

$$\langle x_i \rangle = \sum_{v}^{M} x_{v,i} P_v^{eq} \tag{6.28}$$

where $x_{v,i}$ denotes the protonation or redox form of site i in the charge state $\vec{x}_v$. For small systems, this sum can be evaluated explicitly. For larger systems, Monte-Carlo techniques can be used to determine these probabilities [51, 52].

For a system of interacting sites, the protonation or reduction probabilities $\langle x_i \rangle$ can show a complex shape, thus rendering the assignment of pK_a values or midpoint potentials to individual sites difficult or even meaningless [45, 53–55]. The energy differences between microstates, however, remain well defined and thus form a convenient basis to describe the system. For individual sites in such a complex system, one can however define pH-dependent pK_a values and solution redox potential dependent midpoint potentials [56].

$$pK_i = pH + \frac{1}{\ln 10}\left(\ln\frac{\langle x_i \rangle}{1 - \langle x_i \rangle}\right) \tag{6.29}$$

$$E_i^o = E - \frac{RT}{F}\left(\ln\frac{\langle x_i \rangle}{1 - \langle x_i \rangle}\right) \tag{6.30}$$

These values define properties that are directly related to a free energy difference [56] and are relevant for understanding enzymatic mechanisms.

6.4.3 An Illustrative Example

Titration curves with a non-standard sigmoidal shape can be seen in many protein titration studies, but their interpretation is often very complicated. Sometimes also small molecules show a complex titration behavior. One example is diethylene-triamine-pentaacetate [57–59]. Here, we discuss a fictitious molecule with three groups and each of them can bind a proton. The intrinsic pK_a values and the interaction energies are given in Fig. 6.10a. The resulting microscopic equilibria are given in Fig. 6.10b. The individual titration are given in Fig. 6.10c. Given all the different energy terms, the population of the microscopic and macroscopic states in dependence of pH can be calculated (Fig. 6.10d). Figure 6.10e shows how the individual titration curves and the population of the different microstates contribute to the titration curves of the individual sites.

The titration curve of the central group is unusual because of its non-monotonic shape. In particular between pH 5 and pH 9, the protonation probability increases with increasing pH (i.e. with decreasing proton concentration). The population of the microstates give a physical rational for the unusual, irregular titration behavior of the central group. At high pH (low proton concentration), the protons bind preferable to the central group. Binding the second proton to one of the terminal groups while the central group stays protonated is unfavorable, because these two proton binding sites repel each other. Therefore, when the second proton binds, it is more favorable to deprotonate the central group and to protonate both terminal groups. When the two terminal amines are protonated, the two protons are at a greater distance from each other, and thus they repel each other less as compared to protonating one terminal and the central amine. Finally at very low pH (high proton concentration), all three sites will bind a proton.

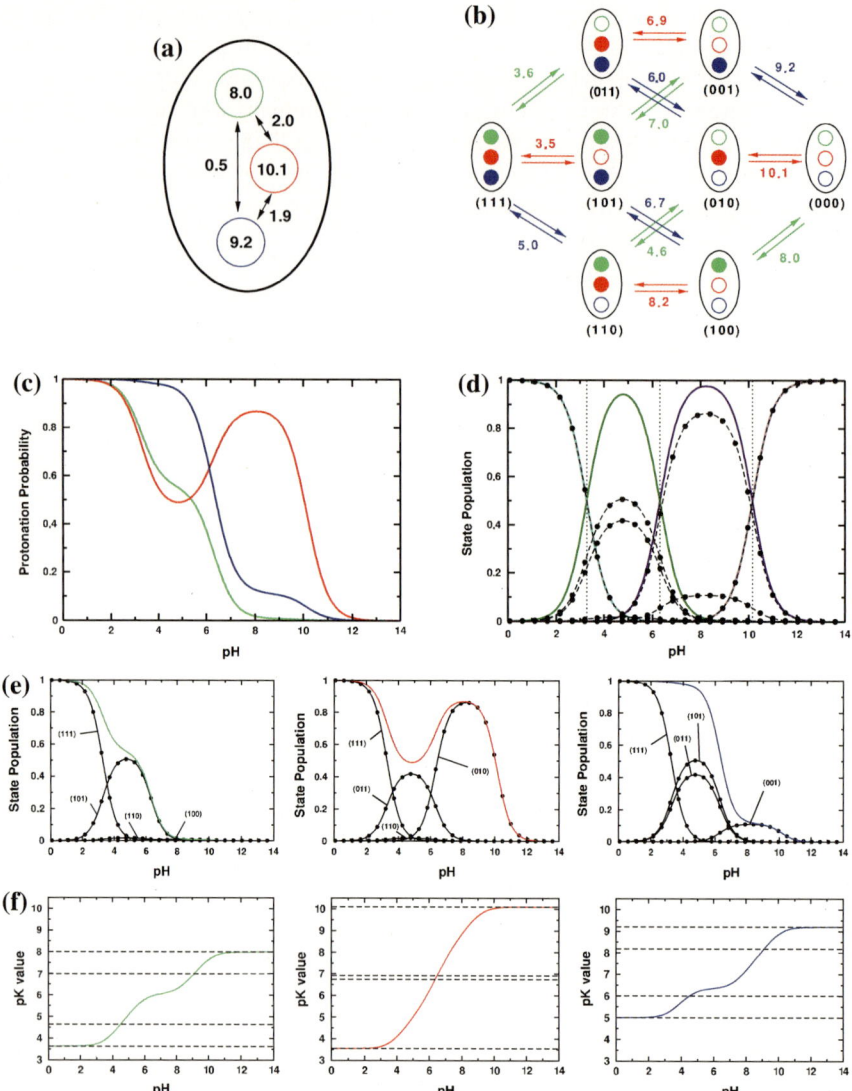

Fig. 6.10 Titration behavior of a fictitious molecule with three titratable groups. **a** Schematic drawing of a fictitious molecule with three titratable groups. The *numbers in the circles* are the intrinsic pK_a values, the numbers at the arrows are the respective interaction energies. All numbers are given in pK_a units. **b** Microscopic protonation equilibria. The numbers at the arrows indicate the microscopic pK_a values. **c** Individual titration curve of the three sites. **d** Populations of the macroscopic (*solid lines*) and microscopic (*lines with circles*) states of the system. The *dotted lines* mark the macroscopic pK_a values. **e** Contributions of the different microstates to the titration curves of the three individual sites. **f** Effective (*solid lines*) and microscopic (*dashed lines*) pK_a values of the three sites

Obviously, it is not easily possible to assign pK_a values to the individual sites. All groups are associated with four microscopic pK_a values (see Fig. 6.10f). However, using Eq. 6.29, it is possible to assign an effective pK_a value to each site (see Fig. 6.10e). It can be seen that the effective pK_a value assumes values that vary within the limits of the microscopic pK_a value.

Even more interesting than the pH-dependence of the pK_a values is the pH-dependence of the protonation energy. The protonation free energy ΔG_{prot} is given by

$$\Delta G_{prot} = RT \ln 10 (pH - pK_a) \qquad (6.31)$$

This equation leads to a linear dependence of the protonation energy for pH-independent pK_a values that are found in the case of isolated titratable groups. However, if the pK_a value is pH-dependent, the pH-dependence of the protonation free energy becomes non-linear.

Even if this example seems to be an extreme example, similarly complicated protonation equilibria caused by charge-charge interactions occur frequently in proteins. A sign of such complications are irregular titration curves. Moreover, even if the titration curves in proteins show apparently a standard sigmoidal shape, the interactions between titratable groups may lead to pH-dependent pK_a values [56]. Such pH-dependent pK_a values may lead to nearly pH-independent protonation energies in a certain pH range and thus may explain why some particular residue can function as proton donor or acceptor over a large pH range allowing catalysis under different pH conditions. Probably for this reason, there are often more protonatable residues in the active site of enzymes than the specific function would require.

6.5 Other Applications of Continuum Electrostatic Calculations

The continuum electrostatic calculations have a great potential for studying the mechanism of enzymes. A step forward in the analysis of enzymatic mechanisms is the combination of continuum electrostatics with quantum chemical calculations as inCOSMO [60] orCPCM [61]. However, it is relatively difficult to model the protein environment appropriately with such methods. An attractive alternative is the self-consistent reaction field method (SCRF) which was developed by Tomasi and coworkers [62] and applied to protein systems by Bashford, Noodlemann and coworkers [63–67]. This method combines quantum chemical calculations with Poisson-Boltzmann calculations and allows to account for the charge distribution within the protein and solvent polarization effects in quantum chemical calculations.

To explore possible mechanisms, it is often required to examine many different possibilities. Sometimes, even many mechanism may be possible at the same time and a single answer may not exist. Such complex reaction schemes can be explored with the help of the microstate model introduced in Sect. 6.4.2.

The kinetics of such reactions can be simulated by a master equation approach. The rate constants which are required for such simulations can be calculated using electrostatic methods [68–71]. Thus, combined with a master equation approach, continuum electrostatics offers also a possibility to access the non-equilibrium behavior of biomolecular systems. In the microstate formalism given by Eqs. 6.25–6.28, charge transfer events are described as transitions between well-defined microstates of a system. The time dependence of the population of each microstate can be simulated using a master equation

$$\frac{d}{dt}P_\nu(t) = \sum_{\mu=1}^{M} k_{\nu\mu}P_\mu(t) - \sum_{\mu=1}^{M} k_{\mu\nu}P_\nu(t) \qquad (6.32)$$

where $P_\nu(t)$ denotes the probability that the system is in charge state ν at time t, $k_{\nu\mu}$ denotes the probability per unit time that the system will change its state from μ to ν. In Eq. 6.32, the first sum includes all the reactions that generate state ν, the second sum includes all the reactions that destroy state ν. The summations run over all possible states μ. In order to restrict the number of states and only consider states that are accessible in a certain energy range, methods like extended Dead End Elimination [72] can be used. Simulating charge transfer by Eq. 6.32 assumes that these processes can be described as a Markovian stochastic process. This assumption implies that the probability of a given charge transfer only depends on the current state of the system and not on the way in which the system has reached this state. The system given by Eq. 6.32 is a system of coupled linear differential equations with constant coefficients, for which an analytical solution exists [69, 70]. Equation 6.32 describes the time evolution of the probability distribution of microstates of the system. For these microstates, energies G_ν and transition probabilities $k_{\nu\mu}$ can be assigned unambiguously. The time-dependent probability of finding a single site in a particular form can be obtained by summing up individual contributions from the time-dependent probabilities $P_\nu(t)$.

$$\langle x_i \rangle(t) = \sum_{\nu}^{M} x_{\nu,i}P_\nu(t) \qquad (6.33)$$

The application of the method outlined about to electron transfer reactions is particularly attractive, since their rates can be estimated using the rate law developed by Moser and Dutton [73, 74] which relies on the Marcus theory [75] and agrees well with experimental data. Mainly three factors govern the rate constants of biological electron transfer reactions: the energy difference between the donor state and the acceptor state, the environmental polarization (reorganization energy), and the electronic coupling between the redox sites. The energy barrier for the transfer process is given in the framework of Marcus theory as

$$\Delta G^{\neq} = \frac{(\Delta G^\circ + \lambda)^2}{4\lambda} \qquad (6.34)$$

where ΔG° is the energy difference between donor and acceptor state and λ is the reorganization energy. The electronic coupling between the redox sites is accounted for by a distance-dependent exponential function $A \exp(-\beta(R - R_{\circ}))$, where R is the edge-to-edge distance between the electron transfer centers, $R_{\circ}$ represents a van der Waals contact distance and A represents an optimal rate.

The free energy ΔG° for a transition between two states v and μ can be calculated within the electrostatic model using Eq. 6.25. The reorganization energy λ contains two contributions, $\lambda = \lambda_o + \lambda_i$, where λ_o is the solvent reorganization energy and λ_i is the inner sphere reorganization energy. λ_o was shown to be accessible to calculations using electrostatic potentials obtained from the solution of the Poisson-Boltzmann equation [75, 76]. The inner sphere reorganization energy λ_i can be estimated by quantum chemical calculations and it is often found to be significantly smaller than the solvent reorganization energy [77–80].

For analyzing a complex charge transfer system, it is of particular interest to follow the flow of charges through the system, i.e., the charge flux. The flux from state v to state μ is determined by the population of state v times the probability per unit time that state v will change into state μ, i.e., by $k_{\mu v}P_v(t)$. The net flux between states μ and v is thus given by

$$J_{v\mu}(t) = k_{v\mu}P_{\mu}(t) - k_{\mu v}P_v(t) \qquad (6.35)$$

The net flux (Eq. 6.35) is positive if there is a net flux from state μ to state v. This flux analysis allows to deduct the reaction mechanism from even very complex reaction schemes [70].

In cases when the number of possible microstates get too large, the differential equation can not be solved analytically anymore. Thus, approximations and simulations need to be applied. One attractive simulation method is the dynamical Monte Carlo Simulation scheme [81], which allows to simulate very complex reaction mechanism such as proton transfer through a protein matrix [82]. Again, the reaction parameter can be obtained from continuum electrostatic calculations. Each simulation trajectory describes one particular reaction path through the possible states of the system. A reaction mechanism can then be inferred from the analysis of many such trajectories. Up to now, this method was only applied to relatively simple systems [82]. However, future application to enzymes which involve chemical transformation, proton and electron transfer as well as conformational changes seem possible making dynamical Monte Carlo simulations a promising future road to analyze enzyme function.

6.6 Conclusion

Electrostatic interactions play a major role in biomolecular systems. In particular, the mechanism of enzyme cannot be understood without the correct evaluation of the effect of all charges involved in the enzyme or in the environment. The methods based on continuum electrostatics are extremely valuable in this task, since they allow the analysis of the electrostatic interactions involving the macromolecular partners in their environment at meaningful time scales. In this review, we have shown how continuum electrostatic methods provide essential information both on the thermodynamics and the kinetics of biological mechanisms. These methods model essential biophysical aspects correctly and allow a computationally efficient calculation of biochemical reactions. We believe that continuum electrostatics has a broad range of applications in biomolecular modeling and the value of the method will become even more obvious when more and larger protein machines will be investigated.

Acknowledgments This work was supported by the DFG Grants UL 174/8 and BO 3578/1.

References

1. Feig M (ed) (2010) Modeling solvent environments. Wiley-VCH Verlag GmbH & Co. KGaA, Weinheim
2. van Gunsteren WF, Bakowies D, Baron R, Chandrasekhar I, Christen M et al (2001) Biomolecular modeling: goals, problems, perspectives. Angew Chem Int Ed Engl 45:4064–4092
3. Honig B, Nicholls A (1995) Classical electrostatics in biology and chemistry. Science 268:1144–1149
4. Richards FM (1977) Areas, volumes, packing and protein structure. Annu Rev Biophys Bioeng 6:151–176
5. Baker NA, Sept D, Joseph S, Holst MJ, McCammon JA (2001) Electrostatics of nanosystems: application to microtubules and the ribosome. Proc Natl Acad Sci USA 98:10037–10041
6. Kirkwood JG (1934) Theory of solutions of molecules containing widely separated charges with special application to zwitterions. J Chem Phys 2:351–361
7. Daune M (1999) Molecular Biophysics. University Press, Oxford
8. Warwicker J, Watson HC (1982) Calculation of the electrostatic potential in the active site cleft due the α-helix dipols. J Mol Biol 186:671–679
9. Bashford D (1997) An object-oriented programming suite for electrostatic effects in biological molecules. In Yutaka I, Rodney RO, John VWR, Marydell T (eds) Scientific computing in object-oriented parallel environments. Springer, Berlin, pp 233–240
10. Im W, Beglov D, Roux B (1998) Continuum solvation model: electrostatic forces from numerical solutions to the Poisson-Boltzmann equation. Comp Phys Comm 111:59–75
11. Boschitsch AH, Fenley MO (2011) A fast and robust Poisson-Boltzmann solver based on adaptive cartesian grids. J Chem Theor Comput 7:1524–1540
12. Holst MJ, Saied F (1993) Multigrid solution of the Poisson-Boltzmann equation. J Comput Chem 14:105–113
13. Holst MJ, Saied F (1995) Numerical-solution of the nonlinear Poisson-Boltzmann equation: developing more robust and efficient methods. J Comput Chem 16:337–364
14. Holst MJ (2001) Adaptive numerical treatment of elliptic systems on manifolds. Adv Comp Math 15:139–191
15. Baker NA, Sept D, Joseph S, Holst MJ, McCammon JA (2001) Electrostatics of nanosystems: application to microtubules and the ribosome. Proc Natl Acad Sci USA 98:10037–10041
16. Sklenar H, Eisenhaber F, Poncin M, Lavery R (1990) Including solvent and counterion effects in the force fields of macromolecular mechanics: the field integrated electrostatic approach (FIESTA). In: David LB, Richard L (eds) Theoretical biochemistry and molecular biophysics, pp 317–335

17. Cortis CM, Friesner RA (1997) Numerical solution of the Poisson-Boltzmann equation using tetrahedral finite-element meshes. J Comp Chem 18:1591–1608
18. Lee B, Richards FM (1971) The interpretation of protein structures: estimation of static accessibility. J Mol Biol 55:379–380
19. Klapper I, Hagstrom R, Fine R, Sharp K, Honig B (1986) Focusing of electric fields in the active site of Cu–Zn superoxide dismutase: effects of ionic strength and amino-acid modification. Proteins 1:47–59
20. Sengupta D, Behera RN, Smith JC, Ullmann GM (2005) The α-helix dipole: screened out? Structure 13:849–855
21. Sengupta D, Meinhold L, Langosch D, Ullmann GM, Smith JC (2005) Energetics of helical-peptide orientations in membranes. Proteins 58:913–922
22. Humphrey W, Dalke A, Schulten K (1996) VMD: visual molecular dynamics. J Mol Graph 14:33–38
23. Bashir Q, Scanu S, Ubbink M (2011) Dynamics in electron transfer protein complexes. FEBS J 278:1391–1400
24. Ullmann GM, Knapp EW, Kostic NM (1997) Computational simulation and analysis of the dynamic association between plastocyanin and cytochrome f. Consequences for the electron-transfer reaction. J Am Chem Soc 119:42–52
25. Ubbink M, Ejdebäck M, Karlson BG, Bendall DS (1998) The structure of the complex of plastocyanin and cytochrome f, determined by paramagnetic NMR and restrained rigid body molecular dynamics. Structure 6:323–335
26. Qin L, Kostic NM (1993) Importance of protein rearrangement in the electron-transfer reaction between the physiological partners cytochrome f and plastocyanin. Biochemistry 32:6073–6080
27. Pearson DC, Gross EL, David E (1996) Electrostatic properties of cytochrome f: implications for docking with plastocyanin. Biophys J 71:64–76
28. De Rienzo F, Gabdoulline RR, Menziani MC, De Benedetti PG, Wade RC (2001) Electrostatic analysis and Brownian dynamics simulation of the association of plastocyanin and cytochrome f. Biophys J 81:3090–3104
29. Metropolis N, Rosenbluth AW, Rosenbluth MN, Teller AH (1953) Equation of state calculation by fast computing machines. J Chem Phys 21:1087–1092
30. Bashir Q, Volkov AN, Ullmann GM, Ubbink M (2010) Visualization of the encounter ensemble of the transient electron transfer complex of cytochrome c and cytochrome c peroxidase. J Am Chem Soc 132:241–247
31. Volkov AN, Bashir Q, Worrall JAR, Ullmann GM, Ubbink M (2010) Shifting the equilibrium between the encounter state and the specific form of a protein complex by interfacial point mutations. J Am Chem Soc 132:11487–11495
32. Kostić NM (1996) Dynamic aspects of electron-transfer reactions in metalloprotein complexes. In: Metal-containing polymeric materials. Plenum Press, New York, pp 491–500
33. Madura JD, Davis ME, Gilson MK, Wade RC, Luty BA et al (1994) Biological applications of electrostatic calculations and Brownian dynamics. Rev Comp Chem 5:229–267
34. Gabdoulline RR, Wade RC (2002) Biomolecular diffusional association. Curr Opin Struct Biol 12:204–213
35. Andrew SM, Thomasson KA, Northrup SH (1993) Simulation of electron-transfer self-exchange in cytochromes c and b_5. J Am Chem Soc 115:5516–5521
36. Northrup SH, Allison SA, McCammon JA (1984) Brownian dynamics simulation of diffusion-influenced bimolecular reactions. J Chem Phys 80:1517–1524
37. Northrup SH (1994) Hydrodynamic motions of large molecules. Curr Opin Struct Biol 4:269–274
38. Pearson DC, Gross EL (1995) The docking of cytochrome f with plastocyanin: three possible complexes. In: Mathis P (ed) Photosynthesis: from light to biosphere, vol II. Kluwer Academic Publishers, New York, pp 729–732
39. Haddadian EJ, Gross EL (2006) A Brownian dynamics study of the effects of cytochrome f structure and deletion of its small domain in interactions with cytochrome c_6 and plastocyanin in *Chlamydomonas reinhardtii*. Biophys J 90:566–577

40. Gross EL, Rosenberg I (2006) A Brownian dynamics study of the interaction of phormidium cytochrome f with various cyanobacterial plastocyanins. Biophys J 90:366–380

41. Haddadian EJ, Gross EL (2005) Brownian dynamics study of cytochrome f interactions with cytochrome c(6) and plastocyanin in *Chlamydomonas reinhardtii* plastocyanin, and cytochrome c(6) mutants. Biophys J 88:2323–2339

42. Ullmann GM, Hauswald M, Jensen A, Kostic NM, Knapp EW (1997) Comparison of the physiologically-equivalent proteins cytochrome c_6 and plastocyanin on the basis of their electrostatic potentials. Tryptophane 63 in cytochrome c_6 may be isofunctional with tyrosine 83 in plastocyanin. Biochemistry 36:16187–16196

43. Ullmann GM, Hauswald M, Jensen A, Knapp EW (2000) Superposition of ferredoxin and flavodoxin using their electrostatic potentials. Implications for their interactions with photosystem I and ferredoxin: NADP reductase. Proteins 38:301–309

44. Hodgkin E, Richards W (1987) Molecular similarity based on electrostatic potential and electric field. Int J Quant Chem Quant Biol Symp 14:105–110

45. Klingen AR, Ullmann GM (2006) Theoretical investigation of the behavior of titratable groups in proteins. Photochem Photobiol Sci 5:588–596

46. Bashford D, Karplus M (1990) pK_as of ionizable groups in proteins: atomic detail from a continuum electrostatic model. Biochemistry 29:10219–10225

47. Ullmann GM, Knapp EW (1999) Electrostatic computations of protonation and redox equilibria in proteins. Eur Biophys J 28:533–551

48. Ullmann GM (2000) The coupling of protonation and reduction in proteins with multiple redox centers: theory, computational method, and application to cytochrome c_3. J Phys Chem B 104:6293–6301

49. Gunner MR, Mao J, Song Y, Kim J (2006) Factors influencing the energetics of electron and proton transfers in proteins. What can be learned from calculations. Biochim Biophys Acta 1757:942–968

50. Nielsen JE, McCammon JA (2003) Calculating pKa values in enzyme active sites. Protein Sci 12:1894–1901

51. Beroza P, Fredkin DR, Okamura MY, Feher G (1991) Protonation of interacting residues in a protein by a Monte Carlo method: application to lysozyme and the photosynthetic reaction center. Proc Natl Acad Sci USA 88:5804–5808

52. Ullmann RT, Ullmann GM (2012) GMCT: a Monte Carlo simulation package for macromolecular receptors. J Comp Chem 33:887–900

53. Ullmann GM (2003) Relations between protonation constants and titration curves in polyprotic acids: a critical view. J Phys Chem B 107:6293–6301

54. Onufriev A, Case DA, Ullmann GM (2001) A novel view on the ph titration of biomolecules. Biochemistry 40:3413–3419

55. Onufriev A, Ullmann GM (2004) Decomposing complex ligand binding into simple components: connections between microscopic and macroscopic models. J Phys Chem B 108:11157–11169

56. Bombarda E, Ullmann GM (2010) pH-dependent pk_a values in proteins-a theoretical analysis of protonation energies with practical consequences for enzymatic reactions. J Phys Chem B 114:1994–2003

57. Kula R, Sawyer D (1964) Protonation studies of anion of diethylenetriaminepentaacetic acid by nuclear magnetic resonance. Inorg Chem 3:458

58. Sudmeier JL, Reilley CN (1964) Nuclear magnetic resonance studies of protonation of polyamine and aminocarboxylate compounds in aqueous solution. Analyt Chem 36:1698–1706

59. Letkeman P (1979) An NMR protonation study of metal diethylenetriaminepentaacetic acid complexes. J Chem Ed 56:348–351

60. Klamt A, Schuurmann G (1993) Cosmo: a new approach to dielectric screening in solvents with explicit expressions for the screening energy and its gradient. J Chem Soc Perkin Trans 2:799–805

61. Cossi M, Rega N, Scalmani G, Barone V (2003) Energies, structures, and electronic properties of molecules in solution with the C-PCM solvation model. J Comput Chem 24:669–681

62. Miertus S, Scrocco E, Tomas J (1981) Electrostatic interaction of a solute with a continuum. A direct utilization of Ab initio molecular potentials for the prevision of solvent effects. Chem Phys 55:117–129

63. Chen JL, Noodleman L, Case D, Bashford D (1994) Incorporating solvation effects into density functional electronic structure calculations. J Phys Chem 98:11059–11068

64. Li J, Fischer CL, Chen JL, Bashford D, Noodleman L (1996) Calculation of redox potentials and pK_a values of hydrated transition metal cations by a combined density functional and continuum dielectric theory. J Phys Chem 96:2855–2866

65. Richardson WH, Peng C, Bashford D, Noodleman L, Case DA (1997) Incorporating solvation effects into density functional theory: calculation of absolute acidities. Int J Quant Chem 61:207–217

66. Li J, Nelson MR, Peng CY, Bashford D, Noodleman L (1998) Incorporating protein environments in density functional theory: a self-consistent reaction field calculation of redox potentials of [2Fe$_2$S] clusters in ferredoxin and phthalate dioxygenase reductase. J Phys Chem A 102:6311–6324

67. Liu T, Han WG, Himo F, Ullmann GM, Bashford D et al (2004) Density functional vertical self-consistent reaction field theory for solvatochromism studies of solvent-sensitive dyes. J Phys Chem B 108:11157–11169

68. Sham YY, Muegge I, Warshel A (1999) Simulating proton translocations in proteins: probing proton transfer pathways in the *Rhodobacter sphaeroides* reaction center. Proteins 36:484–500

69. Ferreira A, Bashford D (2006) Model for proton transport coupled to protein conformational change: application to proton pumping in the bacteriorhodopsin photocycle. J Am Chem Soc 128:16778–16790

70. Becker T, Ullmann RT, Ullmann GM (2007) Simulation of the electron transfer between the tetraheme-subunit and the special pair of the photosynthetic reaction center using a microstate description. J Phys Chem B 111:2957–2968

71. Bombarda E, Ullmann GM (2011) Continuum electrostatic investigations of charge transfer processes in biological molecules using a microstate description. Faraday Discuss 148:173–193

72. Kloppmann E, Ullmann GM, Becker T (2007) An extended dead-end elimination algorithm to determine gap-free lists of low energy states. J Comp Chem 28:2325–2335

73. Moser CC, Keske JM, Warncke K, Farid RS, Dutton PL (1992) Nature of biological electron transfer. Nature 355:796–802

74. Page CC, Moser CC, Chen X, Dutton PL (1999) Natural engineering principles of electron tunneling in biological oxidation–reduction. Nature 402:47–52

75. Marcus RA (1963) Free energy of nonequilibrium polarization systems. II. Homogeneous and electrode systems. J Chem Phys 38:1858–1862

76. Sharp KE (1998) Calculation of electron transfer reorganization energies using the finite difference Poisson Boltzmann model. Biophys J 73:1241–1250

77. Marcus RA, Sutin N (1985) Electron transfer in chemistry and biology. Biochim Biophys Acta 811:265–322

78. Williams RJP (1999) Electron transfer and proton coupling in proteins. J Solid State Chem 145:488–495

79. Olsson MHM, Ryde U, Roos BO (1998) Quantum chemical calculation of the reorganization energy of blue copper proteins. Prot Sci 81:6554–6558

80. Ryde U, Olsson MHM (2001) Structure, strain and reorganization energy of blue copper models in the protein. Int J Quant Chem 81:335–347

81. Gillespie DT (2001) Approximate accelerated stochastic simulation of chemically reacting systems. J Chem Phys 115:1716–1733

82. Till MS, Becker T, Essigke T, Ullmann GM (2008) Simulating the proton transfer in gramicidin A by a sequential dynamical Monte Carlo method. J Phys Chem B 112:13401–13410

Chapter 7
Molecular Mechanics/Coarse-Grained Models

Alejandro Giorgetti and Paolo Carloni

7.1 Molecular Mechanics/Coarse-Grained (MM/CG) Methods

Ligand-protein docking is currently an important tool in drug discovery efforts, indeed, in the last years, structure-based drug design protocols has been the subject of important developments [1–4]. These are well portrayed in the rising number of available protein-ligand docking software programs [5]. Molecular dynamics (MD) simulations are also instruments for determining poses and energetics in these cases. Still a very important challenge needs to be overcome: the prediction of a ligand-protein complex when there is no structural data for the protein and templates of low sequence identity, lower than 30 %, are the only choice for building the model by using homology techniques. Indeed, in bioinformatics based homology models the orientation of the side chains, a key factor in the interaction

A. Giorgetti
Computational Biophysics, German Research School for Simulation Sciences, Jülich, Germany

A. Giorgetti
Department of Biotechnology, University of Verona, Ca' Vignal 1, Verona, Italy

P. Carloni (✉)
Computational Biophysics, German Research School for Simulation Sciences (joint venture of RWTH Aachen University and Forschungszentrum Jülich, Germany), Jülich, Germany
e-mail: p.carloni@grs-sim.de

P. Carloni
Computational Biomedicine, Institute for Advanced Simulation IAS-5, Forschungszentrum Jülich, Jülich, Germany

P. Carloni
Institute of Neuroscience and Medicine INM-9, Forschungszentrum Jülich, Jülich, Germany

© Springer International Publishing Switzerland 2014
G. Náray-Szabó (ed.), *Protein Modelling*, DOI 10.1007/978-3-319-09976-7_7

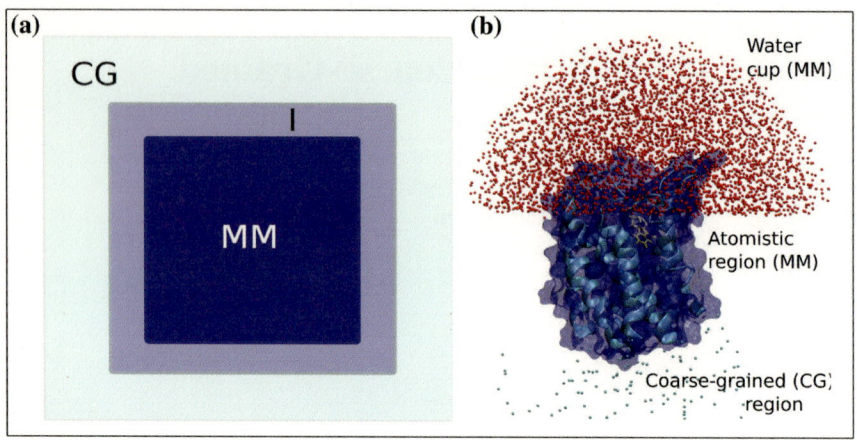

Fig. 7.1 Molecular mechanics/coarse-grained system set-up. **a** Schematic representation of the regions defined in the MM/CG model. The CG, I and MM regions are colored in an increasing *dark blue* tonalities. **b** MM/CG representation of the hTAS2R38 receptor in complex with its agonist. Water molecules are represented as *red points*. The agonist is represented in *yellow*. The protein Cα atoms are represented as *cyan points*. The membrane wall is represented as a *blue surface*

between the ligand and the protein, is not highly accurate, hampering consequently the correct prediction of docking poses obtained by using standard docking algorithms [6]. A way of overcoming these difficulties is to combine homology models and docking procedures with hybrid Molecular Mechanics/Coarse-Grained (MM/CG) approaches [7–16], that is including together, in the same simulation, high- and low-resolution models [7, 9, 17–19] of the system that will be simulated. From one hand a portion of the protein (i.e. the ligand binding site) is described in full atomistic details, while the remaining regions of the protein are described using a coarse-grained representation [10, 11, 20] (Fig. 7.1a). Thus, the accuracy of multiscale models relies upon the methods used for constructing an accurate connection between the boundary of models with different resolutions [21].

We have recently adapted to membrane proteins an MM/CG approach originally developed by us to be used with enzymes. In our case a limited number of water molecules is included in the set up and is confined in a drop next to the binding cavity by introducing repulsive walls into the system [10, 11, 20] (see Fig. 7.1b and below for details).

Thus, the potential energy function in our MM/CG scheme reads:

$$V = E_{MM} + E_I + E_{I/MM} + E_{CG} + E_{CG/I} \tag{7.1}$$

where E_{MM}, E_I and E_{CG} are the potential energy of the atomistic (MM) region, the interface (I) and the coarse grained (CG) regions, respectively. $E_{I/MM}$ and $E_{CG/I}$ describe the interaction energy between I and the MM region and that between the interface and the CG region, respectively. E_{MM}, E_I and $E_{I/MM}$ are described by the GROMOS96 force field [22], whereas E_{CG} and $E_{CG/I}$ are characterized by a Go-like model (Eq. 7.2). $E_{CG/I}$ ensures the integrity and stability of the protein backbone, acting as a structural scaffold. Indeed, the latter term includes the bonded interactions between the CG atoms and the $C\alpha$ atoms in the interface, as well as the non-bonded interactions between CG atoms and the $C\alpha$, $C\beta$ atoms in the interface. The E_{CG} term reads:

$$E_{CG} = \frac{1}{4} \sum_i K_b \left(|R_i - R_{i+1}|^2 - b_{ii+1}^2 \right)^2 + \sum_{i>j} V_0 \left\{ 1 - \exp\left[-B_{ij}\left(|R_i - R_j| - b_{ij} \right) \right] \right\}^2 \tag{7.2}$$

The first term describes the interaction between consecutive CG beads (the $C\alpha$ atoms), where K_b is the force constant and b_{ij} is the equilibrium distance corresponding to the original distances between CG atoms as calculated from the initial structure/model. Non-bonded interactions are taken into account by using a Morse-type potential (second term), here $V_0 = 5.3$ kJ mol^{-1} is the well depth and its modulating coefficient is $B_{ij} = 6/b_{ij}$ nm^{-1}. These two parameters have been already employed in investigating both soluble and membrane proteins [10, 11]. Here, B_{ij} is set to $5 + 6/b_{ij}$ nm^{-1}. This setup ensures the stability of the protein inside its transmembrane site.

The thermal and viscous solvent effects acting on the system are taken into account by using the Langevin approach with a potential of mean force, $V(r_i)$ [23]:

$$m_i \frac{d^2 r_i}{dt^2} = -m_i \gamma_i \frac{dr_i}{dt} + V(r_i) + \eta_i(t) \tag{7.3}$$

where γ_i is the friction coefficient and η_i is a stochastic noise satisfying the relations: $\langle \eta_i(t) \rangle = 0$ and $\langle \eta_i(t) \eta_j(t') \rangle = \delta_{ij}\delta(t - t')2K_B T\gamma_i$; where K_B is the Boltzmann constant and T is the temperature. If the I and MM regions are solvent exposed, the solvent is treated explicitly by using the SPC water model [24]. In the framework of the MM/CG approach: a droplet of water molecules is centered around the MM and I regions. Within this approach, water properties are very similar to those of the bulk water in proximity of the all-atom region, but approaching the drop border located approximately at the interface region, the water density lowers, providing a rough approximation of the bulk behavior [10].

In our approach, the presence of implicit membrane is realized by introducing five repulsive walls ($\varphi_i, i = 1, 2 \ldots 5$) into the system [20]. The five walls, around the protein are described by five corresponding functions using a level-set approach [25]. The region of points $\mathbf{r}$ where all the five $\varphi_i(\mathbf{r})$ are positive characterizes the protein site. The wall i itself is formed by the set of points for which φ_i vanishes. Two planar walls ($\varphi_i, i = 1, 2$) coincide with the height of the heads of membrane lipids. Two hemispheric walls ('outer walls', $\varphi_i, i = 3, 4$) (Fig. 7.1b), capping the extracellular and cytoplasmic ends of the protein, are described by the functions $\varphi_i(\mathbf{r}) = r_i - \|\mathbf{r} - c_{hi}\|$ defined only outside the membrane region. The center c_{hi} of each hemisphere is located at the height of the heads of phospholipids, above/under the center of mass of the protein. The radius r_i of each hemisphere is defined such that the minimum distance between any protein atoms and the wall is 15 Å. This creates a droplet of waters around the MM region similar to Refs. [10, 11]. The membrane wall φ_5 is defined by $\varphi_5(\mathbf{r}) = r_p - \min_j \|\mathbf{r} - c_j\|$, where the distance between the point r and the closest initial position of Cα atoms c_j is computed, and r_p is a distance parameter with a default value 2.0 Å. Additionally, a smoothing technique [20] is applied to avoid discontinuities in the wall.

Boundary potentials $V_i(d)(i = 1, 2 \ldots 5)$ are added to the MM/CG potential energy function. They are defined as functions of a distance d of an atom from the corresponding walls:

$$V_i(d) = \frac{1}{d} \qquad \text{for } i = 1, 2; \tag{7.4}$$

$$V_i(d) = 4\varepsilon \left[\left(\frac{\sigma}{d} \right)^2 - \frac{\sigma}{d} \right] \quad \text{for } i = 3, 4, 5. \tag{7.5}$$

In particular, the potential applied to an atom is the one corresponding to the closest wall φ_i from that atom, i.e. $V_i \left(i : \min_{r'}(\mathbf{r} - \varphi_i(\mathbf{r}')) = d \right)$. $V_i(i = 1, 2)$ is purely repulsive; $V_i(i = 3, 4, 5)$ is a softened Lennard-Jones-type potential; ε is the depth of the potential well; and σ is the finite distance at which the potential $V_i(i = 3, 4, 5)$ is zero. The minimum of the potential is at $d = 2\sigma = r_p$. Waters, Cα atoms of both MM and CG regions, and atoms belonging to external aromatic residues Trp and Tyr are influenced by these potentials. The membrane wall potential V_5 constrains the shape of the protein while providing a good degree of flexibility. This model neither includes electrostatics nor allows distinguishing between different types of bilayers.

The force due to the presence of the wall is derived from the following equations:

$$\overrightarrow{F_i}(r) = -\frac{\partial V_i}{\partial d} \nabla d(r). \tag{7.6}$$

The cut-off distance of the force is set to 7 Å for the repelling walls $V_i(i = 1, 2)$, and to $1.5r_p$ for the outer walls and membrane wall $V_i(i = 3, 4, 5)$. The first value is chosen such that a water molecule cannot pass through this distance during one time step, while the second value guarantees that the force does not affect the MM region. The force is shifted so that it is continuous at the cut-off distance, to avoid a sharp disruption. In addition, it is set to a finite value (1,000 kJ mol^{-1} nm^{-1}) near the wall to prevent too large forces acting on the system.

7.2 Application: G Protein-Coupled Receptors (GPCRs)

GPCRs form the largest membrane-bound receptor family expressed by mammalians (encompassing ca. 4 % of the protein-coding human genome) [26] and are of paramount importance for pharmaceutical intervention (ca. 40 % of currently marketed drugs target GPCRs) [27]. GPCRs are located in the plasma membrane and initiate signaling cascades that allow cells to react to changes within their environment

[28]. Indeed, they transduce signals from the extracellular environment through their interactions with agonists and intracellular heterotrimeric guanine nucleotide-binding proteins (G proteins). All GPCRs share a common three-dimensional (3D) fold that comprises an extracellular N-terminal loop (N-term), followed by seven trans-membrane (TM) α-helices (TM1 to TM7) connected by intracellular, extracellular loops, and an intracellular C-terminal loop (C-term) [29]. The tertiary structure resembles a barrel, with the seven transmembrane helices forming a cavity within the plasma membrane that serves as ligand-binding domain, often covered by the extracellular loop 2 (EL-2). In several cases they can exist as homo- or hetero-dimers or higher-order oligomers during their life cycle in vivo [30].

Recently, there was an explosion in the crystallography of GPCRs [29, 31–33]. On the other hand, while ca. 800 human GPCR' sequences are public, there are actually only 24 unique experimental structures (as reported in the http://blanco.biomol.uci.edu/mpstruc web site) [34]. Only recently one structure belonging to the frizzled/taste2 family [35] and two structures from the secretin family [36, 37] were solved. No structure is still available for the ca. 400 receptors large olfactory receptors sub-branch, as well for the bitter taste receptors [or Taste2 receptors (T2R), which constitute about a half of the frizzled/taste2 family] and for the glutamate, and adhesion families.

The lack of structural information for many of the members of the GPCR family, calls upon computational biology-based structural predictions. The average sequence identity between GPCRs is often below 20 % [38], making target selection and alignment required for homology modeling far from trivial [39–42]. Furthermore, research aimed at elucidating the underlying principles determining the molecular responsiveness range of GPCRs that mediate senses, such as T2R [43, 44] and olfactory [45] receptors, depends on the ability to build reliable models of the interaction sites. Experimental validation is thus crucial for accurate structural characterization (see Refs. [46–51] for some recent examples). A crucial step in understanding specificity and promiscuity in molecular recognition and structure-based design is to identify residues that are important for ligand binding. Several groups have successfully applied homology-based of non-rhodopsin GPCRs structure modeling approaches to ligand-binding elucidation [47, 52–55]. Interestingly, the bigger amount of works based on homology modeling is dedicated to rhodopsin-like GPCRs [56–58] and T2R GPCRs [20, 44, 47, 55, 59]. In these articles, the combination of homology modeling and/or molecular docking with a rapid growing number of protein structures deposited in the PDB database, together with the availability of functional assays, allowed to represent the ligand-binding interactions at an unprecedented level of detail. On the other hand, standard docking procedures on homology modeling, such as those used in Refs. [60, 61], suffer from severe limitations which greatly limit the predictive power of these methods. The limitations in structural predictions include the difficulty in predicting correctly side chains orientations in the binding site [6] and neglecting the presence of explicit solvent [62]. This is particularly important for GPCRs, as water molecules are often found in the binding site of several receptors and were shown to be crucial for the stabilization of the agonists [63, 64]. Unfortunately, the current molecular docking approaches do not use the laws of statistical mechanics, which provide the only way to calculate free energy values.

In our approach, the GPCR's ligand, the binding site and the water molecules around it are treated using an atomistic force field, whilst the protein frame is described at CG level. This methods is not only much cheaper than full atoms MD [20], but it is likely also to be more accurate in the case of homology models with low sequence identity (SI) with the template, exactly like in the case of GPCR [38]. Indeed, if we do not know where side chains are located (which is the case with SI about 20 %), it might actually be better not to include them rather than including them in wrong orientations.

We have tested the predictive power of the MM/CG approach on the crystal structure of human β2 adrenergic receptor (β2AR) (PDB: 2RH1) [65] in complex with its inverse agonist S-Carazolol (S-Car) and its agonist R-Isoprenaline (R-Iso) [20]. For the validation of our approach we have use an all-atom MD simulation performed on the same system by Vanni et al. [66]. The MM/CG simulations were carried out for up to 800 ns. The MM region consisted of 476 and 486 atoms, while the overall system was made of just 4597 and 4587 atoms for β2AR in complex with S-Car and R-Iso, respectively. This allowed us to obtain a 15-fold speedup compared to the all-atom MD simulations of the same system [66]. The structure of the complex between the β2AR and the agonist R-Iso was obtained following the procedure in Ref. [66]. The MM/CG simulations carried out on the β2AR/S-Car and on the β2AR/R-Iso systems allowed to capture the principal features regarding ligand-receptor interactions, in agreement with the corresponding ones with all-atom MD. In a second step of the procedure, with the aim of eliminating bias introduced in the calculations putatively caused by the initial positioning of the ligands and to gain insights into the predictive power of our method, we ran additional simulations in which we have deliberately docked the ligand in a wrong position, with all the main interactions with the residues found in the X-ray structure of hb2-AR/S-Car complex lacking. In these new simulations, the ligand migrates to the correct pose in about 200 ns, again capturing and forming the key interactions. We have then produced a homology model of the same complex using a standard modeling approach that included the use of the MODELLER [67] program. The chosen template for building up the model of the β2AR was the structure of squid rhodopsin (PDB id 2Z73) [68], that share a sequence identity of 20 % with the target protein. After 0.8 μs of MM/CG simulation time, the β2AR structure in complex with S-Car is similar to the X-ray structure (root-mean-square-deviation of the Cα atoms 2 Å). The interactions observed between the ligand and the protein present in the X-ray structure are reproduced also in the MM/CG simulation. Hence, MM/CG simulations on a homology-modeled structure reproduce the ligand pose as in the X-ray structure [65], indicating that this approach can be used in general for ligand/GPCRs complexes.

We have then applied the same procedure to the human TAS2R38 receptor [55] in complex with its agonists phenylthiocarbamide (PTC) (Fig. 7.1b) and propylthiouracil (PROP). To study the interactions with PTC and PROP, the best homology model was then funneled through a standard docking protocol using the information driven Haddock program [55]. We have then selected two representative docking poses. Although these models largely satisfied the existing experimental data

[47], in order to obtain a more accurate description of the binding poses and the specific receptor-ligand interactions, the two models underwent μs-long MM/CG simulations, each at room temperature. MM/CG identified the best binding pose of each agonist. New site-directed mutagenesis experiments were carried out, which confirmed the predicted models. These predictions are consistent with data sets based on more than 20 site-directed mutagenesis and functional calcium imaging experiments of TAS2R38 [55]. The calculations pointed out key interactions between hTAS2R38 and its agonists, which would have been impossible to capture with standard bioinformatics/docking approaches [55]. Another very interesting outcome regards the EL-2 loop, indeed, after MM/CG simulations validated with site-directed mutagenesis experiments, we concluded that the loop conformation may resemble the non-rhodopsin models, because we have shown that the loop does not interact with the ligands.

7.3 Conclusions

In this chapter we have illustrated efforts aimed at characterizing the interactions on ligand-GPCR complexes for which there is a lack of structural information and bioinformatics based structural predictions is challenging. A protocol combining low-resolution homology modeling, docking and MM/CG simulations such as those developed by us shows considerable predictive power. The encouraging results reported here lead us to suggest that in the near future combined methodologies such as those described here may help in structure-based drug design studies.

References

1. Michel J (2014) Current and emerging opportunities for molecular simulations in structure-based drug design. Phys Chem Chem Phys 16:4465–4477
2. Congreve M, Dias JM, Marshall FH (2014) Structure-based drug design for g protein-coupled receptors. Prog Med Chem 53:1–63
3. Klebe G (2013) Protein modeling and structure-based drug design. In: Klebe G (ed) Drug design. Springer, Berlin, pp 429–448
4. Rastelli G (2013) Emerging topics in structure-based virtual screening. Pharm Res 30:1458–1463
5. Sousa SF, Ribeiro AJ, Coimbra JT, Neves RP, Martins SA, Moorthy NS, Fernandes PA, Ramos MJ (2013) Protein-ligand docking in the new millennium—a retrospective of 10 years in the field. Curr Med Chem 20:2296–2314
6. Eswar N, Webb B, Marti-Renom MA, Madhusudhan MS, Eramian D, Shen MY, Pieper U, Sali A (2007) Comparative protein structure modeling using modeller. Curr Protoc Protein Sci 5–6
7. Villa E, Balaeff A, Mahadevan L, Schulten K (2004) Multiscale method for simulating protein-DNA complexes. Multiscale Model Simul 2:527–553
8. Shi Q, Izvekov S, Voth GA (2006) Mixed atomistic and coarse-grained molecular dynamics: simulation of a membrane-bound ion channel. J Phys Chem B 110:15045–15048

9. Villa E, Balaeff A, Schulten K (2005) Structural dynamics of the lac repressor, DNA complex revealed by a multiscale simulation. Proc Natl Acad Sci USA 102:6783–6788
10. Neri M, Anselmi C, Cascella M, Maritan A, Carloni P (2005) Coarse-grained model of proteins incorporating atomistic detail of the active site. Phys Rev Letters 95:218102
11. Neri M, Baaden M, Carnevale V, Anselmi C, Maritan A, Carloni P (2008) Microseconds dynamics simulations of the outer-membrane protease T. Biophys J 94:71–78
12. Kalli AC, Campbell ID, Sansom MS (2011) Multiscale simulations suggest a mechanism for integrin inside-out activation. Proc Natl Acad Sci USA 108:11890–11895
13. Messer BM, Roca M, Chu ZT, Vicatos S, Kilshtain AV, Warshel A (2010) Multiscale simulations of protein landscapes: using coarse-grained models as reference potentials to full explicit models. Proteins 78:1212–1227
14. Rzepiela AJ, Louhivuori M, Peter C, Marrink SJ (2011) Hybrid simulations: combining atomistic and coarse-grained force fields using virtual sites. Phys Chem Chem Phys 13:10437–10448
15. Wassenaar TA, Ingólfsson HI, Prieß M, Marrink SJ, Schäfer LV (2013) Mixing martini: electrostatic coupling in hybrid atomistic–coarse-grained biomolecular simulations. J Phys Chem B 117:3516–3530
16. Han W, Schulten K (2012) Further optimization of a hybrid united-atom and coarse-grained force field for folding simulations: improved backbone hydration and interactions between charged side chains. J Chem Theory Comput 8:4413–4424
17. Lyman E, Ytreberg FM, Zuckerman DM (2006) Resolution exchange simulation. Phys Rev Lett 96:028105
18. Lyman E, Zuckerman DM (2006) Resolution exchange simulation with incremental coarsening. J Chem Theory Comput 2:656–666
19. Liu P, Voth GA (2007) Smart resolution replica exchange: an efficient algorithm for exploring complex energy landscapes. J Chem Phys 126(4):045106
20. Leguebe M, Nguyen C, Capece L, Hoang Z, Giorgetti A, Carloni P (2012) Hybrid molecular mechanics/coarse-grained simulations for structural prediction of G-protein coupled receptor/ligand complexes. PLoS ONE 7:e47332
21. Ayton GS, Noid WG, Voth GA (2007) Multiscale modeling of biomolecular systems: in serial and in parallel. Curr Opin Str Biol 17:192–198
22. Scott WRP, Hunenberger PH, Tironi IG, Mark AE, Billeter SR, Fennen J, Torda AE, Huber T, Kruger P, van Gunsteren WF (1999) The GROMOS biomolecular simulation program package. J Phys Chem A 103:3596–3607
23. Nadler W, Brunger AT, Schulten K, Karplus M (1987) Molecular and stochastic dynamics of proteins. Proc Natl Acad Sci USA 84:7933–7937
24. Berweger CD, van Gunsteren WF, Müller-Plathe F (1995) Force field parametrization by weak coupling. Re-engineering SPC water. Chem Phys Lett 232:429–436
25. Osher S, Sethian JA (1988) Fronts propagating with curvature-dependent speed—algorithms based on hamilton-jacobi formulations. J Comput Phys 79:12–49
26. Schoneberg T, Schulz A, Biebermann H, Hermsdorf T, Rompler H, Sangkuhl K (2004) Mutant G-protein-coupled receptors as a cause of human diseases. Pharmacol Ther 104:173–206
27. Overington JP, Al-Lazikani B, Hopkins AL (2006) How many drug targets are there? Nat Rev Drug Discov 5:993–996
28. Audet M, Bouvier M (2012) Restructuring G-protein-coupled receptor activation. Cell 151:14–23
29. Venkatakrishnan AJ, Deupi X, Lebon G, Tate CG, Schertler GF, Babu MM (2013) Molecular signatures of G-protein-coupled receptors. Nature 494:185–194
30. Gurevich VV, Gurevich EV (2008) GPCR monomers and oligomers: it takes all kinds. Trends Neurosci 31:74–81
31. Rosenbaum DM, Rasmussen SG, Kobilka BK (2009) The structure and function of G-protein-coupled receptors. Nature 459:356–363
32. Topiol S, Sabio M (2009) X-ray structure breakthroughs in the GPCR transmembrane region. Biochem Pharmacol 78:11–20
33. Sprang SR (2011) Cell signalling: binding the receptor at both ends. Nature 469:172–173

34. Bjarnadottir TK, Gloriam DE, Hellstrand SH, Kristiansson H, Fredriksson R, Schioth HB (2006) Comprehensive repertoire and phylogenetic analysis of the G protein-coupled receptors in human and mouse. Genomics 88:263–273
35. Wang C, Wu H, Katritch V, Han GW, Huang XP, Liu W, Siu FY, Roth BL, Cherezov V, Stevens RC (2013) Structure of the human smoothened receptor bound to an antitumour agent. Nature. doi:10.1038/nature12167
36. Hollenstein K, Kean J, Bortolato A, Cheng RK, Dore AS, Jazayeri A, Cooke RM, Weir M, Marshall FH (2013) Structure of class B GPCR corticotropin-releasing factor receptor 1. Nature 499:438–443
37. Siu FY, He M, de Graaf C, Han GW, Yang D, Zhang Z, Zhou C, Xu Q, Wacker D, Joseph JS, Liu W, Lau J, Cherezov V, Katritch V, Wang MW, Stevens RC (2013) Structure of the human glucagon class B G-protein-coupled receptor. Nature 499:444–449
38. Rayan A (2010) New vistas in GPCR 3D structure prediction. J Mol Model 16:183–191
39. Giorgetti A, Raimondo D, Miele AE, Tramontano A (2005) Evaluating the usefulness of protein structure models for molecular replacement. Bioinformatics 21 Suppl 2: ii72–ii76
40. Lupieri P, Nguyen CH, Bafghi ZG, Giorgetti A, Carloni P (2009) Computational molecular biology approaches to ligand-target interactions. Hfsp J 3:228–239
41. Tramontano A, Cozzetto D, Giorgetti A, Raimondo D (2008) The assessment of methods for protein structure prediction. Methods Mol Biol 413:43–57
42. Kufareva I, Rueda M, Katritch V, Stevens RC, Abagyan R (2011) Status of GPCR modeling and docking as reflected by community-wide GPCR Dock 2010 assessment. Structure 19:1108–1126
43. Cui M, Jiang P, Maillet E, Max M, Margolskee RF, Osman R (2006) The heterodimeric sweet taste receptor has multiple potential ligand binding sites. Curr Pharm Des 12:4591–4600
44. Brockhoff A, Behrens M, Niv MY, Meyerhof W (2010) Structural requirements of bitter taste receptor activation. Proc Natl Acad Sci USA 107:11110–11115
45. Reisert J (2010) Origin of basal activity in mammalian olfactory receptor neurons. J Gen physiol 136:529–540
46. Khafizov K, Anselmi C, Menini A, Carloni P (2007) Ligand specificity of odorant receptors. J Mol Model 13:401–409
47. Biarnes X, Marchiori A, Giorgetti A, Lanzara C, Gasparini P, Carloni P, Born S, Brockhoff A, Behrens M, Meyerhof W (2010) Insights into the binding of Phenyltiocarbamide (PTC) agonist to its target human TAS2R38 bitter receptor. PLoS ONE 5:e12394
48. Carlsson J, Coleman RG, Setola V, Irwin JJ, Fan H, Schlessinger A, Sali A, Roth BL, Shoichet BK (2011) Ligand discovery from a dopamine D3 receptor homology model and crystal structure. Nat Chem Biol 7:769–778
49. Levit A, Barak D, Behrens M, Meyerhof W, Niv M (2012) Homology model-assisted elucidation of binding sites in GPCRs. In: Membrane protein structure and dynamics. Humana Press, New York, pp 179–205
50. Mobarec JC, Sanchez R, Filizola M (2009) Modern homology modeling of G-protein coupled receptors: which structural template to use? J Med Chem 52:5207–5216
51. Yarnitzky T, Levit A, Niv MY (2010) Homology modeling of G-protein-coupled receptors with X-ray structures on the rise. Curr Opin in Drug Disc 13:317–325
52. Petrel C, Kessler A, Dauban P, Dodd RH, Rognan D, Ruat M (2004) Positive and negative allosteric modulators of the Ca^{2+}-sensing receptor interact within overlapping but not identical binding sites in the transmembrane domain. J Biol Chem 279:18990–18997
53. de Graaf C, Rognan D (2009) Customizing G Protein-coupled receptor models for structure-based virtual screening. Curr Pharm Des 15:4026–4048
54. Bhattacharya S, Subramanian G, Hall S, Lin J, Laoui A, Vaidehi N (2010) Allosteric antagonist binding sites in class B GPCRs: corticotropin receptor 1. J Comput Aided Mol Des 24:659–674
55. Marchiori A, Capece L, Giorgetti A, Gasparini P, Behrens M, Carloni P, Meyerhof W (2013) Coarse-grained/molecular mechanics of the TAS2R38 bitter taste receptor: experimentally-validated detailed structural prediction of agonist binding. PLoS ONE 8(5):e64675
56. Niv M, Skrabanek L, Filizola M, Weinstein H (2006) Modeling activated states of GPCRs: the rhodopsin template. J Comput Aided Mol Des 20:437–448

57. Niv MY, Filizola M (2008) Influence of oligomerization on the dynamics of G-protein coupled receptors as assessed by normal mode analysis. Proteins 71:575–586
58. Ivanov AA, Barak D, Jacobson KA (2009) Evaluation of Homology Modeling of G-Protein-Coupled Receptors in Light of the A2A Adenosine Receptor Crystallographic Structure. J Med Chem 52:3284–3292
59. Slack JP, Brockhoff A, Batram C, Menzel S, Sonnabend C, Born S, Galindo MM, Kohl S, Thalmann S, Ostopovici-Halip L, Simons CT, Ungureanu I, Duineveld K, Bologa CG, Behrens M, Furrer S, Oprea TI, Meyerhof W (2010) Modulation of bitter taste perception by a small molecule hTAS2R antagonist. Curr Biol 20:1104–1109
60. Garcia-Perez J, Rueda P, Alcami J, Rognan D, Arenzana-Seisdedos F, Lagane B, Kellenberger E (2011) Allosteric model of maraviroc binding to CC chemokine receptor 5 (CCR5). J Biol Chem 286:33409–33421
61. Kothandan G, Gadhe CG, Cho SJ (2012) Structural insights from binding poses of CCR2 and CCR5 with clinically important antagonists: a combined in silico study. PLoS ONE 7:e32864
62. Camacho CJ (2005) Modeling side-chains using molecular dynamics improve recognition of binding region in CAPRI targets. Proteins 60:245–251
63. Angel TE, Chance MR, Palczewski K (2009) Conserved waters mediate structural and functional activation of family A (rhodopsin-like) G protein-coupled receptors. Proc Natl Acad Sci USA 106:8555–8560
64. Nygaard R, Valentin-Hansen L, Mokrosinski J, Frimurer TM, Schwartz TW (2010) Conserved water-mediated hydrogen bond network between TM-I, -II, -VI, and -VII in 7TM receptor activation. J Biol Chem 285:19625–19636
65. Cherezov V, Rosenbaum DM, Hanson MA, Rasmussen SG, Thian FS, Kobilka TS, Choi HJ, Kuhn P, Weis WI, Kobilka BK, Stevens RC (2007) High-resolution crystal structure of an engineered human beta2-adrenergic G protein-coupled receptor. Science 318:1258–1265
66. Vanni S, Neri M, Tavernelli I, Rothlisberger U (2011) Predicting novel binding modes of agonists to b adrenergic receptors using all-atom molecular dynamics simulations. PLoS Comput Biol 7:e1001053
67. Eswar N, Webb B, Marti-Renom MA, Madhusudhan MS, Eramian D, Shen MY, Pieper U, Sali A (2006) Comparative protein structure modeling using modeller. Curr Protoc Bioinform 5–6
68. Murakami M, Kouyama T (2008) Crystal structure of squid rhodopsin. Nature 453:363–367

Chapter 8
Modelling the Dynamic Architecture of Biomaterials Using Continuum Mechanics

Robin Oliver, Robin A. Richardson, Ben Hanson, Katherine Kendrick, Daniel J. Read, Oliver G. Harlen and Sarah A. Harris

8.1 Length and Time Scales in Biomolecular Simulation

Many cellular processes are inherently multi-scale, and can span several computational regimes. For example, the action of the muscle protein myosin involves quantum mechanics at the shortest time and length-scales, as the coupling of ATP hydrolysis to the conformational changes driving the motor result from the enzyme catalysed breaking and reformation of covalent bonds. Motion at the macroscopic level is then achieved through the co-ordinated action of many myosin molecules. The success of techniques such as X-ray crystallography and Nuclear Magnetic Resonance (NMR) in providing biomolecular structures (the

R. Oliver
Sheffield University, Hicks Building Hounsfield Road, Sheffield S3 7RH, UK
e-mail: robin.oliver@sheffield.ac.uk

R.A. Richardson · B. Hanson · K. Kendrick · S.A. Harris (✉)
School of Physics and Astronomy, E C Stoner Building, University of Leeds, Leeds LS2 9JT, UK
e-mail: s.a.harris@leeds.ac.uk

B. Hanson
e-mail: py09bh@leeds.ac.uk

K. Kendrick
e-mail: py10kek@leeds.ac.uk

D.J. Read · O.G. Harlen
School of Mathematics, University of Leeds, Leeds LS2 9JT, UK
e-mail: d.j.read@leeds.ac.uk

O.G. Harlen
e-mail: o.g.harlen@leeds.ac.uk

© Springer International Publishing Switzerland 2014
G. Náray-Szabó (ed.), *Protein Modelling*, DOI 10.1007/978-3-319-09976-7_8

protein database contained 98,900 structures at the time of writing) has stimulated the simulation community to provide computational methods to comprehensively explore the dynamics of these molecules. For example, the MoDEL database, which captures the set of nonhomologous cytoplasmic proteins currently present in the PDB, contains 1,700 atomistic molecular dynamics (MD) simulations of proteins of at least 10 ns in duration [1]. Since these atomistic MD calculations rely on MD forcefields which are parameterised using quantum chemical calculations, quantum and molecular mechanical models have developed somewhat in tandem. Recent improvements in experimental techniques such as cryo-Electron tomography, ion mobility mass spectrometry and NMR have now made it possible to study "the molecular sociology of the cell" [2] in which complex biomolecular assemblies and single molecules are located in crowded macromolecular environments that are far closer to those present in vivo. For these assemblies however, there is often insufficient experimental information to construct an atomistic model. The Electron Microscopy Database (EMDB) currently contains 2,267 three-dimensional EM density maps and grew by a record 544 EMDB maps in 2013 [3]. These density maps describe the overall shape of a biomolecule, biomolecular complex or super-macromolecular structure. The relevant time and length scales are often too large to be computationally accessible using atomistic MD, especially for these larger structures, and so coarse-grained simulation techniques are being increasingly employed as more low resolution experimental data becomes available.

8.2 Mesoscale Simulation Methods

Since the computational expense of calculations at the quantum and atomistic levels places strict limits on the system sizes and time-scales that can be explored, accessing the mesoscale is only possible with coarse-grained simulation. A common scheme for coarse-graining reduces the computational expense of the calculation by combining groups of atoms. For example the Martini forcefield combines four heavy atoms into a single bead [4]. Since these calculations are based on equivalent physical principles to atomistic MD simulation and have been reviewed recently [5], we focus on alternative techniques that are arguably less established for biomolecular simulation. Elastic network models are introduced in Sect. 8.2.1, Brownian dynamics in Sect. 8.2.2, dissipative particle dynamics (DPD) in Sect. 8.2.3 and lattice Boltzmann methods in Sect. 8.2.4. The application of DPD and lattice Boltzmann calculations to fluid flows in biology has been the subject of a comprehensive recent review [6]. Continuum mechanics methods are then introduced in Sect. 8.2.5. In Sect. 8.3, we present a detailed account of fluctuating finite element analysis (FFEA), since this newer approach to mesoscale biomolecular modelling has not been reviewed elsewhere.

8.2.1 Elastic Network Models

Elastic Network Models (ENM) treat a protein as a series of interconnected harmonic springs. A matrix (known as the Hessian matrix) is constructed, containing the spring constants that define these bonded interactions. Normal mode analysis can then be performed, which involves diagonalising this spring constant matrix to obtain a set of eigenvectors, which define structural changes within the protein associated with the modes, and their corresponding eigenvalues, which give the amplitudes of these motions. While in principle all atoms can be included in a normal mode calculation, the difficulty of assigning appropriate spring constants and obtaining a sufficiently robust local minimum means that these calculations normally use coarse-grained protein models. Typically, the structure of the protein of interest is downloaded from the Protein Data Bank (PDB), and all of the atoms except the backbone carbon atoms are discounted. Then of the remaining atoms, all within a certain distance of one another are connected via springs [7]. The cutoff distance is selected by the user, but is typically around 10 Å, as shown in Fig. 8.1.

Elastic network models provide an efficient computational method for finding the global motions of proteins since they do not require extensive simulation. Several online servers are now available to take PDB files and calculate the global motions such as AD-ENM [8], elNémo [9] and Hingeprot [10], and ENM have been the subject of several recent reviews [11] and perspectives [12]. Recent successes using this technique for biomolecular modelling include the demonstration

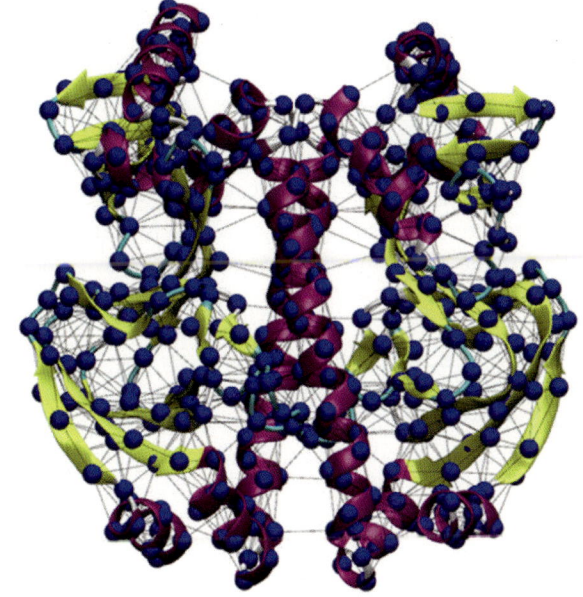

Fig. 8.1 An elastic network model of a Catabolite activator protein (CAP) dimer. The network obtained by connecting α carbon atoms is shown in *grey*, α carbons are shown as *blue* spheres. Image reproduced from McLeish et al. [7]

that these simplified network models provide global motions of the bovine pancreatic trypsin inhibitor protein that broadly agree with the results of atomistic MD calculations performed over millisecond timescales using the Anton supercomputer [13]; the finding that low frequency motions in proteins can modulate allosteric interactions in the absence of conformational changes [14]; the demonstration that the binding of CO_2 to the connexin hemichannel restricts its motion and might therefore explain its function as a CO_2 sensor [15] and finally the observation that central pore dilation is a high frequency mode of the nuclear pore complex [16], leading to the conclusion that it is a non-favourable mechanical process unless the correct cargo is present, which is to be expected for selective transport processes.

8.2.2 Brownian Dynamics

Brownian dynamics captures the diffusion of an object immersed in a fluid using the Langevin equation [17, 18]:

$$m\frac{d^2\mathbf{x}}{dt^2} = -\lambda\frac{d\mathbf{x}}{dt} + \mathbf{E}(t) + \mathbf{f}(t), \tag{8.1}$$

where m is the particle mass, $\mathbf{x}$ is the particle position, λ is the viscous drag coefficient, $\mathbf{E}(t)$ is the interaction force (e.g. due to some external background potential) and $\mathbf{f}(t)$ is the thermal force. The Langevin equation can be viewed as a generalisation of Newton's equation $F = ma$ to include random thermal fluctuations. The thermal noise $\mathbf{f}(t)$ continuously injects energy to the simulation, whereas the viscous friction term due to the fluid viscosity dissipates this thermal excitation. At equilibrium, the thermal noise must be chosen so that the average energy provided by the thermal forces is balanced by viscous energy dissipation. This law is known as the fluctuation-dissipation theorem. At equilibrium, since the thermal force has no net direction [19]:

$$\langle f(t) \rangle = 0, \tag{8.2}$$

where $\langle \ldots \rangle$ denotes an ensemble average.

The fluctuation-dissipation theorem is obtained by considering the variance of the thermal noise (for a complete derivation see Kubo [19]):

$$\langle f(t)f(t') \rangle = 2\lambda k_B T \delta(t - t'), \tag{8.3}$$

where $\delta(t)$ is the Dirac delta function. The fluctuation-dissipation relation effectively defines the strength of the thermal noise and sets the energy balance of the system. However, momentum conservation is not achieved in Brownian dynamics. Momentum is randomly imparted by the thermal noise term and is therefore not conserved. While Brownian dynamics does provide a representation of stochastic thermal noise at the local level, it represents solvent interactions through a local friction and so does not capture hydrodynamic interactions, which can introduce long range dynamical coupling through interactions with the solvent environment.

In spite of this approximation, Brownian dynamics simulations have proven particularly useful for investigating the behaviour of collections of interacting

Fig. 8.2 Brownian dynamics simulation by McGuffee and Elcock [20] of the bacterial cytoplasm using rigid protein structures downloaded from the PDB

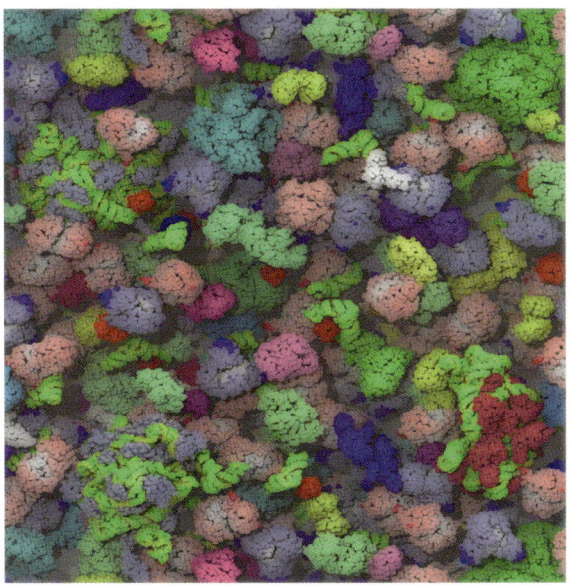

biomolecules at the mesoscale. In a pioneering series of simulations, Elcock and McGuffee used Brownian dynamics to simulate the crowded cytoplasmic environment over microsecond timescales [20], as shown in Fig. 8.2. The proteins were approximated as rigid units and simple local Stokes drag was used to represent solvent interactions. Ongoing calculations aim to include long range hydrodynamics, in spite of the increase in computational complexity [21]. Recent applications of Brownian dynamics to crowded environments have shown that the shape of the proteins acting as crowding agents has a large effect on translation and rotational diffusion coefficients [22]. Brownian dynamics simulations have also been used to study DNA compaction; a recent study of bacterial DNA interacting with the H-NS protein showed that the compacted DNA conformations obtained were very sensitive to the protein/DNA interaction energies, and that deposition on a mica surface, as is common for AFM experiments, can significantly perturb the behaviour of the biopolymer [23]. Moreover, Brownian dynamics simulations of chromatin applied in conjunction with high speed live cell fluorescent imaging have shown that chromatin dynamics can be described using the simple Rouse polymer model [24], in which the motion of the chromosome is dominated by the crowded environment of the nucleus [25].

8.2.3 Dissipative Particle Dynamics

Dissipative particle dynamics (DPD) is a mesoscale simulation method that reproduces hydrodynamics as well as thermodynamics by ensuring both momentum and energy conservation. The method includes three types of force: viscous dissipation, thermal noise and particle repulsion [26]. Crucially for momentum conservation, all of these are applied as inter-particle forces. In DPD, the force on the ith particle is given by:

$$\mathbf{f}_i = \sum_{j \neq i} (\mathbf{F}_{ij}^D + \mathbf{F}_{ij}^C + \mathbf{F}_{ij}^T), \tag{8.4}$$

where $\mathbf{F}_{ij}^D$ is the viscous dissipation force, $\mathbf{F}_{ij}^C$ is the repulsive force and $\mathbf{F}_{ij}^T$ is the thermal force of particle j on particle i. Momentum conservation is achieved by ensuring that the three force tensors $\mathbf{F}_{ij}^D$, $\mathbf{F}_{ij}^C$ and $\mathbf{F}_{ij}^T$ are anti-symmetric under permutation of the indices [26]. Thus, momentum gained by one particle is lost by its counterpart. The viscous force takes the form,

$$\mathbf{F}_{ij}^D = -\gamma \omega (\hat{\mathbf{r}}_{ij} \cdot \mathbf{v}_{ij}) \hat{\mathbf{r}}_{ij}, \tag{8.5}$$

where γ is related to the viscosity, ω is a weight function that depends only on the distance between particles, $\hat{\mathbf{r}}_{ij}$ is the unit vector corresponding to the displacement $\mathbf{r}_{ij} = \mathbf{r}_j - \mathbf{r}_i$, and similarly $\mathbf{v}_{ij} = \mathbf{v}_j - \mathbf{v}_i$ is the velocity difference between particles j and i.

The fluctuation-dissipation relation couples the dissipative term $\mathbf{F}_{ij}^D$ with $\mathbf{F}_{ij}^T$ [26] so that the average amount of energy drained by the viscous terms is equal to that added by the thermal term at thermal equilibrium, as described in detail by Español and Warren [27]. The resultant thermal noise term is:

$$\mathbf{F}_{ij}^T = \sigma \omega^{\frac{1}{2}} \boldsymbol{\theta}_{ij}(t) \hat{\mathbf{r}}_{ij}, \tag{8.6}$$

where ω is the same weight function as in Eq. (8.5) and $\sigma = (2k_B T \gamma)^{\frac{1}{2}}$, which relates the thermal noise to the viscosity via γ. To complete the coupling, $\boldsymbol{\theta}_{ij}(t)$ is a matrix of stochastic processes obeying the statistics:

$$\langle \theta_{ij}(t) \rangle = \mathbf{0} \tag{8.7}$$

$$\langle \theta_{ij}(t) \theta_{kl}(t') \rangle = (\delta_{ik} \delta_{jl} + \delta_{il} \delta_{jk}) \delta(t - t') \tag{8.8}$$

The first average states that the average thermal force on a particle is zero while the second term effectively ensures that the matrix $\theta_{ij}(t)$ is symmetric.

The final term in used in DPD, $\mathbf{F}_{ij}^C$, is a soft repulsive force. The strength of this interaction can be tuned for different particle types to set their relative affinities for one another, and provides the opportunity to describe chemical specificity for the different particle species in a simulation. A detailed recipe for how to select appropriate parameters to describe the strength of the interaction using Flory Huggins theory is presented by Groot and Warren [26]. Figure 8.3 shows that a suitable choice of different soft repulsive forces for the different chemical species within a DPD simulation can lead to phase separation of immiscible particles.

The main advantage of DPD is that the use of soft potentials allows large timesteps to be taken [26], while it is still possible to retain some chemical detail through an appropriate choice of repulsive inter-particle potentials. Boundary

Fig. 8.3 Phase separation of immiscible DPD particles in solution

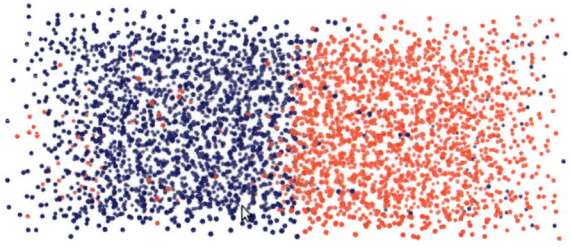

conditions that impose hard walls or even an external flow can also be used [28]. Moreover, DPD allows for an arbitrary level of coarse-graining through the choice of what each "particle" in the system represents; for example, a DPD particle may represent an atom, a group of atoms or a large colloid particle. DPD is therefore sufficiently versatile that it has been applied to multiple length and timescales. For example, DPD has been used to model the next generation of drug delivery vehicles and their transport across lipid membranes [6, 29]. Peng et al. constructed a DPD model of the lipid bilayer and cytoskeleton in an entire red blood cell. These calculations showed that the explicit inclusion of the elastic interaction between the bilayer and the cytoskeleton was necessary to obtain agreement with the thermal fluctuations within red blood cells observed experimentally [30]. However, DPD is computationally inefficient compared to lattice Boltzmann calculations (described below), as the force calculations entail a double sum over all particles at each timestep.

8.2.4 Lattice Boltzmann Methods

While MD, DPD and Brownian dynamics all consider physical objects interacting through Newtonian physics, lattice Boltzmann techniques are based on kinetic theory, evolving probability distributions of particle position and velocity, rather than the corresponding dynamical variables, using the Boltzmann equation [31]. The technical details underlying lattice Boltzmann methods applied to complex fluid flow are described in the review by Aidun and Clausen [32]. The probability distribution functions in the lattice Boltzmann method represent average occupancies of discrete velocity states. To illustrate this, it is useful to firstly discuss lattice gas automata [31, 33].

Figure 8.4 shows a two dimensional lattice gas model for a hexagonal lattice. The lattice gas is populated by particles that move from one node to another along connected grid lines. The model is therefore discrete in both space and time; in a single timestep a particle can only move to a neighbouring node. For the hexagonal grid shown in Fig. 8.4, each node in the lattice gas model can be occupied by a maximum of six particles, since every node has six neighbours and can therefore support six particles with velocities that point in the direction of all neighbouring nodes. Taking a timestep in the lattice gas method requires two consecutive

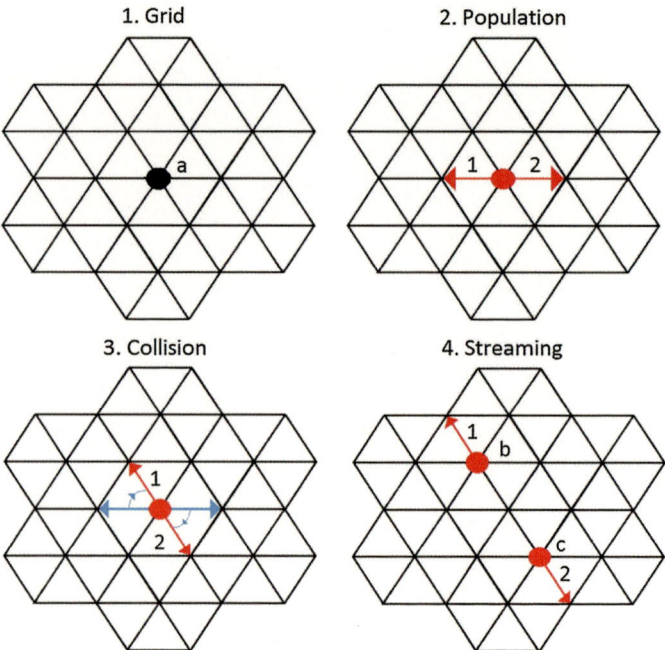

Fig. 8.4 **1** The hexagonal grid of a lattice gas model, with the central node highlighted. **2**. Two particles (*1* and *2*) are shown occupying the central node, with velocities indicated by the *red arrows*. **3**. As two particles occupy the same node, a collision must occur. The original velocities of particles *1* and *2* on the central node are shown in *blue*; the new post collision velocities of particles *1* and *2* are shown in *red*. **4**. Post collision, the particles are streamed onto new nodes so that particle *1* moves from node *a* onto node *b* and particle *2* moves onto node *c* whilst retaining the same post collision velocities

operations to be performed on all nodes: collision and streaming. The collision operation changes the velocities of particles if they occupy the same node and the streaming operation translates particles to neighbouring nodes following their direction of travel. As the collisions preserve both particle number and momentum, fluid mechanics is recovered at length and time scales much larger than the discretisation. However, this can cause convergence to the correct distributions to be slow. The lattice Boltzmann method was developed to improve the computational efficiency of the calculations [31, 33].

The lattice Boltzmann model replaces discrete particle occupancies on each node with their average value [33]. While the streaming step remains the same, the procedure for collisions needs to be generalised to account for the fact that the occupancies are now averaged. This is achieved through the Bhatnagar-Gross-Krook (BGK) relaxation term [31, 33], which can be derived directly from the Boltzmann equation [34]. The BGK relaxation term calculates the equilibrium distribution of average occupancies for a node while preserving particle number and momentum. The average occupancies then relax toward this equilibrium within a given time scale. The equation of motion for the lattice Boltzmann method is then given by [34]:

$$f_i(\mathbf{x} + c\mathbf{e}_i \Delta t, t + \Delta t) - f_i(\mathbf{x}, t) = -\frac{f_i(\mathbf{x}, t) - f_i^{eq}(\mathbf{x}, t)}{\tau}, \tag{8.9}$$

where f_i is an average occupancy number on a node, e_i are the directions allowed on the grid, c is the discretised speed and τ is the relaxation time to equilibrium.

The lattice Boltzmann method is being increasingly used as an alternative to continuum methods for solving time-dependent macroscale fluid mechanics problems on complex domains. The lattice Boltzmann method is easier to implement on large numbers of processors because the collision step is entirely local, and the boundary conditions are also easy to specify [35]. The lattice Boltzmann method has been widely applied to model flows within biological systems, as described in the review by Mills et al. [6]. This is mainly due to the ease of boundary condition application, such that blood flow through vessels and even at junctions can be modelled [36]. For example, Zhang et al. [37, 38] used lattice Boltzmann to simulate the aggregation of red blood cells. In the absence of flow, the red blood cells formed cylindrical stacks in order to maximise surface area contact. However, performing the calculations in the presence of a shear flow (flow can be applied in LB by simply adapting the lattice edge boundary conditions) resulted in disaggregation of the cellular stacks, as is observed experimentally.

The primary limitation of the lattice Boltzmann method arises from the fact that it is a discrete method and that the average occupancy numbers propagate at a single speed throughout the grid, which is the effective sound speed. Consequently it cannot represent a truly incompressible fluid. There is also no thermal noise, although a number of approaches to include stochastic behaviour have been implemented [39, 40].

8.2.5 Continuum Models

Continuum models are most commonly associated with macroscale simulations in the athermal limit such as those found in civil, mechanical or aeronautical engineering applications. A continuum model describes a physical system using locally averaged variables such as velocity, density, stress, strain, temperature, concentration, composition, or charge, which are considered to be (usually smooth) functions of position, with dynamics often governed by partial differential equations. To describe fluid flow or material transport the two relevant equations are the continuity equation and the momentum equation [41]. The continuity equation is as follows:

$$\frac{\partial \rho}{\partial t} + \nabla \cdot (\rho \mathbf{v}) = 0, \tag{8.10}$$

where ρ is the density of the continuum and $\mathbf{v}$ the fluid velocity. This equation enforces mass conservation; the change in density as a function of time is equal to the amount of material flowing into or out of an (infinitesimally) small unit volume around the point. Similarly the change in momentum is given by [41]:

$$\rho \frac{D\mathbf{v}}{Dt} = \nabla \cdot \sigma + \mathbf{f}, \tag{8.11}$$

where $\frac{D}{Dt} = \frac{\partial}{\partial t} + \mathbf{v} \cdot \nabla$ is the material derivative (the time derivative in the frame moving with the fluid velocity), σ is the stress within the continuum and $\mathbf{f}$ an external force per unit volume.

The momentum equation expresses Newton's second law for a continuum mechanics system. The material derivative gives the rate of change of velocity of a segment of the material as it moves under flow. This acceleration is produced by a combination of external forces, and material stress. The stress tensor σ (in three dimensions) is a symmetric 3 by 3 matrix [41] which describes the internal forces within the continuum material. The component σ_{ij} is the force per unit area, in direction i, acting across a surface with normal direction j.

The stress in a continuum material is obtained from the constitutive equation, which includes parameters for material properties such as the viscosity or elasticity. Many different constitutive models exist, the simplest being linear elasticity or Hooke's law [42], $\sigma_S = E\varepsilon_S$, where E is the Young's modulus and σ_S and ε_S are the stress and strain in the material respectively. If the material has viscous properties, then an internal stress is induced due to the rate of change of strain, $\sigma_D = \eta \frac{\partial \varepsilon_D}{\partial t}$, where η is the viscosity co-efficient. Other material models such as Maxwell or Kelvin-Voigt [43] are used for viscoelastic materials, in which both internal viscosity and elasticity need to be considered. The choice of the appropriate constitutive model depends on the types of forces the material will experience, and whether the material will permanently deform under the action of the force.

The Maxwell model treats the continuum as an elastic spring and viscous damper connected in series [44], as shown in Fig. 8.5. The stress in the spring and damper are therefore equal, $\sigma = \sigma_S = \sigma_D$, and the strains sum to give the total strain, $\varepsilon = \varepsilon_S + \varepsilon_D$. By considering the time derivative of the strain we can see how it is related to the stress:

$$\frac{d\varepsilon}{dt} = \frac{\sigma}{\eta} + \frac{1}{E}\frac{d\sigma}{dt} \tag{8.12}$$

In the Kelvin-Voigt model the spring and damper components are connected in parallel rather than series [44], as shown in Fig. 8.6, which means that this time the *strains* are equal, $\varepsilon = \varepsilon_S = \varepsilon_D$, and the stresses sum to give the total, $\sigma = \sigma_S + \sigma_D$. Following similar analysis to the Maxwell model, we obtain:

$$\frac{d\varepsilon}{dt} = \frac{\sigma}{\eta} - \frac{E}{\eta}\varepsilon \tag{8.13}$$

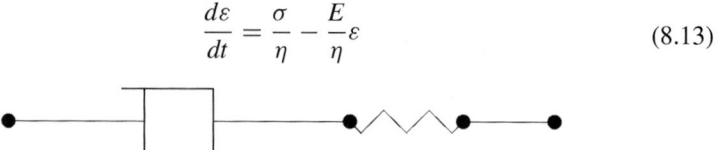

Fig. 8.5 The Maxwell model for visco-elastic media consists of a viscous damper (on the *left*) in series with an elastic spring (on the *right*)

Fig. 8.6 The Kelvin-Voigt Model for visco-elastic media is comprised of a viscous damper (on the *bottom*) connected in parallel with an elastic spring (on the *top*)

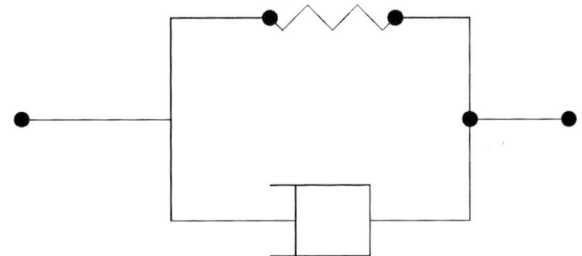

When a constant stress σ_0 is applied to a Kelvin-Voigt material at time $t = 0$, the solution of Eq. (8.13) is:

$$\varepsilon(t) = \frac{\sigma_0}{E}\left(1 - e^{-\frac{E}{\eta}t}\right) \tag{8.14}$$

Equation 8.14 shows that as $t \to \infty$, the viscous effect decays away exponentially and we are left with Hooke's Law, which indicates that the material only experiences the effects of elasticity. Therefore, a deformed Kelvin-Voigt material will always return to its starting configuration over a sufficiently long time scale. In contrast, when a constant strain is applied to a Maxwell material, an analogous solution but for Eq. (8.12) shows that the induced stress will exponentially decay to zero causing the continuum to relax into a new stable state. Therefore, permanent deformation is possible with the Maxwell model.

Numerical methods are employed to solve the partial differential equations involved in continuum mechanics, of which there are three common methods; finite difference [45], finite volume [46] and finite element methods [47]. Finite difference methods store the values of the physical variables at discrete points on a grid with spatial derivatives approximated from differences in the values of quantities on neighbouring nodes. The computational grid must be constructed using an orthogonal lattice, which makes it difficult to represent complex boundary shapes. This problem is resolved by finite volume and finite element methods, which are geometrically more flexible. In the finite volume method the computational domain is divided into small volumes (typically cuboids) and the continuum equations are converted into integrals over the surface of these volumes involving fluxes between volumes. The finite element method involves a similar subdivision of the domain into simple shapes typically hexahedra or tetrahedra, known as finite elements. The continuum equations are converted into volume integrals, which can be computed locally in each element.

Finite element analysis (FEA) is routinely used for biomedical applications at macroscopic length-scales, such as whole heart modelling [24]. Continuum mechanics has been applied to model the deposition of rigid nanoparticles under flow [48], cell motility [49] and blood flow in the aorta [50] over length scales from hundreds of nanometers and time scales of microseconds (for the nanoparticles), up to micrometers and minutes (for the cell motility simulations). Bathe [51] has employed a continuum mechanics model to extract the normal modes of proteins, and has developed an online database of finite element structures constructed from EMDB maps [52]. However, until recently the application of continuum methods to the nanoscale has been limited because of the difficulties of including thermal fluctuations in the calculations. This has led to the development of fluctuating finite element analysis (FFEA), which specifically includes thermal noise within a continuum mechanics framework.

8.3 Fluctuating Finite Element Analysis

Fluctuating finite element analysis (FFEA) is an alternative scheme for mesoscale biological simulations which models the dynamics of globular proteins using a continuum mechanics model that includes thermal fluctuations. In its current form, FFEA treats a protein as a viscoelastic material using a Kelvin-Voigt model, thus giving the protein a permanent equilibrium structure about which it fluctuates [53]. The continuum equations can then be solved numerically using finite element analysis. Since the finite element method can handle complex geometries with relative ease, this makes it particularly useful for the treatment of the dynamics of globular proteins. The thermal noise, which is implemented as a stochastic component of stress, is derived analytically by solving the fluctuation-dissipation relation within the finite element approximation [53], thus ensuring that the computed dynamical trajectories are consistent with the requirements of classical thermodynamics.

FFEA treats a globular macromolecule as a continuous medium of density ρ subject to thermal noise, viscous dissipation and elasticity. The equation of motion for the system (using index notation together with summation convention) is the momentum equation:

$$\rho \left(\frac{\partial u_i}{\partial t} + u_j \frac{\partial u_i}{\partial x_j} \right) = \frac{\partial \sigma_{ij}}{\partial x_j}, \tag{8.15}$$

where $\left(\frac{\partial u_i}{\partial t} + u_j \frac{\partial u_i}{\partial x_j} \right)$ is the total time derivative of the velocity vector field in the Lagrangian frame of the material. The stress σ_{ij} can be sub-divided into three contributions:

σ_{ij}^v, σ_{ij}^e and σ_{ij}^t are the stresses due to viscosity, conservative elastic forces and ther-

$$\sigma_{ij} = \sigma_{ij}^v + \sigma_{ij}^e + \sigma_{ij}^t \tag{8.16}$$

mal fluctuations respectively. Since the viscous and elastic stresses are summed in this model, this corresponds to a Kelvin-Voigt material [54].

The viscous stress σ_{ij}^v is assumed to be linear [55], and is written as:

$$\sigma_{ij}^v = \mu \left(\frac{\partial u_i}{\partial x_j} + \frac{\partial u_j}{\partial x_i} \right) + \lambda \frac{\partial u_m}{\partial x_m} \delta_{ij}, \tag{8.17}$$

where μ is the shear viscosity and λ is the second coefficient of viscosity, giving a bulk viscosity $\mu_{bulk} = \lambda + \frac{2}{3}\mu$. Physically, Eq. (8.17) means that the material resists the build up of velocity gradients within it.

The elasticity in the FFEA model has been derived from a strain energy functional [56] such that the material is hyperelastic. This means that the stored elastic energy is fully recoverable, even if the elastic forces are non-linear in the material deformation. In general, the elastic forces in continuum mechanics are framed in terms of the deformation gradient tensor F_{ij}. The deformation gradient tensor F_{ij} measures the material deformation from its initial configuration to a new configuration so that $F_{ij} = \frac{\partial x_i}{\partial X_j}$, where $\mathbf{x}(\mathbf{X}, t)$ is the current position of material initially located at $\mathbf{X}$. In our current FFEA treatment [53] the elastic stress in the material is written as:

$$\sigma_{ij}^e = \frac{G}{\det(F)} F_{ik} F^T{}_{kj} + B(\det(F) - \alpha)\delta_{ij}. \tag{8.18}$$

where G is the shear modulus, $K = B - \frac{G}{3}$ is the bulk modulus and $\alpha = 1 + \frac{G}{B}$. From the bulk and shear moduli, the Young's Modulus of the material is given by $E = \frac{9KG}{3K+G}$, as described in detail by Oliver et al. [53].

In addition to the viscous and elastic stresses defined above, a fluctuating thermal stress σ_{ij}^t is added to ensure that the motion of the model is consistent with thermodynamic equilibrium. We detail our formulation of the thermal stress in Sect. 8.3.3, but before doing so it is convenient to discuss the finite element discretisation of the continuum model.

8.3.1 Finite Element Formulation Within FFEA

Here we provide a short overview of the finite element method, with emphasis on how it has been implemented within FFEA. For more information on the formalism and general method, we offer the following references for introductory [47] and advanced [57] topics.

In general, the finite element method works by dividing the material into 3D elements, called finite elements, that are geometrically simple [58]. Typically, in three dimensions, tetrahedral or hexahedral elements are chosen, as these can be used to cover domains with complicated geometries with a relatively small error in the total shape and volume at the surface. The idea is then to find an approximate solution of our equation of motion, Eq. (8.15), within each element using functions constructed from linear combinations of basis functions $\phi_1(\mathbf{x}), \ldots, \phi_n(\mathbf{x})$, so that the material velocity is approximated as:

$$\mathbf{u}(\mathbf{x}) = \sum_{\alpha=1}^{n} \mathbf{v}_\alpha \phi_\alpha(\mathbf{x}). \tag{8.19}$$

where α is the node index. These nodes are the vertices of the finite elements. The basis functions ϕ_α are typically chosen as polynomials over each element with a value of 1 at a given element node (α) and a value of zero at all other nodes. The velocity $\mathbf{v}_\alpha$ can be interpreted as the material velocity at node α. The velocity at all other points in space is a polynomial interpolation between these nodal values, dependant on the order of our basis function.

The key step in the finite element method is to construct a "weak-form" of Eq. (8.15), which is formed by integrating Eq. (8.15) together with a set of weight functions, w_α:

$$\int_V w_\alpha(\mathbf{x}) \left[\rho \left(\frac{\partial u_i}{\partial t} + u_j \frac{\partial u_i}{\partial x_j} \right) - \frac{\partial \sigma_{ij}}{\partial x_j} \right] dV = 0. \tag{8.20}$$

This means that we are finding solutions where the error in Eq. (8.15) is "orthogonal" to the weight functions. We then perform integration by parts on the stress term to move the derivative onto the weight function,

$$\int_V \rho w_\alpha(x) \frac{\partial u_i}{\partial t} dV = - \int_V \frac{\partial w_\alpha(x)}{\partial x_j} \sigma_{ij} dV - \int_V \rho w_\alpha(x) u_j \frac{\partial u_i}{\partial x_j} dV \tag{8.21}$$

This is known as the "weak form" of Eq. (8.15). One of advantage of this representation is that we no longer need to calculate the derivative of the stress, which allows us to use a stress that is discontinuous between elements. In the Galerkin formulation of the finite element method [47], the weight functions are chosen to be the same as the basis functions ϕ_α, this reduces the system to a set of linear algebraic equations (one equation for each weight function) which can then be solved numerically for the nodal velocity values. The final result is a set of equations which can be written in matrix-vector form as follows:

$$M_{pq} \frac{\partial v_q}{\partial t} = N_p - K_{pq} v_q - \nabla_p U(\mathbf{x}), \tag{8.22}$$

where M_{pq} is a mass matrix that describes the distribution of mass within the continuum, v_p is a velocity vector, N_p is the fluctuating thermal force vector, K_{pq} is a viscosity matrix and $\nabla_p U(\mathbf{x})$ is the elastic force vector. Once we solve this equation, we obtain the nodal velocity values which can be interpolated throughout the system if required. We now discuss the properties of the fluctuating force.

8.3.2 Thermal Noise and the Fluctuation-Dissipation Relation

As in other models such as Brownian dynamics or DPD, the thermal noise in FFEA must be coupled to the viscosity through a fluctuation-dissipation relation, giving rise to a fluctuating stress tensor.

The fluctuation-dissipation relation concerns energy balance at thermal equilibrium. On average the amount of energy dissipated from the system by the viscosity must be balanced by the energy input from the thermal noise. As the elastic term is conservative, it does not contribute to this energy balance. Thus, we expect the fluctuation-dissipation relation for the FFEA system to include terms from the viscosity matrix and the noise vector. By considering the average change in kinetic energy over a discrete timestep Δt, Oliver et al. [53] showed that the noise vector must have the following statistics:

$$\langle N_p \rangle = 0, \tag{8.23}$$

$$\langle N_p N_q \rangle = \frac{k_B T}{\Delta t}(K_{pq} + K_{qp}). \tag{8.24}$$

These equations are analogous to the fluctuation-dissipation relations for both Brownian dynamics and DPD. Specifically, the second moment average of the thermal noise is correlated to the temperature, system time step, and the viscous dissipation. The only difference is that in Brownian dynamics and DPD the dissipation is via a frictional force whilst in FFEA the dissipation is via a viscous stress; this results in a matrix for the correlations in thermal forces on the nodes of each element. For first-order tetrahedral finite elements (first order basis functions), these correlated forces can be cast in terms of a thermal stress tensor within a single element, of the form:

$$\sigma_{ij}^t = \left(\frac{2k_B T}{V \Delta t}\right)^{\frac{1}{2}} \left(\mu^{\frac{1}{2}} X_{ij} + \lambda^{\frac{1}{2}} X^0 \delta_{ij}\right) \tag{8.25}$$

where V is the element volume. In applying Eq. (8.25), seven random numbers of unit variance must be chosen at each timestep, for each element in the model. These numbers are X^0 and the 6 independent elements of the symmetric tensor X_{ij}. This result matches the mathematics of a much earlier description for the thermal fluctuations of continuum fluids by Landau and Lifshitz [59].

To validate that the fluctuation-dissipation relation is correct within the continuum model, Oliver et al. [53] extracted the probability distribution of the nodal velocities for several nodes of a cylindrical mesh and showed that these distributions were in agreement with the expected Gaussian distribution of velocities at each node, with covariance $\langle v_p v_q \rangle = k_b T M_{pq}^{-1}$. Temporal convergence was established by noting that the average kinetic and potential energies converged to the correct equipartition values with decreasing simulation timestep. Spatial convergence is more subtle; with increasing spatial resolution, the number of degrees of freedom increases and so the kinetic and potential energies also increase accordingly ($\frac{k_B T}{2}$ per quadratic degree of freedom). However, the large scale fluctuations in the deformation of the model were quantitatively shown to converge with increasing mesh resolution. To show that the FFEA model explores conformational space in the correct manner, the thermal deformations of idealised beams [60] with various cross sections were simulated, and it was shown using Fourier analysis that the computed results matched the predictions for the amplitudes of fluctuations of the normal modes. This confirmed that the conformational space explored by the beams is consistent with Boltzmann statistics.

8.3.3 Performing an FFEA Simulation

An FFEA calculation requires low resolution structural information about the overall shape of the biomolecule, such as an EMBD electron density map, or a SAXS envelope. If available, an FFEA simulation can also be performed using a conventional atomistic structure from the PDB, although only the global shape of the biomolecule would be used in the calculation. The following procedure is used to perform an FFEA calculation (see Fig. 8.7; software tools are given in bold).

1. Obtain information on the overall shape of the biomolecule (e.g. an electron density map from the EMDB).
2. Convert the structure to a finite element surface mesh:

Chimera—Visualise the density map at different density levels and export a .vrml surface map at the required level.
FFEA_ tools—Parse a .vrml file to remove overlapping and repeated elements.
FFEA_ tools—Convert from a density map to a .surf surface file at the required level.

3. Coarse grain the surface mesh:

FFEA_ tools—Coarse-grain the surface to remove small finite elements. (To be physically realistic each element must contain several atoms).

4. Create a volume mesh from the surface mesh:

Netgen/Tetgen—Import .surf/.vrml file and construct a volumetric mesh file, .vol, which is used to generate input files for the FFEA simulation.

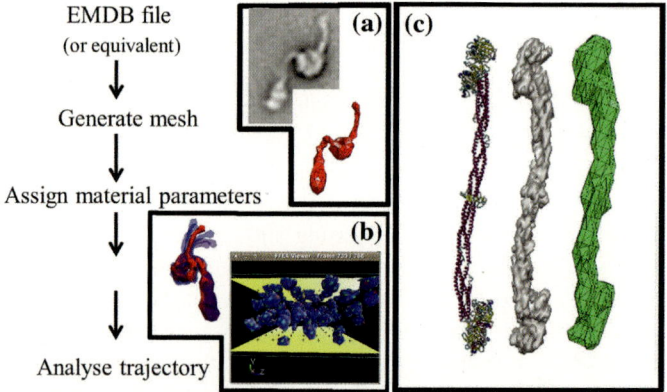

Fig. 8.7 A flowchart of the meshing process. **a** Shows a negative stain image of axonemal dynein [61], converted into a volumetric mesh. **b** Shows superposed frames from an FFEA trajectory of an individual dynein molecule, and a screenshot of a simulation of interacting myoglobin molecules. **c** Illustrates the meshing process starting from a PDB file, converting first to a density map and then creating the volume mesh. This mesh is for human fibrinogen [62], a protein involved in blood clotting

5. Set the material parameters for the structure and the surroundings.
6. Set the characteristic energies of the relevant biomolecular interactions:

FFEA_ tools—Converts a .vol file, together with simulation parameters, into a set of input files required by FFEA.

7. Run the simulation for long enough to obtain sufficient statistics:

FFEA—Runs the simulation with the above as input files.

'FFEA_ tools' is a toolkit developed in house to streamline the production of the relevant input files for an FFEA simulation. It contains file conversion scripts and input file creation and manipulation tools, written in Python and C++.

From an EMDB density map an isosurface (surface of constant density) can be constructed using isolines, to define a volume. The Chimera visualisation tool [63] can be used to view a change in the density level of the isolines, which usually has a recommended value on the EMDB. VRML surface files exported by Chimera have been known to cause fatal errors with volumetric meshing software, such as Netgen [64], due to overlapping surface elements and repeated edges. FFEA_ tools contains scripts to parse VRML files and clean them of errors, or create alternative surface files straight from the density map. Once a clean surface mesh is established, further processing is required to ensure that the object does not contain finite elements that are too small. The timestep required to maintain the stability of the numerical integration scheme depends on the length of the smallest element, so larger elements are computationally more efficient. Consequently, the mesh is repeatedly coarse-grained before being parsed to a finite element mesh generator like Netgen or Tetgen, which creates the final 3D volume mesh.

The 3D finite element mesh consists of tetrahedra that together make up the shape of the biomolecule. For each element, it is necessary to assign the density, bulk and shear viscous moduli for the protein and the solvent, as well as the bulk and shear elastic moduli for the protein. Typically, we have used values for the viscosity that are comparable with that of water and values for the elasticity similar to those of low density polyethylene. These values agree with experimentally determined values for the material properties of biomolecules by AFM [65]. While in principle the material parameters can be assigned to each element independently, giving a highly inhomogeneous biomolecule, in practice it has been sufficient to use at most two different values of Young's modulus to describe a single macromolecule. The current FFEA model includes a local friction term, with associated Brownian noise, to model the interaction with the surrounding solvent. However, long-range hydrodynamic interactions mediated by the solvent are currently neglected.

FFEA simulations performed using viscosities representative of water require a long time to converge to thermodynamic equilibrium, because biomolecular dynamics are heavily overdamped. These strong viscous interactions with the solvent enormously decrease the rate of exploration of conformational space by biomolecules, and set the length of the relevant timescales in biomolecular systems. However, if only thermodynamic quantities are required, the exploration of conformational space can be accelerated by reducing the value of the external viscous drag to that which gives critical damping, although clearly this results in trajectory timescales that are not physically meaningful.

The code has been parallelised for shared memory, and can therefore take advantage of parallel processing. The mathematical structure of FFEA, which keeps the calculation of the stress local to an element, reduces the communication costs meaning that FFEA simulations parallelise efficiently.

Figure 8.8 shows the setup of an FFEA simulation of the measles virus nucleoprotein capsid. The nucleocapsid is made up of an ensemble of identical protein subunits arranged in a helical structure. By duplicating a section of a nucleocapsid extracted from an EM density map [66], and taking advantage of the periodicity of the helical structure, a longer mesh was produced. This mesh had dimensions 20 nm by 138 nm, representing 330 protein subunits. The simulations contained 45,322 finite elements, and required 1,000 h to obtain 250 ns of dynamics running in parallel on 8 processors. Figure 8.8 also shows the low frequency modes extracted from the FFEA simulation trajectory using Principal Components Analysis [67]. The low frequency dynamics of the cylindrical virus capsid consists of global bending (modes 1, 2, 3, 6 and 7), twisting (modes 4 and 8) and stretching (mode 5). These observed modes agree with those expected for a cylindrical elastic object suggest that 250 ns is a sufficiently long trajectory for the lowest frequency harmonics of the molecule to be captured under critically damped conditions.

A reduced representation of the nucleocapsid was created after the simulation by extracting the trajectories of one node at the centre of each subunit, resulting in a simplified helix. We assumed that nodes central to each subunit would accurately represent their net motion allowing us to neglect motions local to each subunit. This allowed fast analysis on the motion of the overall capsid and enabled the calculation of idealised motion vectors. Taking the dot product of the idealised motion vectors with the calculated eigenvectors allowed the modes to be analysed in a quantitative fashion, and for the results to be compared with those calculated analytically for a perfect elastic cylinder. The results from the long mesh showed a stronger correlation with the idealised beam bending modes than when the simulation was run with the original smaller section of nucleocapsid, as the modes produced by the smaller section are more affected by end effects, which results in the modes appearing as combinations of the ideal twisting and bending modes.

For the case of systems with many interacting molecules, the current FFEA code has also been parallelised to enable each molecule to be placed onto individual processors. Figure 8.9 shows an FFEA simulation of 128 interacting myoglobin proteins (shown in blue) located between a pair of adhesive surfaces (in yellow), with parallel boundary conditions in the xy-plane. This simulation was performed to inform the design of protein biosensors for nanotechnological applications. We have used FFEA to determine how the binding and unbinding rates of myoglobin to the surface are affected by the material properties of the protein. For individual myoglobins, we have shown that halving the Young's modulus increases the binding constant by a factor of around 4, due to the fact that softer molecules can deform to achieve a higher contact area with the surface.

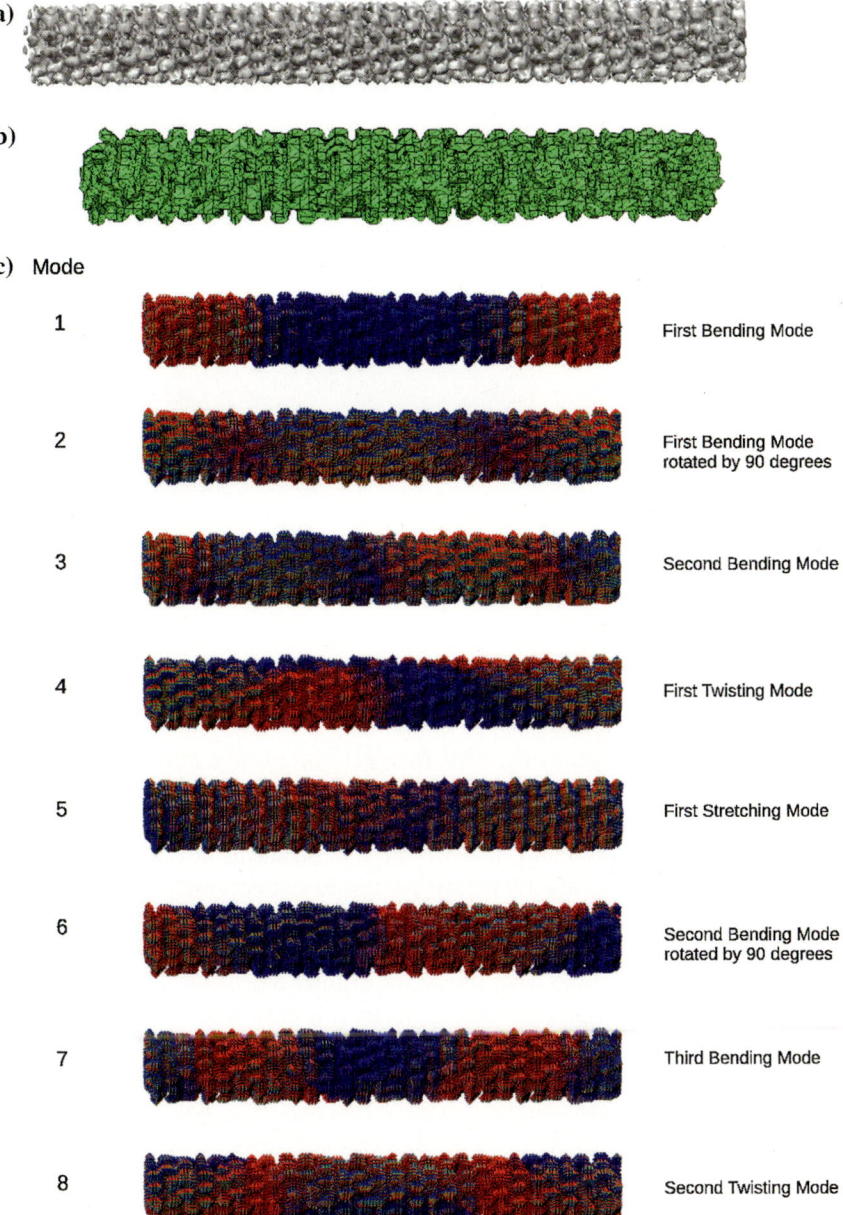

Fig. 8.8 FFEA simulations of the measles virus nucleocapsid. **a** Shows the EMBD map, **b** shows the corresponding finite element mesh created and **c** shows the first 8 dynamical modes of the capsid extracted from the FFEA trajectory with Principal Components Analysis

Fig. 8.9 A screenshot
of interacting myoglobin
molecules

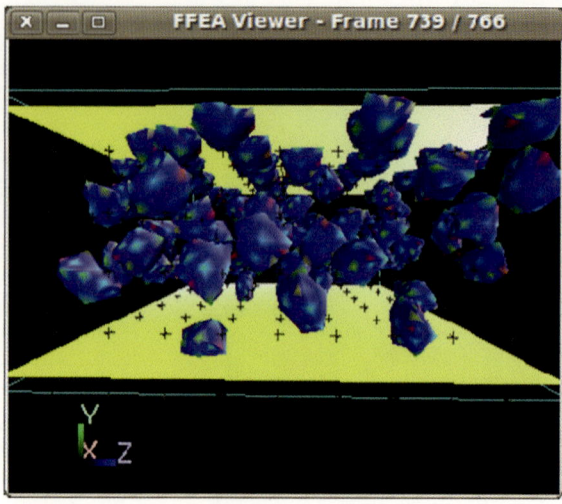

8.3.4 Prospects for Biomolecular Modelling at the Mesoscale

Biomolecular simulation at mesoscopic time and length scales will become increasingly important in deepening our understanding of the role of dynamics in biomolecular function due to the growth of lower resolution experimental structural databases, such as the EMDB. Most importantly, structural biophysics has now developed techniques to look at biomolecules not just in isolation, as is the case for those stored in the PDB, but in a biological environment that is more similar to the in vivo system [2]. Since these experiments describe biomolecular structures, environments and biological processes that occur over time and length scales inaccessible to quantum or atomistic simulation, coarse-grained approaches will need to be continuously developed and improved. However, given the multi-scale nature of some of the most important cellular processes, such as the action of the ribosome or the muscle protein myosin, it is clear that no single simulation method will be capable of spanning such diverse computational regimes. This calls for a multi-scale approach capable of combining complementary simulation techniques within a single calculation.

References

1. Meyer T et al (2010) MoDEL (molecular dynamics extended library): a database of atomistic molecular dynamics trajectories. Structure 18(11):1399–1409
2. Robinson CV, Sali A, Baumeister W (2007) The molecular sociology of the cell. Nature 450(7172):973–982
3. www.emdatabank.org. Emdb deposition and annotation statistics: Emdatabank. http://www.emdatabank.org/dpstn_annot_stats.html, April 2014

4. Marrink SJ, Tielman DP (2013) Perspective on the martini model. Chem Soc Rev 42(16):6801–6822
5. Tozzini V (2010) Minimalist models for proteins: a comparative analysis. Q Rev Biophys 43(3):333–371
6. Mills ZG, Mao W, Alexeev A (2013) Mesoscale modeling: solving complex flows in biology and biotechnology. Trends Biotechnol 31(7):426–434
7. McLeish TC, Rodgers TL, Wilson MR (2013) Allostery without conformation change: modelling protein dynamics at multiple scales. Phys Biol 10(5):056004
8. Zheng W, Liao JC, Brooks BR, Doniach S (2007) Toward the mechanism of dynamical couplings and translocation in hepatitis c virus ns3 helicase using elastic network model. Proteins Struct Funct Bioinf 67(4):886–896
9. Suhre K, Sanejouand YH (2004) Elnémo: a normal mode web server for protein movement analysis and the generation of templates for molecular replacement. Nucleic Acids Res 32(2):W610–W614
10. Emekli U, SchneidmanDuhovny D, Wolfson HJ, Nussinov R, Haliloglu T (2008) Hingeprot: automated prediction of hinges in protein structures. Proteins Struct Funct Bioinf 70(4):1219–1227
11. Bahar I, Lezon TR, Bakan A et al (2010) Global dynamics of proteins: bridging between structure and function. Ann Rev Biophys 39:23–42
12. Noid WG (2013) Perspective: coarse-grained models for biomolecular systems. J Chem Phys 139(9)
13. Gur M, Zomot E, Bahar I (2013) Global motions exhibited by proteins in micro- to milliseconds simulations concur with anisotropic network model predictions. J Chem Phys 139(12)
14. Rodgers TL, Townsend PD, Burnell D et al (2013) Modulation of global low-frequency motions underlies allosteric regulation: demonstration in CRP/FNR family transcription factors. PLOS Biol 11(9)
15. Meigh L et al (2013) CO_2 directly modulates connexin 26 by formation of carbamate bridges between subunits. Elife 2:e01213
16. Kin D, Nguyen C, Bathe M (2010) Conformational dynamics of supramolecular protein assemblies. J Struc Biol 173:261–270
17. Ermak DL, Buckholz H (1980) Numerical integration of the Langevin equation: Monte carlo simulation. J Comp Phys 35(2):168–182
18. Chen JC, Kim AS (2004) Brownian dynamics, molecular dynamics, and monte carlo modelling of colloidal systems. Adv Colloid Interface Sci 112(1):159–173
19. Kubo R (1966) The fluctuation-dissipation theorem. Rep Prog Phys 29(1)
20. McGuffee SR, Elcock AH (2010) Diffusion, crowding and protein stability in a dynamic molecular model of the bacterial cytoplasm. PLOS Comp Biol 6(3)
21. Frembgen-Kesner T, Elcock AH (2010) Absolute protein-protein association rate constants from flexible, coarse-grained brownian dynamics simulations: the role of intermolecular hydrodynamic interactions in barnase-barstar association. Biophys J 99(9):L75–L77
22. Balbo J, Mereghetti P, Herten D, Wade RC (2013) The shape of protein crowders is a major determinant of protein diffusion. Biophys J 104(7):1576–1584
23. Joyeux M, Vreede J (2013) A model of h-ns mediated compaction of bacterial DNA. Biophys J 104(7):1615–1622
24. Doi M, Edwards SF (1998) The theory of polymer dynamics. Oxford University Press, Oxford
25. Hajjoul H, Mathon J, Ranchon H et al (2013) High-throughput chromatin motion tracking in living yeast reveals the flexibility of the fiber throughout the genome. Genome Res 23(11):1829–1838
26. Groot RD, Warren PB (1997) Dissipative particle dynamics: bridging the gap between atomistic and mesoscopic simulation. J Chem Phys 107(11):4423
27. Español P, Warren PB (1995) Statistical mechanics of dissipative particle dynamics. Europhys Lett 30(4):191
28. Pivkin IV, Karniadakis GE (2005) A new method to impose no-slip boundary conditions in dissipative particle dynamics. J Comp Phys 207(1):114–128

29. Liu F, Wu D, Kamm RD et al (2013) Analysis of nanoprobe penetration through a lipid bilayer. Biochim Biophys Acta Biomembr 1828(8):1667–1673
30. Peng Z, Li X, Pivkin I, Dao M, Karniadakis G, Suresh S (2013) Lipid bilayer and cytoskeletal interactions in a red blood cell. Proc Natl Acad Sci USA 110(33):13356–13361
31. Succi S (2001) The lattice Boltzmann equation: for fluid dynamics and beyond. Oxford University Press, Oxford
32. Aidun C, Clausen J (2010) Lattice-boltzmann method for complex flows. Ann Rev Fluid Mech 42:439–472
33. Chen S, Doolen GD (1998) Lattice boltzmann method for fluid flows. Ann Rev Fluid Mech 30(1):329–364
34. He X, Luo LS (1997) Theory of the lattice boltzmann method: from the boltzmann equation to the lattice boltzmann equation. Phys Rev E 56(6):6811
35. Zou Q, He X (1997) On pressure and velocity boundary conditions for the lattice boltzmann BGK model. Phys Fluids 9(6):1591–1598
36. Yin X, Thomas T, Zhang J (2013) Multiple red blood cell flows through microvascular bifurcations: cell free layer, cell trajectory, and hematocrit separation. Microvasc Res 89:47–56
37. Zhang J, Johnson PC, Popel AS (2008) Red blood cell aggregation and dissociation in shear flows simulated by lattice boltzmann method. J Biomech 41(1):47–55
38. Liu Y, Zhang L, Wang X, Liu WK (2004) Coupling of navierstokes equations with protein molecular dynamics and its application to hemodynamics. Int J Numer Methods Fluids 46:1237–1252
39. Adhikari R, Stratford K, Cates ME, Wagner AJ (2005) Fluctuating lattice boltzmann. Europhys Lett 71(3):473
40. Gross M, Adhikari R, Cates ME, Varnik F (2010) Thermal fluctuations in the lattice boltzmann method for nonideal fluids. Phys Rev E 82(5)
41. Fung YC (1977) A first course in continuum mechanics. Prentice-Hall Inc., Englewood Cliffs
42. Sokolnikoff IS, Specht RD (1956) Mathematical theory of elasticity, vol 83. McGraw-Hill, New York
43. Eringen AC (1980) Mechanics of continua. Robert E. Krieger Publishing Co, Malabar
44. Reddy JN (2013) An Introduction to Continuum Mechanics. Cambridge University Press, Cambridge
45. Fadlun EA, Verzicco R, Orlandi P, Mohd-Yusof J (2000) Combined immersed-boundary finite-difference methods for three-dimensional complex flow simulations. J Comp Phys 161(1):35–60
46. Versteeg HK, Malalasekera W (2007) An introduction to computational fluid dynamics: the finite volume method. Pearson Education, Delhi
47. Reddy JN (1993) An introduction to the finite element method. McGraw-Hill, New York
48. Shah S, Liu Y, Hu W, Gao J (2011) Modeling particle shape-dependent dynamics in nanomedicine. J Nanosci Nanotech 11(2):919
49. Gracheva ME, Othmer HG (2004) A continuum model of motility in ameboid cells. Bull Math Biol 66(1):167–193
50. De Hart J, Baaijens FPT, Peters GWM, Schreurs PJG (2003) A computational fluid-structure interaction analysis of a fiber-reinforced stentless aortic valve. J Biomech 36(5):699–712
51. Bathe M (2008) A finite element framework for computation of protein normal modes and mechanical response. Proteins 70:1595–1609
52. Kim D, Altschuler J, Strong C, McGill G, Bathe M (2011) Conformational dynamics data bank: a database for conformational dynamics of proteins and supramolecular protein assemblies. Nucleic Acids Res 39:451–455
53. Oliver RC, Read DJ, Harlen OG, Harris SA (2013) A stochastic finite element model for the dynamics of globular macromolecules. J Comp Phys 239:147–165
54. Meyers MA, Chawla KK (2009) Mechanical behavior of materials. Cambridge University Press, Cambridge
55. Lai WM, Rubin DH, Rubin D, Krempl E (2009) Introduction to continuum mechanics. Butterworth-Heinemann, Oxford

56. Bower AF (2011) Applied mechanics of solids. CRC Press, Boca Raton
57. Ross CTF (1998) Advanced applied finite element methods. Woodhead Publishing, Cambridge
58. Dhatt G, Lefrançois E, Touzot G (2012) Finite element method. Wiley, New York
59. Landau LD, Lifshitz EM (1959) Fluid mechanics: course of theoretical physics, vol 6. Pergamon Press, New York
60. Gere JM (2004) Mechanics of materials, 6th edn. Brookes/Cole
61. Burgess SA, Walker ML, Sakakibara H, Knight PJ, Oiwa K (2003) Dynein structure and powerstroke. Nature 421(6924):715–718
62. Kollman JM, Pandi L, Sawaya MR, Riley M, Doolittle RF (2009) Crystal structure of human fibrinogen. Biochemistry 48(18):3877–3886
63. Pettersen EF, Goddard TD, Huang CC, Couch GS, Greenblatt DM, Meng EC, Ferrin TE (2004) UCSF chimera—a visualization system for exploratory research and analysis. J Comp Chem 25(13):1605–1612
64. Schöberl J (1997) Netgen an advancing front 2d/3d-mesh generator based on abstract rules. Comput Vis Sci 1(1):41–52
65. Kurland NE, Drira Z, Yadavalli VK (2012) Measurement of nanomechanical properties of biomolecules using atomic force microscopy. Micron 43(2):116–128
66. Desfossee A, Goret G, Estrozi LF, Ruigrok RWH, Gutsche I (2011) Nucleo-protein-rna orientation in the measles virus nucleocapsid by three-dimensional electron microscopy. J Virology 85(3):1391–1395
67. Meyer T, Ferrer-Costa C, Perez A, Rueda M, Bidon-Chanal A, Luque F, Laughton CA, Orozco M (2006) Essential dynamics: a tool for efficient trajectory compression and management. J Chem Theory Comput 2(2):251–258

Chapter 9
Structure Prediction of Transmembrane Proteins

Gábor E. Tusnády and Dániel Kozma

9.1 Introduction

In this chapter we discuss the various structural aspects of transmembrane proteins (TMPs) and survey the tasks and methods needed for modeling their structure. The structure prediction of TMPs from the pure amino acid sequence translated from genome projects may go through the following steps: (i) remove annotated or predicted cleavable parts (transit sequences, signal peptides); (ii) determination of the protein type (TMP or not); (iii) localization of TM segments within the amino acid sequence (topography prediction, 2D prediction) and the soluble parts of the protein relative to the membrane (topology prediction, 2.5D prediction); (iv) modeling the tertiary structure (3D) of membrane embedded protein parts which, depending on the amino acid similarity to the available relatives whose structure are already solved, may be based on homology modeling; may use the advantage of threading or may be de novo predictions including the contact prediction of amino acids of TM segments; (v) prediction of oligomerization propensity; (vi) finding the orientation in the membrane. In the following sections we guide the reader through these consecutive steps (Fig. 9.1) on how to derive the biologically active form of an unknown TMP purely computationally.

G.E. Tusnády (✉) · D. Kozma
Institute of Enzymology, Research Centre for Natural Sciences, Hungarian Academy of Sciences, P.O. Box 7, Budapest 1518, Hungary
e-mail: tusnady.gabor@ttk.mta.hu

© Springer International Publishing Switzerland 2014
G. Náray-Szabó (ed.), *Protein Modelling*, DOI 10.1007/978-3-319-09976-7_9

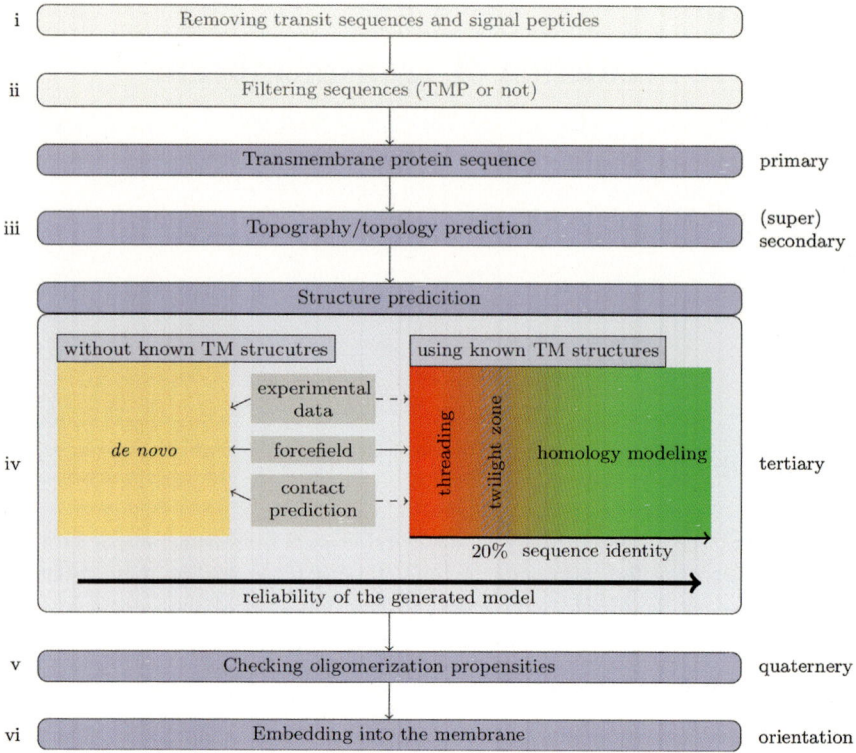

Fig. 9.1 Summary of the prediction pipeline of transmembrane proteins

9.2 Structural Aspects of Transmembrane Proteins

The lipid bilayer is an amphipathic slab with hydrophilic surfaces and a hydrophobic core region, from where the water molecules are excluded. Therefore the membrane segments of the polypeptide chains must adopt structures where all hydrogen donor and acceptor atoms are bound intramolecularly. This constraint leads to the formation of α-helical bundles and β-barrel secondary structures that are the most common secondary structures in the membrane spanning regions of TMPs. Therefore, based on the secondary structure of protein segments in the membrane regions, TMP can be classified into two main groups: α-helical and β-barrel. All plasma-membrane proteins are α-helical bundles with a large conformational variation, which is partly due to the water molecules penetrated into the membrane regions of the TMPs [161] forming water-filled cavities which makes the hydropathy-based topology predictions more difficult as well.

Very rarely, coil regions can be found in the membrane-embedded structure parts, mostly in re-entrant regions (that enter and exit on the same side of the membrane) or at kinks, where the translational symmetry breaks. Secondary structures do not terminate necessarily at the membrane water interface; sometimes these penetrate to the hydrophilic water phase. Often, on the membrane—water barrier interfacial helices (α-helices laying close and approximately parallel to the membrane surface) can be found, which have various (but not fully understood) functional roles, e.g. gating regulation and co-factor shielding [104].

While α-helical TMPs exist in all super-kingdoms, β-barrel TMPs can be found only in bacterial porins and in the inner membrane of mitochondria of eukaryote cells. For a long time it was believed that β-barrel TMPs always have even number of strands and in the range between 8 and 22, but this is refuted by the recently solved structure of the voltage-dependent anion channel (VDAC) [9] and the translocation domain of bacterial usher proteins [113, 150] containing 19 and 24 transmembrane (TM) β-strands, respectively.

The number of TM segments in α-helical TMPs range from 1 up to 24 (sodium channel protein type 2, α-subunit), but regarding their number in autonomous protein domains the highest known number is 15. Genome-wide analyses showed that distribution of the number of TM segments in α-helical TMPs is not random, proteins with 6 and 12 transmembrane helices (TMHs), such as small-molecule transporters, sugar transporters and ABC transporters, are predominant in uni-cellular organisms [3, 28, 67, 73, 123, 157]. In contrast, proteins with 7 TMHs are frequent in worms and human due to the high abundance of G-protein coupled receptors (GPCRs) [100]. Partly due to this abundance, the seven-helix membrane protein family members are the most important current drug targets.

9.3 Estimated Size of the Structure Space of Transmembrane Proteins

For globular/soluble proteins the total number of distinct globular folds that exist in nature is predicted to be a rather limited number [23], probably no more than 10,000 [70, 162], regardless of the astronomical number of the possible combination of structural elements. In TMPs, due to the physical constraints imposed by the lipid bilayer the number of possible folds is much smaller. Most of the TMPs adhere to one principal topology, involving one or more α-helices arranged parallel to each other and oriented about perpendicular with respect to the membrane plane. For β-barrel TMPs, they have a smaller structural diversity than α-helical ones. The short loops between helices constrain the possible folds of TMHs, therefore conformation space can be sampled effectively for small numbers of helices, and there are only about 30 possible folds for a TMP with three transmembrane helices (TMHs) [15]. However, the number of combinatorially possible folds was shown to increase exponentially with the number of TMHs to 1.5 million folds for seven helices, studies have showed that increasing number of membrane regions does not mean the exponential expansion of the fold space. Moreover structures with 8 or more transmembrane helices have less different architectures which reuse elements of folds with 3 or 4 helices [100]. Therefore the size of the fold space cannot be predicted based on combinatorial considerations. As an upper estimate of the number of different structures, the number of protein families can be used that are identifiable based on sequence similarity alone. Obviously this is a rough approximation, but provides a definite and reliable upper limit.

Liu et al. [87] showed in a study of 26 proteomes that there are about 10 times more soluble protein families than membrane protein families. Oberai et al. [107] set up a numerical experiment to estimate the number of distinct TMP folds. They found that any given residue has an 80 % chance to fall into one of about 500 families and observed a significant decrease in the number of members between the first and the second 20 most populous families. These results indicate that there are only a few very large and many very small families of membrane proteins, similarly to soluble proteins. The largest families are populated by various signaling proteins (e.g. GPCRs) and channels (e.g. potassium channels) [24, 48, 129], different transporters (secondary transporters and the ABC transporter family [32, 128, 132]), and TMPs involved in energy production (cytochrome b and NADH ubiquinone oxidoreductases) [12, 39]. As a consequence of the rapid fall-of and the asymptotic tail of the family size distribution, Oberai et al. [107] concluded that 670 families will cover 80 % of the structured sequence space but 1,720 families are needed to cover 90 % of the structured sequence space for all extant polytopic membrane proteins. These numbers are still an upper limit, as in SCOP [2] hierarchy a family is a subset of a fold. Assuming that the distribution of folds over families is similar to the one of soluble proteins and applying a stretched exponential model [44], Oberai et al. [107] estimated that only 550 folds cover 90 % and 300 folds cover 80 % of membrane protein structured sequence space. Finally, taking into account the physical constraint that stem from the membrane bilayer environment, they expect this is still an overestimate. Currently about only a hundred distinct (good quality, X-ray) transmembrane folds are known from various organisms. Known TMP structures by now make possible to create model for 26 % of the human α-helical transmembrane proteome using homology modeling (see Sect. 9.4.2.1), this ratio could be increased up to 56 % with 100 more new evenly selected and determined structures [115].

Another interesting paper discusses the number of different helix-helix contact architectures as a function of the number of transmembrane segments [100]. They developed a method for predicting helical interaction graphs and found that membrane proteins with 8 and more helices have significantly fewer arrangements than proteins with up to 7 helices. The most striking cases are transporter proteins with either 8 or 11 transmembrane helices, which according to Neumann et al., all seem to share a common helix interaction pattern. It was observed that TMPs with 10, 12, 14 membrane segments have significantly more distinct interaction graph than TMPs with 11, 13 or 15. This implies a hypothesis, TMPs with more than 8 TM segments may originated from TMPs with 5, 6 and 7 membrane regions that themselves are distributed over many different helix interaction clusters. While odd number of regions cannot stem from gene duplication, this could be an acceptable explanation for the phenomena described above [100].

9.4 Predicting Different Levels of Structures

9.4.1 Topography and Topology Prediction

Starting structure modeling from the amino acid sequence the first task is to check the presence of signal peptides and to decide whether it codes a globular or transmembrane protein. Here we refer some recent reviews, where these problems are discussed [146].

As a second step, one has to locate membrane spanning segments within the sequence. The information refers the location of the membrane spanning regions within the sequence is called topography. While in the case of helical TMPs the transmembrane segments are formed by 15–20 hydrophobic amino acids, in case of the β-barrel TMPs the length of the TM segments are shorter and only every second amino acid has to be hydrophobic making their topography prediction harder. In this section we do not discuss topography prediction of β-barrel TMPs; instead we focus on helical transmembrane segment prediction.

Earlier topography prediction methods [35, 74] explored the fact that membrane spanning segments are more hydrophobic than other parts of the protein chain. These segments can be identified by averaging the hydrophobicity of the amino acids within a sliding window over the sequence investigated. Other statistical approaches, like the Dense Alignment Surface (DAS) algorithm [27] overcomes the difficulties caused by the different hydrophobicity scales by a special alignment procedure [26], where the unrelated TMPs recognize each other without applying any hydrophobicity scales. Later it was shown, that in the case of a properly chosen hydrophobicity scale, accuracy of topography prediction can be as high as of the best state-of-the-art prediction methods [11].

For topology prediction the next step is orienting the membrane spanning segments from outside to inside or vice versa. This is equivalent to localize the sequence segments between membrane spanning segments alternatively inside or outside. The difference between topography and topology is that topology refers the location of the non-membrane segments as well. However, there are only a few properties of TMPs that help this task. The first and most prevalent such feature of TMPs is that the positively charged amino acids are more abundant on the cytosolic part of polypeptide chain, than on the extra-cytosolic ones (positive-inside rule) [137, 151]. Most topology predictions apply this rule after the topography prediction to choose the more likely from the two possible models [126]. Some prediction methods, such as TOPPRED [137, 153], utilize this rule both for topography and for topology prediction, by generating several models with certain and possible transmembrane segments, and choosing the model where the differences of the number of lysines and arginines were the highest between the even and odd loops. The MEMSAT method [60] incorporates the positive-inside rule indirectly by maximizing the sum of log-likelihoods of amino acid preferences taken from various structural parts of membrane proteins in a model recognition approach.

By increasing the number of TMPs whose topology were experimentally proven, machine learning algorithms like hidden Markov model (HMM) [119], support vector machine (SVM) [25] and artificial neural network (ANN) [94] can provide high prediction accuracies due to the fact, that the amino acid compositions of the various structural parts of TMPs are specific and machine learning algorithms are capable of learning these compositions during supervised learning [139]. Novel machine learning methods report higher and higher prediction accuracies due to the continuously growing and more reliable training sets and combining various techniques (e.g. using SVM or ANN for residue prediction and HMM for segment identification [149]). However, as these methods usually operate with parameter sets that are hard or near-impossible to integrate biochemically we cannot learn from these methods about the topology forming rules of TMPs. Moreover, to predict the topology of novel TMPs were never seen earlier by the machine learning methods, these methods may need to be retrained.

Replacing supervised learning technique by unsupervised one for HMMs, the training phase can be eliminated and the dependence on the training set can be avoided as well. Methods, such as HMMTOP, therefore do not need to be retrained from time to time. The success of unsupervised learning is based on the fact that a polypeptide chain of a TMP goes through various spaces of a cell with different physico-chemical properties (hydrophobic, polar, negatively charged, water-lipid interface etc.), therefore, the amino acid compositions of the TMP segments will be different in each type of regions. We do not need to know and as a consequence the constructed method does not need to learn these characteristic amino acid compositions to successfully predict the topology of TMPs. According to the law of maximal probability, these structural parts can be identified by segmenting in a way, that the amino acid compositions of the various structural parts show maximal divergence. This partitioning can be found by hidden Markov models.

There are two additional possibilities to increase the prediction accuracy of topology predictions. The first one is the utilization of consensus prediction methods. In addition to getting better predictions, using the results of several prediction methods allows us to estimate the reliability of the predicted topology as well. The consensus approach was also applied to predict partial membrane topologies, i.e. the part of the sequence where the majority of the applied methods agree. The other technique to increase the prediction accuracy is the use of constrained prediction methods. These can be used if there is/are one or more experimental data about the topology and prediction method can handle these data as constraints and not only to filter results that agree with the given experimental data. Thus, given a constraint (e.g. the N-terminus is inside), a constrained prediction method gives a prediction that satisfies this criterion. In a HMM based method this is achieved by the modification of the Baum-Welch and Viterbi algorithms. The first such application was HMMTOP2 [145]. Later the two other HMM based methods, TMHMM and Phobius were also modified to include this feature [62, 139]. The mathematical details of the necessary modification can be found in Ref. [6]. The optimal placement of constraints was also investigated, and it was shown that the accuracy can be increased by 10 % if the N- or C-terminal of the polypeptide chain is constrained in the above mentioned way, and 20 % is the maximum obtainable increase if one of each loop or tail residue in turn is fixed to its experimentally annotated location

[120]. Constraints can be either experimental results or bioinformatical evidence. In the first molecular biology experiments transposons were used to create random chimera proteins [52], later more specific molecular biology techniques were applied to investigate the topology of TMPs of interest (for review of these techniques see [144, 148]). The continuous development of biotechnology allows scientist to analyse the topology of all TMPs in an organism. In the topology analyses of *E. coli* and *S. cerevisiae*, the results of C-terminal fusion proteins were applied as constraints [28, 31, 67, 68, 120]. Recently high through-put techniques became available, where the surface of a living cell is labeled by chemical agents and the labeled peptides are investigated by coupled analytical technique after purification and degradation [13, 46, 93, 101]. In TOPDB more than 4,500 experimental results were collected for ~1,500 TMPs, and these constraints were applied to make constrained topology predictions for the ~1,500 TMPs. Regarding bioinformatical approaches, locations of compartment specific domains and sequence motifs can also be used as constraints. Such domains and motifs were collected into the TOPDOM database [144] from various databases such as SMART [45, 84], Pfam [36] and Prosite [136] for the purpose of constrained prediction.

9.4.2 Tertiary Structure Prediction of Transmembrane Proteins

Despite of the theoretical and computational difficulties, during the last two decades scientists have developed valuable methods to approximately model the tertiary structure of TMPs. Predicting TMP structures, at first, seems to be a relatively easy problem compared to understanding soluble protein structures. The fact that many TMPs share similar folds even with marginal sequence identities [43, 129] proves that TMPs are more structurally conserved than globular proteins. This is due to the strict conformational constraints that come from the membrane lipid bilayer, which dramatically decreases the size of the conformational space. However, the presence of an additional environment may cause previously unforeseen difficulties.

There are three main strategies to solve the tertiary structure of unknown TMP sequences. Homology modeling can be used when there is a sequential homologue with sequence identity greater than 20 %. In the case when no sequential homologue is available but (ideally) all folds are known, one can use threading methods to select the best packing of the query sequence. When neither sequential relative nor all folds are captured solely, de novo methods are still usable. It is worth mentioning that the order of this enumeration reflects the reliability of the methods as well (Fig. 9.1). Therefore it's not surprising that—due to the fundamentally unfeasible sampling of the whole structure space while looking for native structures—de novo methods are at the end of this list.

In the following sections we go through these three main families of transmembrane tertiary structure prediction strategies, namely comparative modeling, threading and de novo methods.

9.4.2.1 Comparative Modeling Techniques

Comparative modeling (also known as homology modeling) is a structure build-ing strategy for unknown protein structures, which can be used when at least one sequential homologue with known structure is available for a given query TMP. The 3D structure of the sequentially homologue protein are used as a template (or target). Once the template has been selected and an alignment is generated between the tem-plate and target sequences, the non-conserved residues are replaced and insertions (regions with no template structure) are modeled as loop regions using de novo meth-ods [117]. It is important to note that as for globular proteins, the accuracy of a homol-ogy model is strongly dependent on the identity between the two sequences [38].

While this technique basically relies on sequence alignments, at first, we have to declare the sequence identity level from where two TMPs can be considered as structural homologue. It was shown for globular proteins [125] that proteins with 30 % sequence identity the probability of sharing the same fold is ~90 % (below 25 % identity this probability drops to 10 %), in alignments longer than 80 resi-dues. Although the application of this well-known fact has become second nature for researchers in the case of globular proteins, shedding light on the twilight zone (where structural similarity starts to diverge rapidly as sequential identity decreases) of TMPs is only a recent improvement [38, 108]. This lagging is due to the difficulties in experimental structure determination methods [75] applied and its consequence, the relatively small number of known transmembrane structures.

In a recent study [108], sequence–structure relation was analyzed using TMP structures with resolution <4 Å. It was found for the membrane region of TMPs that at >35 % sequence identity the structure RMSDs (RMSD—Root Mean Square Deviation) were 0.89 ± 0.43 Å and 0.80 ± 0.32 Å for α-helical and β-barrel mem-brane proteins, respectively. In addition, at 20–30 % sequence identity RMSDs increased—as expected—to 1.59 ± 0.55 Å and 1.30 ± 0.35 Å. According to expec-tations, TMPs show lower RMSD values than globular proteins, as structure in the membrane region is more conserved or restricted than in the non-membrane regions. Consequently, in the case of membrane regions of TMPs it is possible to use structures even with low sequence identity (<20 %) for comparative modeling. Moreover, β-barrel architecture seems much more robust to sequence variations. They found that sequence–structure similarity is generally independent of the num-ber of membrane regions. The authors [108] concluded that functional mechanisms are preserved by high structural conservation and their functional specificity is mainly determined by the variable solvent-exposed regions.

Although homology modeling of globular proteins is a tried-and-true technique to predict 3D structure of query sequences having a sequential homologue, but in the field of TMPs this approach is in its infancy. There are some examples for modeling GPCR receptors [4, 43], but there isn't any fully automated, membrane protein spe-cific method. Other, non-specific methods [122] are used as well, but the constraints imposed by the membrane are not utilized in the modeling, and the applied scoring functions designed for globular proteins might lead to distorted models.

Neglecting the scoring function and other technical details, a typical template-based modeling protocol can be briefly described in the following steps. At first, query protein searched against a related database containing TMP sequences with known structure and one or more homologue templates are selected based on their sequential identity. Next, query sequence is aligned to all template sequences. These steps are usually merged and performed together, while most methods for detecting templates rely on the production of sequence alignments. As known, the primary criteria of database search algorithms is speed, therefore alignments resulted in database searches may not be as accurate as alignment produced by non-searching techniques. However, these kinds of algorithms are widely used to detect, and to generate alignment for homologue templates from database, e.g. PSI-BLAST [1, 131] and HHsearch [138]. The alignment of the target to template sequence(s) is the most important step of the whole procedure. Aligning transmembrane sequences used to be a long-standing unsolved problem, but by now numerous TMP sequence aligner methods have been developed, e.g. AlignMe [140] and MP-T [53]. According to Forrest et al. [38], comparative modeling of TMPs has been estimated to obtain accuracy as high as that of soluble proteins if the alignment for TMPs achieves the accuracy of its soluble protein counterpart.

The last step is the coordinate generation based on the alignment. For predicting the conformation of loop regions one can use Loopy [163], which is one of the fastest or PLOP [57], which is one of the most accurate techniques. FREAD [22] uses environment specific scoring parameters to improve the sampling for their loop structure prediction algorithm. RAPPER [29] and FALCm4 [81] rely on fine-grained residue-specific φ/ψ propensity tables for conformational sampling. Recently a coarse-grained method for loop prediction [90] was also developed of which computational time scales better than others, while the accuracy was preserved.

To highly increase the accuracy of the final structure, a genetic algorithm developed by John et al. [58] can be used to iteratively build better alignment for distant homologues. This method builds target-template alignments and structure models, and after assessing generated models, the alignments of the best models are used for generating further alignments.

Here we sketched the basic principles of the homology modeling techniques, in the following we review some recent methods based on comparative modeling.

A web server for homology modeling of TMPs named Memoir [34] is a pipeline utilizing iMembrane [65], a membrane protein annotator using CGDB [21] coarse-grained database; MP-T [53] target-template aligner; Medeller [66], a coordinate generator and FREAD [22], a loop modeler. Memoir does not search for a homologue template, therefore it needs this as an input parameter and does not provide any information on the reliability of the resulting structures.

A novel method, GPCRM [79] is developed for GPCR membrane protein structure predictions with averaging of multiple template structures and profile-profile comparison. It also utilizes two distinct loop modeling techniques: Modeller [154] and Rosetta [124] and excluding models with lipid penetrated loops.

At the border of homology and de novo modeling, the SWISS-MODEL [4] 7TM interface is developed for the modeling of TMPs with 7 transmembrane helices. SWISS-MODEL 7TM performs homology modeling on experimental and theoretical templates; to use this server user needs to provide the location of TMHs in the query sequence and also a template.

9.4.2.2 Threading Algorithms

Threading becomes very useful in cases when a sequence does not have any sequential relative with a known structure. This is a common scenario in the case of TMPs, where only a highly restricted number of TMPs show significant sequence identity to any known structure. As a consequence, homology modeling techniques discussed above have serious limitations which can be bridged using threading. Nevertheless, for building an efficient and reliable pipeline first we need a representative structure set of the conformational space of TMPs.

As discussed in Sect. 9.3, only a very small ratio, about the one fifth of the TMP structure space is known. Therefore, an efficient threading algorithm must not only find the structure with the lowest energy, but it has to discriminate native and the 'most-stable' decoy structures as well. These structure assessing algorithms are discussed in Sect. 9.4.2.4.

Due to the significant physico-chemical differences between soluble and membrane proteins, threading methods developed for globular proteins cannot be used directly, however a few methods have been customized for TMPs.

TASSER [165] is a two-step method that threads the sequence onto parts of solved protein structures and then refines the resulting template. The method was validated on a set of 38 non-homologue TMP structures, a little fewer than half of which have the RMSDs less than 6.5 Å compared to the native structure, but in the other cases RMSDs are in excess of 10 Å. It was used systematically to predict human GPCRs and these seemed consistent with experimental data. However, when there was no significant sequential relative, it was uncertain if the results represent the native structure.

A recent method, TMFR [158] is a sequence based fold recognition algorithm and has the accuracy of 49.2 and 82.2 % for α-helical and β-barrel TMPs, respectively. It utilizes topological features which improve the fold recognition [49] and can accurately align the target sequence to the template structure and generate reliable alignment raw scores to evaluate the structural similarity between the target and template. This provides practically only a sequence alignment. Therefore, algorithm traces back structure prediction problem to something akin to homology modeling.

However, this type of approximation widen the horizon of TMP structure prediction significantly, unfortunately the lack of structural representatives limits the usability of threading henceforward. In the next subsection, de novo methods are discussed which try to get over these difficulties.

9.4.2.3 De Novo Methods

De novo modeling does not use homologue proteins of known structures to predict the structure of an unknown protein. For an effective de novo structure prediction method there are two crucial requirements: accurate energetic representation of a protein structure and an efficient sampling of conformational space [82]. While structural space expands rapidly with the sequence length, these methods are mainly applicable for small soluble proteins [16], not for TMPs, which are often large structures [158]. Although methods do not have restrictions on the number of known structures as homology modeling or threading do. However, as combinatorial approaches these require large amounts of computing time which often cannot be run on a single desktop computer, hence reducing their availability for structural biologists.

Contact Aided Structure Prediction

Contact prediction methods originates from the article, written by Göbel et al. [42], that describes how one could infer spatial information from multiple sequence alignments. This concept is based on the observation that the structure is more conserved than the sequence. Therefore, if a residue fulfilling structurally important role in a protein mutates, than another spatially close residue has to change to preserve both the structure and the function. Later it turned out that this assumption is a poor approximation of real proteins and their evolutionary processes.

Contact prediction methods can be classified into two main categories, namely local and global methods. The first one contains the 'classical' correlated mutation algorithms (CMA), which could be subdivided into further subcategories. To extract spatially close residues from multiple sequence alignments simple covariance analysis with various substitution matrices [42, 109]), χ^2-test [64], information theoretic approaches [33], machine-learning [20, 103, 118], alignment perturbation (SCA [88], ELSC [30]), probabilistic and empirical matrix methods or formal language [155, 156] were used. Further on, consensus methods are developed [41], which did not succeed to significantly overcome the performance of the previous methods (Acc. ~10 %, see Ref. [55]). CASP10 [95] confirmed the need for the development of contact prediction methods. On a test set of newly identified structures the best algorithm performed at an accuracy of ~30 %. Machine learning methods (PROFcon [118], CMAPpro [20], MEMPACK [106], PhyCMAP [159]) generally outperform others based on statistical considerations. Despite of the better predictor abilities, machine learning based approaches make their results difficult to interpret biophysically. In addition, the lack of a physical model makes the limits of their usability ill-defined. A recent study [47] showed that using three representatively selected contact prediction methods, there is no such linear combination of selected local techniques which could reach a satisfiable performance level. In addition, when a consensus method was trained and tested on only two, ABC-B and ABC-C protein families, despite of a nearly over parameterized model, these techniques could not reach a satisfying performance limit.

The main problem is that the observable correlations among sites do not stem from spatial closure purely. Atchley et al. [5] formalized the sources of these correlations, that—apart from structural constraints—could came from phylogenetic noise, function and higher-order statistical non-independence of positions. In addition, random noise or uneven sampling could bias measured correlations as well. The orders of magnitude of these factors are investigated by Noivirt et al. [102];

they found that correlation from structural, functional and phylogenetic constraints are in the same order of magnitude. Therefore, using background co-evolution signal correction [33] proved to be a valuable tool to reduce phylogenetic noise and to increase precision of methods significantly albeit even these precisions remains low.

Even if we neglect the disturbingly high correlation from functional and phylogenetic sources, there still remains a significant problem, namely disentangling direct and indirect interactions [17]. Global methods can take it into account with estimating joint probabilities of multiple residues. For reliable statistics huge number of sequences is required, which is a limiting factor still could not be overcame yet. Burger and van Nimwegen [17] had developed a Bayesian-network based method, which can take into account that the probability for residues to be in contact depends on their primary sequence separation and that highly conserved residues tend to participate in a larger number of contacts [17]. With this or other methods using maximum-entropy model [77, 97], sparse inverse covariance estimating [59] approaches could break through the barriers set by indirect contacts and multiple correlations. This network-based conceptional change of view and the increased size of sequence families result in significant performance gain.

Another possibility for calculating structural constraints is the prediction of helical interaction only, instead of predicting directly residue—residue contacts. It is an easier task than identifying all individual residue contacts in an α-helical TMP. However, this does not give any information on the orientation of helices and all the helices are treated as perpendicular to the membrane plane [40, 100]. Using propensity estimation techniques, e.g. lipid exposure predictors, the precision of helix-helix interaction and orientation estimations can be improved [92, 103, 116]. It has been known for a while that the tilted orientation of transmembrane helices is a principal compensation mechanism for hydrophobic mismatch [111]. Nevertheless, spanning regions are not necessarily straight: kinked or bended helices exist as well [76, 152], which complicates helical contact prediction even further.

These methods are valuable tools in themselves, which help to get closer to the biological understanding of TMP structures, functions and their mutational processes, but unfortunately they are still not as trustworthy as e.g. topology prediction techniques.

It is worth to mention, contact predictions cannot be used to directly reconstruct the 3D structure of proteins [110] not even using perfect predictor, not even for TMPs. This is due to their contradictory results originates from the oligomerization or conformational changes of the studied proteins. In the case of oligomers we would need to distinguish between intra- and interchain contacts. Another problem arises from multiple conformations of proteins, as in the case of the open and closed conformations of the *E. coli* GlpT or human OCTN1 [96]. When neither conformation change nor oligomerization has an influence on the inspected protein structure, theoretically an essential set of structure determining residue contacts is enough to replicate the 3D structure [130].

If we approximate the given problem from a reverse way, we could use predicted contacts as constraints in simulated annealing simulations [54, 91] or to aid the separation of native like TMP folds from decoys [92, 103, 127]. Obviously, one could use experimental techniques, such as e.g. NMR, to earn useful structural constraints to build TMP model structures, but in this section we discussed this problem from the computational point of view only.

Forcefield-Based Approaches

There are many forcefield-based methods for determining the 3D structure of proteins; here we review some of those that were developed for TMPs. Given the structural simplicity observed in the β-barrel conformational space, structure prediction methods focused on estimating structures of α-helical TMPs. This imbalance will be observable in this paragraph as well.

In the early studies, such as Fleishman and Ben-Tal [37] residue environment preferences were used to predict the likely arrangement of transmembrane helices, and they were able to predict the native structure of TMP glycophorin A. Ledesma et al. [80] suggesting a model for the uncoupling protein 1 (UCP1), utilizing a computational docking method. Chen and Chen [19] used a lattice model of membrane proteins with a composite energy function to study their folding dynamics and native structures in Monte Carlo simulations. This model successfully predicts the seven helix bundle structure of sensory rhodopsin I by employing a three-stage folding. FILM [112] was developed for predicting small TMP structures based on assembling super-secondary segments taken from a protein structure library. The native structure is searched by simulated annealing. The main limitation of FILM is that the potential function is not able to reproduce the compactness of transmembrane bundles.

RosettaMembrane [164] (a derivation of Rosetta [124]) uses an all-atom physical model to describe intra-protein and protein-solvent interactions in the membrane environment. The surrounding environment is divided into 5 layers: water-exposed, polar, interface, outer and inner hydrophobic in both directions of the membrane core. Here a log-likelihood pair and environment potential were used, which penalizes steric overlap but favours packing density like characteristics of membrane proteins and strands. The method was tested on 12 membrane proteins with known structure, The length of the query sequences were between 51 and 145, which was predicted with RMSD <4 Å. However, mainly due to the technically unfeasible sampling of the conformational space this method performs poorly for large and complex proteins, independent on being soluble or transmembrane. In a newer version [7] of RosettaMembrane, experimental and predicted constraints were used to aid structure prediction. A great advantage of this method is that it can take into account cofactors, which could significantly modify structures.

BCL::MP-Fold [160] uses a three layer (solution, transition, membrane) implicit membrane representation with transition regions and a knowledge-based potential derived using Bayes' theorem and the inverse-Boltzmann relation. The final score resulted as the linear combination of many energy terms with optimized weights. The search for the native structure starts from randomly placed helices oriented perpendicular to the membrane plane. As a next step, folding is performed with simulated annealing [63].

As GPCRs are the most common drug targets, specific methods for predicting GPCR structures, such as MembStruk [85, 147] and PREDICT [135] have been developed as well. Both methods provide full-atom models for GPCRs on the basis of physico-chemical principles. In the PREDICT algorithm a concept of structural decoys is employed to ensure that the algorithm identifies the correct structure and to avoid trapping in a local minimum.

3D-SPOT [99] is a template-free method utilizing a statistical mechanical model [98] and an empirical potential function; TMSIP [56] is to predict the 3D structure of a given β-barrel TMP. While this method is based on physical interactions and does not require template structures, it can be applied for predicting structures of novel folds. The method performs well; in a blind test it was able to generate accurate structures of the transmembrane regions with a median main chain RMSD of 3.9 Å, on a set of 23 proteins.

9.4.2.4 Validating Predicted 3D Structures

Several methods have been developed for judging the reliability of predicted structures and identifying erroneous regions. Methods like PROCHECK [78] can be used for TMPs as well, because it takes into account only fundamental properties, namely the ψ/φ backbone torsion angles. There are various attempts to develop a measure, like the normalized QMEAN score [10] for soluble proteins, describing absolute quality of each structure for membrane proteins too. Phatak et al. [114] described a method filtering near-native structures from decoys using low-complexity Support Vector Regression models for predicting relative lipid accessibility (RLA). The quality assessment is based on the consistency of the predicted and observed RLA profiles. ProQM [121] utilizes SVM and membrane protein specific features, tested on GPCR structures. As it turned out, this method is capable of disentangling correct models from incorrect ones and has the ability to identify poor quality regions. IQ (Interaction-based Quality assessment) [50] incorporates four types of inter-residue interactions and achieves high prediction power on the independently constructed dataset (GPCR Dock 2008 (206 models), GPCR Dock 2010 (284 models), and HOMEP (92 models)). However, further validation of this method is needed. Recent results suggest that among conformations very dissimilar to native structures, this scoring function cannot correctly identify the best one. This is largely understandable since this scoring function relies primarily on the number of hydrophobic interactions. Lots of incorrectly formed hydrophobic interactions in decoy conformations could bias the IQ value (Li Zhijun personal communication).

9.4.3 Quaternary Structures of Membrane Proteins

As it was discussed earlier, genome-wide analysis of domain combinations of helical membrane proteins revealed that α-helical TMPs exist mostly as single domains. Oligomerization within the membrane may could be the general mechanism for membrane proteins to gain new biological functions [83, 86, 87]. Therefore, discovering principles of oligomer formation of TMPs is needed to understand theirs functions and to gain new therapeutic strategies.

For globular proteins there are various methods to predict oligomerization propensities. PQS [72] and PISA [51] can identify the biologically active oligomer from X-ray structures. However, these methods cannot be used for TMPs, therefore in the PDBTM [71, 142, 143] database a simple homology search was used for predicting oligomeric state of potentially existing novel transmembrane structures, independent on their type (α-helical or β-barrel).

Bowie [69] and coworkers predicted the structure of α-helical TMP oligomers (glycophorin A and M2 proton channel) using knowledge of the oligomer symmetry. They used a simple softened van der Waals potential and Monte Carlo minimization to pack ideal α-helices. Bordner [14] have developed a method to predict biding sites of TMPs using a Random Forest classifier, trained on residue type distributions and evolutionary conservation for individual surface residues, followed by spatial averaging of the residue scores. Random Forest predictions were first made for individual surface residues and then the resulting scores of nearby residues were averaged in order to arrive at the final prediction score. Docking based approaches for predicting oligomerization has been developed as well [18]. In a recent review, the suitability of some widely-used docking algorithms for modeling complexes of α-helical TMPs was studied and the dependence of the docking performance on the protein features discussed as well [61].

Although α-helical TMPs pose a greater challenge, the oligomerization state of β-barrel membrane proteins can be accurately predicted computationally [98]. Based on the TMSIP [56] empirical potential function and the reduced conformational state model, extensive and contiguous weakly stable regions in many β-barrel membrane proteins seem to be an indicator of oligomerization propensities of β-barrel membrane proteins. Furthermore, as structural information is not essential for such predictions, the oligomerization state can also be predicted quite successful even when only sequence information is considered [98].

As it was discussed, there are various methods to model the quaternary structure of a TMP if it is a homomer or all the different subunits are known. Cases when the other subunits are unknown cannot be solved yet.

9.5 Orientation of Membrane Proteins in the Lipid Bilayer

Neither monomeric nor oligomeric TMPs do not exist alone without the amphiphilic membrane bilayer. By removing the hydrophobic environment the native structure breaks down. For experimental structure determination special handle with detergent is needed to extract TMP from the membrane and to preserve its native structure. Accordingly during experimental structure determination of TMPs, the information on the orientation disappears. While this information is essential for understanding the biological function and the mechanism of action of TMPs, experimental methods cannot recover it and thus has to be defined using computational techniques. Orientation and burial of TMPs are very important e.g. for drug design to identify accessible parts of TMPs.

There are various attempts to predict orientation and burial of TMPs. One of the first methods was IMPALA [8], which uses amino acid propensities. TMDET [141] algorithm utilizes a geometrical algorithm to locate the most probable orientation of the given TMP in the membrane slab. OPM [89] applies a more sophisticated description of the problem, but does not outperform TMDET significantly. Senes et al. [134] have developed an empirical low-resolution potential called E_z, for protein insertion in the lipid membrane. A recent paper describes a method for predicting membrane protein orientation using a knowledge-based statistical potential [105, 133].

References

1. Altschul SF, Madden TL, Schäffer AA, Zhang J, Zhang Z, Miller W, Lipman DJ (1997) Gapped BLAST and PSI-BLAST: a new generation of protein database search programs. Nucleic Acids Res 25(17):3389–3402
2. Andreeva A, Howorth D, Chandonia J-M, Brenner SE, Hubbard TJP, Chothia C, Murzin AG (2008) Data growth and its impact on the SCOP database: new developments. Nucleic Acids Res 36:D419–D425
3. Arai M, Ikeda M, Shimizu T (2003) Comprehensive analysis of transmembrane topologies in prokaryotic genomes. Gene 304:77–86
4. Arnold K, Bordoli L, Kopp J, Schwede T (2006) The SWISS-MODEL workspace: a web-based environment for protein structure homology modelling. Bioinformatics 22(2):195–201
5. Atchley WR, Wollenberg KR, Fitch WM, Terhalle W, Dress AW (2000) Correlations among amino acid sites in bHLH protein domains: an information theoretic analysis. Mol Biol Evol 17(1):164–178
6. Bagos PG, Liakopoulos TD, Hamodrakas SJ (2006) Algorithms for incorporating prior topological information in HMMs: application to transmembrane proteins. BMC Bioinform 7:189
7. Barth P, Wallner B, Baker D (2009) Prediction of membrane protein structures with complex topologies using limited constraints. Proc Natl Acad Sci USA 106(5):1409–1414
8. Basyn F, Spies B, Bouffioux O, Thomas A, Brasseur R (2003) Insertion of X-ray structures of proteins in membranes. J Mol Graph Model 22(1):11–21
9. Bayrhuber M, Meins T, Habeck M, Becker S, Giller K, Villinger S, Vonrhein C, Griesinger C, Zweckstetter M, Zeth K (2008) Structure of the human voltage-dependent anion channel. Proc Natl Acad Sci USA 105(40):15370–15375
10. Benkert P, Biasini M, Schwede T (2011) Toward the estimation of the absolute quality of individual protein structure models. Bioinformatics 27(3):343–350
11. Bernsel A, Viklund HK, Falk J, Lindahl E, von Heijne G, Elofsson A (2008) Prediction of membrane-protein topology from first principles. Proc Natl Acad Sci USA 105(20):7177–7181
12. Berry EA, Guergova-Kuras M, Huang L-S, Crofts AR (2000) Structure and function of Cytochrome bc complexes. Annu Rev Biochem 69(1):1005–1075
13. Biller L, Matthiesen J, Kühne V, Lotter H, Handal G, Nozaki T, Saito-Nakano Y, Schümann M, Roeder T, Tannich E, Krause E, Bruchhaus I (2014) The cell surface proteome of Entamoeba histolytica. Mol Cell Proteomics MCP 13(1):132–144
14. Bordner AJ (2009) Predicting protein–protein binding sites in membrane proteins. BMC Bioinform 10:312
15. Bowie JU (1999) Helix-bundle membrane protein fold templates. Protein Sci 8(12):2711–2719

16. Bradley P, Misura KMS, Baker D (2005) Toward high-resolution de novo structure prediction for small proteins. Science 309(5742):1868–1871
17. Burger L, van Nimwegen E (2010) Disentangling direct from indirect co-evolution of residues in protein alignments. PLoS Comput Biol 6(1):e1000633
18. Casciari D, Seeber M, Fanelli F (2006) Quaternary structure predictions of transmembrane proteins starting from the monomer: a docking-based approach. BMC Bioinform 7(1):340
19. Chen C-M, Chen C-C (2003) Computer simulations of membrane protein folding: structure and dynamics. Biophys J 84(3):1902–1908
20. Cheng J, Baldi P (2007) Improved residue contact prediction using support vector machines and a large feature set. BMC Bioinform 8:113
21. Chetwynd AP, Scott KA, Mokrab Y, Sansom MSP (2008) CGDB: a database of membrane protein/lipid interactions by coarse-grained molecular dynamics simulations. Mol Membr Biol 25(8):662–669
22. Choi Y, Deane CM (2010) FREAD revisited: accurate loop structure prediction using a database search algorithm. Proteins 78(6):1431–1440
23. Chothia C (1992) Proteins. One thousand families for the molecular biologist. Nature 357(6379):543–544
24. Chou K-C, Elrod DW (2002) Bioinformatical analysis of G-protein-coupled receptors. J Proteome Res 1(5):429–433
25. Cortes C, Vapnik V (1995) Support-vector networks. Mach Learn 20(3):273–297
26. Cserzö M, Bernassau JM, Simon I, Maigret B (1994) New alignment strategy for transmembrane proteins. J Mol Biol 243(3):388–396
27. Cserzö M, Wallin E, Simon I, von Heijne G, Elofsson A (1997) Prediction of transmembrane alpha-helices in prokaryotic membrane proteins: the dense alignment surface method. Protein Eng 10(6):673–676
28. Daley DO, Rapp M, Granseth E, Melén K, Drew D, von Heijne G (2005) Global topology analysis of the *Escherichia coli* inner membrane proteome. Science 308(5726):1321–1323
29. de Bakker PIW, DePristo MA, Burke DF, Blundell TL (2003) Ab initio construction of polypeptide fragments: accuracy of loop decoy discrimination by an all-atom statistical potential and the AMBER force field with the generalized born solvation model. Proteins 51(1):21–40
30. Dekker JP, Fodor A, Aldrich RW, Yellen G (2004) A perturbation-based method for calculating explicit likelihood of evolutionary co-variance in multiple sequence alignments. Bioinformatics 20(10):1565–1572
31. Drew D, Sjöstrand D, Nilsson J, Urbig T, Chin C-N, de Gier J-W, von Heijne G (2002) Rapid topology mapping of *Escherichia coli* inner-membrane proteins by prediction and PhoA/GFP fusion analysis. Proc Natl Acad Sci USA 99(5):2690–2695
32. Driessen AJ, Rosen BP, Konings WN (2000) Diversity of transport mechanisms: common structural principles. Trends Biochem Sci 25(8):397–401
33. Dunn SD, Wahl LM, Gloor GB (2008) Mutual information without the influence of phylogeny or entropy dramatically improves residue contact prediction. Bioinformatics 24(3):333–340
34. Ebejer J-P, Hill JR, Kelm S, Shi J, Deane CM (2013) Memoir: template-based structure prediction for membrane proteins. Nucleic Acids Res 41(Web Server issue):W379–W383
35. Eisenberg D, Schwarz E, Komaromy M, Wall R (1984) Analysis of membrane and surface protein sequences with the hydrophobic moment plot. J Mol Biol 179(1):125–142
36. Finn RD, Mistry J, Schuster-Böckler B, Griffiths-Jones S, Hollich V, Lassmann T, Moxon S, Marshall M, Khanna A, Durbin R, Eddy SR, Sonnhammer ELL, Bateman A (2006) Pfam: clans, web tools and services. Nucleic Acids Res 34:D247–D251
37. Fleishman SJ, Ben-Tal N (2002) A novel scoring function for predicting the conformations of tightly packed pairs of transmembrane α-helices. J Mol Biol 321(2):363–378
38. Forrest LR, Tang CL, Honig B (2006) On the accuracy of homology modeling and sequence alignment methods applied to membrane proteins. Biophys J 91(2):508–517
39. Friedrich T, Böttcher B (2004) The gross structure of the respiratory complex I: a Lego system. Biochim Biophys Acta Bioenerg 1608(1):1–9

40. Fuchs A, Kirschner A, Frishman D (2009) Prediction of helix–helix contacts and interacting helices in polytopic membrane proteins using neural networks. Proteins 74(4):857–871
41. Fuchs A, Martin-Galiano AJ, Kalman M, Fleishman S, Ben-Tal N, Frishman D (2007) Co-evolving residues in membrane proteins. Bioinformatics 23(24):3312–3319
42. Göbel U, Sander C, Schneider R, Valencia A (1994) Correlated mutations and residue contacts in proteins. Proteins 18(4):309–317
43. Gonzalez A, Cordom A, Caltabiano G, Pardo L (2012) Impact of helix irregularities on sequence alignment and homology modeling of G protein-coupled receptors. ChemBioChem 13(10):1393–1399
44. Govindarajan S, Recabarren R, Goldstein RA (1999) Estimating the total number of protein folds. Proteins 35(4):408–414
45. Granseth E, Daley DO, Rapp M, Melén K, von Heijne G (2005) Experimentally constrained topology models for 51,208 bacterial inner membrane proteins. J Mol Biol 352(3):489–494
46. Gu B, Zhang J, Wang W, Mo L, Zhou Y, Chen L, Liu Y, Zhang M (2010) Global expression of cell surface proteins in embryonic stem cells. PLoS One 5(12):e15795
47. Gulyás-Kovács A (2012) Integrated analysis of residue coevolution and protein structure in ABC transporters. PLoS One 7(5):e36546
48. Harte R, Ouzounis CA (2002) Genome-wide detection and family clustering of ion channels. FEBS Lett 514(2–3):129–134
49. Hedman M, Deloof H, Von Heijne G, Elofsson A (2002) Improved detection of homologous membrane proteins by inclusion of information from topology predictions. Protein Sci 11(3):652–658
50. Heim AJ, Li Z (2012) Developing a high-quality scoring function for membrane protein structures based on specific inter-residue interactions. J Comput Aided Mol Des 26(3):301–309
51. Henrick K (1998) PQS: a protein quaternary structure file server. Trends Biochem Sci 23(9):358–361
52. Herrero M, de Lorenzo V, Neilands JB (1988) Nucleotide sequence of the iucD gene of the pColV-K30 aerobactin operon and topology of its product studied with phoA and lacZ gene fusions. J Bacteriol 170(1):56–64
53. Hill JR, Deane CM (2013) MP-T: improving membrane protein alignment for structure prediction. Bioinformatics 29(1):54–61
54. Hopf TA, Colwell LJ, Sheridan R, Rost B, Sander C, Marks DS (2012) Three-dimensional structures of membrane proteins from genomic sequencing. Cell 149(7):1607–1621
55. Horner DS, Pirovano W, Pesole G (2008) Correlated substitution analysis and the prediction of amino acid structural contacts. Brief Bioinform 9(1):46–56
56. Jackups R, Liang J (2005) Interstrand pairing patterns in beta-barrel membrane proteins: the positive-outside rule, aromatic rescue, and strand registration prediction. J Mol Biol 354(4):979–993
57. Jacobson MP, Pincus DL, Rapp CS, Day TJF, Honig B, Shaw DE, Friesner RA (2004) A hierarchical approach to all-atom protein loop prediction. Proteins 55(2):351–367
58. John B, Sali A (2003) Comparative protein structure modeling by iterative alignment, model building and model assessment. Nucleic Acids Res 31(14):3982–3992
59. Jones DT, Buchan DWA, Cozzetto D, Pontil M (2011) PSICOV: precise structural contact prediction using sparse inverse covariance estimation on large multiple sequence alignments. Bioinformatics 28(2):184–190
60. Jones DT, Taylor WR, Thornton JM (1994) A model recognition approach to the prediction of all-helical membrane protein structure and topology. Biochemistry 33(10):3038–3049
61. Kaczor AA, Selent J, Sanz F, Pastor M (2013) Modeling complexes of transmembrane proteins: systematic analysis of protein–protein docking tools. Mol Inform 32(8):717–733
62. Käll L, Krogh A, Sonnhammer ELL (2004) A combined transmembrane topology and signal peptide prediction method. J Mol Biol 338(5):1027–1036

63. Karakaş M, Woetzel N, Staritzbichler R, Alexander N, Weiner BE, Meiler J (2012) BCL:fold–de novo prediction of complex and large protein topologies by assembly of secondary structure elements. PLoS One 7(11):e49240
64. Kass I, Horovitz A (2002) Mapping pathways of allosteric communication in GroEL by analysis of correlated mutations. Proteins 48(4):611–617
65. Kelm S, Shi J, Deane CM (2009) iMembrane: homology-based membrane-insertion of proteins. Bioinformatics 25(8):1086–1088
66. Kelm S, Shi J, Deane CM (2010) MEDELLER: homology-based coordinate generation for membrane proteins. Bioinformatics 26(22):2833–2840
67. Kim H, Melén K, Osterberg M, von Heijne G (2006) A global topology map of the *Saccharomyces cerevisiae* membrane proteome. Proc Natl Acad Sci USA 103(30):11142–11147
68. Kim H, Melén K, von Heijne G (2003) Topology models for 37 *Saccharomyces cerevisiae* membrane proteins based on C-terminal reporter fusions and predictions. J Biol Chem 278(12):10208–10213
69. Kim S, Chamberlain AK, Bowie JU (2003) A simple method for modeling transmembrane helix oligomers. J Mol Biol 329(4):831–840
70. Koonin EV, Wolf YI, Karev GP (2002) The structure of the protein universe and genome evolution. Nature 420(6912):218–223
71. Kozma D, Simon I, Tusnády GE (2012) PDBTM: protein data bank of transmembrane proteins after 8 years. Nucleic Acids Res 41(D1):D524–D529
72. Krissinel E, Henrick K (2007) Inference of macromolecular assemblies from crystalline state. J Mol Biol 372(3):774–797
73. Krogh A, Larsson B, von Heijne G, Sonnhammer EL (2001) Predicting transmembrane protein topology with a hidden Markov model: application to complete genomes. J Mol Biol 305(3):567–580
74. Kyte J, Doolittle R (1982) A simple method for displaying the hydropathic character of a protein. J Mol Biol 157(1):105–132
75. Lacapère J-J, Pebay-Peyroula E, Neumann J-M, Etchebest C (2007) Determining membrane protein structures: still a challenge! Trends Biochem Sci 32(6):259–270
76. Langelaan DN, Wieczorek M, Blouin C, Rainey JK (2010) Improved helix and kink characterization in membrane proteins allows evaluation of kink sequence predictors. J Chem Inf Model 50(12):2213–2220
77. Lapedes A, Giraud B, Jarzynski C (2012) Using sequence alignments to predict protein structure and stability with high accuracy. arXiv:1207.2484
78. Laskowski R, Macarthur M, Moss D, Thornton J (1993) PROCHECK: a program to check the stereochemical quality of protein structures. J Appl Cryst 26:283–291
79. Latek D, Pasznik P, Carlomagno T, Filipek S (2013) Towards improved quality of GPCR models by usage of multiple templates and profile-profile comparison. PLoS One 8(2):e56742
80. Ledesma A, de Lacoba MG, Arechaga I, Rial E (2002) Modeling the transmembrane arrangement of the uncoupling protein UCP1 and topological considerations of the nucleotide-binding site. J Bioenerg Biomembr 34(6):473–486
81. Lee J, Lee D, Park H, Coutsias EA, Seok C (2010) Protein loop modeling by using fragment assembly and analytical loop closure. Proteins 78(16):3428–3436
82. Lee J, Lee J, Sasaki TN, Sasai M, Seok C, Lee J (2011) De novo protein structure prediction by dynamic fragment assembly and conformational space annealing. Proteins 79(8):2403–2417
83. Lehnert U, Xia Y, Royce TE, Goh C-S, Liu Y, Senes A, Yu H, Zhang ZL, Engelman DM, Gerstein M (2004) Computational analysis of membrane proteins: genomic occurrence, structure prediction and helix interactions. Q Rev Biophys 37(2):121–146
84. Letunic I, Copley RR, Schmidt S, Ciccarelli FD, Doerks T, Schultz J, Ponting CP, Bork P (2004) SMART 4.0: towards genomic data integration. Nucleic Acids Res 32:D142–D144
85. Li Y, Goddard WA (2008) Prediction of structure of G-protein coupled receptors and of bound ligands, with applications for drug design. Pac Symp Biocomput 344–353

86. Liang J, Naveed H, Jimenez-Morales D, Adamian L, Lin M (2012) Computational studies of membrane proteins: models and predictions for biological understanding. Biochim Biophys Acta 1818(4):927–941

87. Liu Y, Gerstein M, Engelman DM (2004) Transmembrane protein domains rarely use covalent domain recombination as an evolutionary mechanism. Proc Natl Acad Sci USA 101(10):3495–3497

88. Lockless SW, Ranganathan R (1999) Evolutionarily conserved pathways of energetic connectivity in protein families. Science 286(5438):295–299

89. Lomize MA, Lomize AL, Pogozheva ID, Mosberg HI (2006) OPM: orientations of proteins in membranes database. Bioinformatics 22(5):623–625

90. Macdonald JT, Kelley LA, Freemont PS (2013) Validating a coarse-grained potential energy function through protein loop modelling. PLoS One 8(6):e65770

91. Marks DS, Colwell LJ, Sheridan R, Hopf TA, Pagnani A, Zecchina R, Sander C (2011) Protein 3D structure computed from evolutionary sequence variation. PLoS One 6(12):e28766

92. Miller CS, Eisenberg D (2008) Using inferred residue contacts to distinguish between correct and incorrect protein models. Bioinformatics 24(14):1575–1582

93. Mindaye ST, Ra M, Lo Surdo J, Bauer SR, Alterman MA (2013) Improved proteomic profiling of the cell surface of culture-expanded human bone marrow multipotent stromal cells. J Proteomics 78:1–14

94. Minsky M, Seymour P (1969) Perceptrons. MIT Press, Oxford

95. Monastyrskyy B, D'Andrea D, Fidelis K, Tramontano A, Kryshtafovych A (2014) Evaluation of residue-residue contact prediction in CASP10. Proteins 82(Suppl 2):138–153

96. Morcos F, Jana B, Hwa T, Onuchic JN (2013) Coevolutionary signals across protein lineages help capture multiple protein conformations. Proc Natl Acad Sci USA 110(51):20533–20538

97. Morcos F, Pagnani A, Lunt B, Bertolino A, Marks DS, Sander C, Zecchina R, Onuchic JN, Hwa T, Weigt M (2011) Direct-coupling analysis of residue coevolution captures native contacts across many protein families. Proc Natl Acad Sci USA 108(49):E1293–E1301

98. Naveed H, Jackups R, Liang J (2009) Predicting weakly stable regions, oligomerization state, and protein–protein interfaces in transmembrane domains of outer membrane proteins. Proc Natl Acad Sci USA 106(31):12735–12740

99. Naveed H, Xu Y, Jackups R, Liang J (2012) Predicting three-dimensional structures of transmembrane domains of β-barrel membrane proteins. J Am Chem Soc 134(3):1775–1781

100. Neumann S, Fuchs A, Hummel B, Frishman D (2013) Classification of α-helical membrane proteins using predicted helix architectures. PLoS One 8(10):e77491

101. Niehage C, Steenblock C, Pursche T, Bornhäuser M, Corbeil D, Hoflack B (2011) The cell surface proteome of human mesenchymal stromal cells. PLoS One 6(5):e20399

102. Noivirt O, Eisenstein M, Horovitz A (2005) Detection and reduction of evolutionary noise in correlated mutation analysis. Protein Eng Des Sel 18(5):247–253

103. Nugent T, Jones DT (2010) Predicting transmembrane helix packing arrangements using residue contacts and a force-directed algorithm. PLoS Comput Biol 6(3):e1000714

104. Nugent T, Jones DT (2011) Membrane protein structural bioinformatics. J Struct Biol 179(3):327–337

105. Nugent T, Jones DT (2013) Membrane protein orientation and refinement using a knowledge-based statistical potential. BMC Bioinform 14(1):276

106. Nugent T, Ward S, Jones DT (2011) The MEMPACK alpha-helical transmembrane protein structure prediction server. Bioinformatics 27(10):1438–1439

107. Oberai A, Ihm Y, Kim S, Bowie JU (2006) A limited universe of membrane protein families and folds. Protein Sci 15(7):1723–1734

108. Olivella M, Gonzalez A, Pardo L, Deupi X (2013) Relation between sequence and structure in membrane proteins. Bioinformatics 29(13):1589–1592

109. Olmea O, Rost B, Valencia A (1999) Effective use of sequence correlation and conservation in fold recognition. J Mol Biol 293(5):1221–1239

110. Ortiz AR, Kolinski A, Rotkiewicz P, Ilkowski B, Skolnick J (1999) Ab initio folding of proteins using restraints derived from evolutionary information. Proteins Struct Funct Genet 37(S3):177–185

111. Park SH, Opella SJ (2005) Tilt angle of a trans-membrane helix is determined by hydrophobic mismatch. J Mol Biol 350(2):310–318
112. Pellegrini-Calace M, Carotti A, Jones DT (2003) Folding in lipid membranes (FILM): a novel method for the prediction of small membrane protein 3D structures. Proteins 50(4):537–545
113. Phan G, Remaut H, Wang T, Allen WJ, Pirker KF, Lebedev A, Henderson NS, Geibel S, Volkan E, Yan J, Kunze MBA, Pinkner JS, Ford B, Kay CWM, Li H, Hultgren SJ, Thanassi DG, Waksman G (2011) Crystal structure of the FimD usher bound to its cognate FimC-FimH substrate. Nature 474(7349):49–53
114. Phatak M, Adamczak R, Cao B, Wagner M, Meller J (2011) Solvent and lipid accessibility prediction as a basis for model quality assessment in soluble and membrane proteins. Curr Protein Pept Sci 12(6):563–573
115. Pieper U, Schlessinger A, Kloppmann E, Chang GA, Chou JJ, Dumont ME, Fox BG, Fromme P, Hendrickson WA, Malkowski MG, Rees DC, Stokes DL, Stowell MHB, Wiener MC, Rost B, Stroud RM, Stevens RC, Sali A (2013) Coordinating the impact of structural genomics on the human α-helical transmembrane proteome. Nat Struct Mol Biol 20(2):135–138
116. Pilpel Y, Ben-Tal N, Lancet D (1999) kPROT: a knowledge-based scale for the propensity of residue orientation in transmembrane segments. Application to membrane protein structure prediction. J Mol Biol 294(4):921–935
117. Punta M, Forrest LR, Bigelow H, Kernytsky A, Liu J, Rost B (2007) Membrane protein prediction methods. Methods 41(4):460–474
118. Punta M, Rost B (2005) PROFcon: novel prediction of long-range contacts. Bioinformatics 21(13):2960–2968
119. Rabiner L (1989) A tutorial on hidden Markov models and selected applications in speech recognition. Proc IEEE 77(2):257–286
120. Rapp M, Drew D, Daley DO, Nilsson J, Carvalho T, Melén K, De Gier J-W, Von Heijne G (2004) Experimentally based topology models for E. coli inner membrane proteins. Protein Sci 13(4):937–945
121. Ray A, Lindahl E, Wallner B (2010) Model quality assessment for membrane proteins. Bioinformatics 26(24):3067–3074
122. Reddy CS, Vijayasarathy K, Srinivas E, Sastry GM, Sastry GN (2006) Homology modeling of membrane proteins: a critical assessment. Comput Biol Chem 30(2):120–126
123. Remm M, Sonnhammer E (2000) Classification of transmembrane protein families in the Caenorhabditis elegans genome and identification of human orthologs. Genome Res 10(11):1679–1689
124. Rohl CA, Strauss CEM, Misura KMS, Baker D (2004) Protein structure prediction using Rosetta. Methods Enzymol 383:66–93
125. Rost B (1999) Twilight zone of protein sequence alignments. Protein Eng Des Sel 12(2):85–94
126. Rost B, Fariselli P, Casadio R (1996) Topology prediction for helical transmembrane proteins at 86 % accuracy. Protein Sci 5(8):1704–1718
127. Sadowski MI, Maksimiak K, Taylor WR (2011) Direct correlation analysis improves fold recognition. Comput Biol Chem 35(5):323–332
128. Saier MJ, Beatty J, Goffeau A, Harley K, Heijne W, Huang S, Jack D, Jähn P, Lew K, Liu J, Pao S, Paulsen I, Tseng T, Virk P (1999) The major facilitator superfamily. J Mol Microbiol Biotechnol 1(2):257–279
129. Sansom MS, Shrivastava IH, Bright JN, Tate J, Capener CE, Biggin PC (2002) Potassium channels: structures, models, simulations. Biochim Biophys Acta Biomembr 1565(2):294–307
130. Sathyapriya R, Duarte JM, Stehr H, Filippis I, Lappe M (2009) Defining an essence of structure determining residue contacts in proteins. PLoS Comput Biol 5(12):e1000584
131. Schäffer AA, Aravind L, Madden TL, Shavirin S, Spouge JL, Wolf YI, Koonin EV, Altschul SF (2001) Improving the accuracy of PSI-BLAST protein database searches with composition-based statistics and other refinements. Nucleic Acids Res 29(14):2994–3005
132. Schmitt L (2002) Structure and mechanism of ABC transporters. Curr Opin Struct Biol 12(6):754–760

133. Schramm CA, Hannigan BT, Donald JE, Keasar C, Saven JG, Degrado WF, Samish I (2012) Knowledge-based potential for positioning membrane-associated structures and assessing residue-specific energetic contributions. Structure 20(5):924–935

134. Senes A, Chadi DC, Law PB, Walters RFS, Nanda V, Degrado WF (2007) E(z), a depth-dependent potential for assessing the energies of insertion of amino acid side-chains into membranes: derivation and applications to determining the orientation of transmembrane and interfacial helices. J Mol Biol 366(2):436–448

135. Shacham S, Marantz Y, Bar-Haim S, Kalid O, Warshaviak D, Avisar N, Inbal B, Heifetz A, Fichman M, Topf M, Naor Z, Noiman S, Becker OM (2004) PREDICT modeling and in-silico screening for G-protein coupled receptors. Proteins 57(1):51–86

136. Sigrist CJA, de Castro E, Cerutti L, Cuche BA, Hulo N, Bridge A, Bougueleret L, Xenarios I (2013) New and continuing developments at PROSITE. Nucleic Acids Res 41:D344–D347

137. Sipos L, von Heijne G (1993) Predicting the topology of eukaryotic membrane proteins. Eur J Biochem/FEBS 213(3):1333–1340

138. Söding J (2005) Protein homology detection by HMM-HMM comparison. Bioinformatics 21(7):951–960

139. Sonnhammer EL, von Heijne G, Krogh A (1998) A hidden Markov model for predicting transmembrane helices in protein sequences. Proc Int Conf Intell Syst Mol Biol 6:175–182

140. Stamm M, Staritzbichler R, Khafizov K, Forrest LR (2013) Alignment of helical membrane protein sequences using AlignMe. PLoS One 8(3):e57731

141. Tusnády GE, Dosztányi Z, Simon I (2004) Transmembrane proteins in the protein data bank: identification and classification. Bioinformatics 20(17):2964–2972

142. Tusnády GE, Dosztányi Z, Simon I (2005) PDB_TM: selection and membrane localization of transmembrane proteins in the protein data bank. Nucleic Acids Res 33:D275–D278

143. Tusnády GE, Dosztányi Z, Simon I (2005) TMDET: web server for detecting transmembrane regions of proteins by using their 3D coordinates. Bioinformatics 21(7):1276–1277

144. Tusnády GE, Kalmár L, Hegyi H, Tompa P, Simon I (2008) TOPDOM: database of domains and motifs with conservative location in transmembrane proteins. Bioinformatics 24(12):1469–1470

145. Tusnády GE, Simon I (2001) The HMMTOP transmembrane topology prediction server. Bioinformatics 17(9):849–850

146. Tusnády GE, Simon I (2010) Topology prediction of helical transmembrane proteins: how far have we reached? Curr Protein Pept Sci 11(7):550–561

147. Vaidehi N, Floriano WB, Trabanino R, Hall SE, Freddolino P, Choi EJ, Zamanakos G, Goddard WA (2002) Prediction of structure and function of G protein-coupled receptors. Proc Natl Acad Sci USA 99(20):12622–12627

148. van Geest M, Lolkema JS (2000) Membrane topology and insertion of membrane proteins: search for topogenic signals. Microbiol Mol Biol Rev 64(1):13–33

149. Viklund HK, Elofsson A (2008) OCTOPUS: improving topology prediction by two-track ANN-based preference scores and an extended topological grammar. Bioinformatics 24(15):1662–1668

150. Volkan E, Kalas V, Pinkner JS, Dodson KW, Henderson NS, Pham T, Waksman G, Delcour AH, Thanassi DG, Hultgren SJ (2013) Molecular basis of usher pore gating in Escherichia coli pilus biogenesis. Proc Natl Acad Sci USA 110(51):20741–20746

151. von Heijne G (1986) The distribution of positively charged residues in bacterial inner membrane proteins correlates with the trans-membrane topology. EMBO J 5(11):3021–3027

152. von Heijne G (1991) Proline kinks in transmembrane alpha-helices. J Mol Biol 218(3):499–503

153. von Heijne G (1992) Membrane protein structure prediction. Hydrophobicity analysis and the positive-inside rule. J Mol Biol 225(2):487–494

154. Šali A, Blundell TL (1993) Comparative protein modelling by satisfaction of spatial restraints. J Mol Biol 234(3):779–815

155. Waldispühl J, Berger B, Clote P, Steyaert J-M (2006) Predicting transmembrane beta-barrels and interstrand residue interactions from sequence. Proteins 65(1):61–74

156. Waldispühl J, Steyaert J-M (2005) Modeling and predicting all-α transmembrane proteins including helix–helix pairing. Theor Comput Sci 335(1):67–92
157. Wallin E, von Heijne G (1998) Genome-wide analysis of integral membrane proteins from eubacterial, archaean, and eukaryotic organisms. Protein Sci 7(4):1029–1038
158. Wang H, He Z, Zhang C, Zhang L, Xu D (2013) Transmembrane protein alignment and fold recognition based on predicted topology. PLoS One 8(7):e69744
159. Wang Z, Xu J (2013) Predicting protein contact map using evolutionary and physical constraints by integer programming. Bioinformatics 29(13):i266–i273
160. Weiner BE, Woetzel N, Karakaş M, Alexander N, Meiler J (2013) BCL:MP-fold: folding membrane proteins through assembly of transmembrane helices. Structure 21(7):1107–1117
161. White SH (2009) Biophysical dissection of membrane proteins. Nature 459(7245):344–346
162. Wolf YI, Grishin NV, Koonin EV (2000) Estimating the number of protein folds and families from complete genome data. J Mol Biol 299(4):897–905
163. Xiang Z, Soto CS, Honig B (2002) Evaluating conformational free energies: the colony energy and its application to the problem of loop prediction. Proc Natl Acad Sci USA 99(11):7432–7437
164. Yarov-Yarovoy V, Schonbrun J, Baker D (2006) Multipass membrane protein structure prediction using Rosetta. Proteins 62(4):1010–1025
165. Zhang Y, Devries ME, Skolnick J (2006) Structure modeling of all identified G protein-coupled receptors in the human genome. PLoS Comput Biol 2(3):e13

Chapter 10
Dynamics of Small, Folded Proteins

Petra Rovó, Dóra K. Menyhárd, Gábor Náray-Szabó and András Perczel

10.1 Introduction

While the three-dimensional structure of biological macromolecules provides a wealth of information about their architecture, organization and reactive sites it is evident that just the knowledge of their static coordinates is not enough to precisely describe their mode of action. Thus protein dynamics (in other words internal mobility), the timescale of these motions and all the important states a protein explores during its folding process should also be scrutinized in order to understand its kinetics and thermodynamics [1–3].

Proteins are not static entities but rather dynamic, flexible molecules where the internal mobility significantly affects their biological function. In solution proteins constantly fold and unfold occupying an astronomical large number of conformational states called micro states. In living cells they also get synthesized and degraded thus they exist in a well-balanced steady-state condition. Protein plasticity allows fluctuation around its average three-dimensional structure. This fluctuation occurs over a wide range of time-scales, from bond vibration on the picosecond timescale to the folding process, which may take minutes or even hours. Protein dynamics influences several functions, such as catalytic activity of enzymes [4], ligand recognition [5], signalling and regulation [6], as well as stability [7]. Furthermore, dynamics affects protein folding, aggregation, as well as

P. Rovó · G. Náray-Szabó · A. Perczel (✉)
Laboratory of Structural Chemistry and Biology, Institute of Chemistry, Eötvös Loránd University, Pázmány Péter st. 1A, Budapest 1117, Hungary
e-mail: perczel@chem.eltc.hu

D.K. Menyhárd · A. Perczel
MTA-ELTE Protein Modelling Research Group, Pázmány Péter st. 1A, Budapest 1117, Hungary

© Springer International Publishing Switzerland 2014
G. Náray-Szabó (ed.), *Protein Modelling*, DOI 10.1007/978-3-319-09976-7_10

ligand binding [8]. Fluctuation enables some proteins to perform multiple distinct functions and it is important for the evolution of novel functions [9].

Protein plasticity can be assessed at the atomic level by two major techniques, computational molecular dynamics (MD) and nuclear magnetic resonance (NMR) spectroscopy. MD calculations provide protein properties on the basis of explicit structural models, while NMR spectroscopy can study experimentally a wide range of time-scales that a protein explores throughout its lifetime in a comprehensive, site-specific and non-invasive manner. These two techniques complement each other and therefore they are important tools in understanding the details of dynamic processes in proteins. In this chapter first we give an overview on their most important aspects, then we will discuss two case studies, the Trp cage miniprotein and podocin.

10.2 Molecular Dynamics of Peptides and Proteins

A versatile computational tool for the study of protein movements is molecular dynamics, a method, which is based on a simple empirical energy expression for a molecular system. Since quantum mechanical calculations are not tractable for such large systems, the much simpler molecular mechanics (MM) or force field methods are applied. Molecules are treated as group of atoms, each represented by a point mass, which are connected by bonds, represented by strings. Forces acting along bonds are described by simple mathematical formulae, which are adapted from classical physics. The total energy of the system is then calculated as the sum of some simple terms:

$$E_{MM} = V_{stretch} + V_{bend} + V_{torsion} + V_{outofplane} + V_{nonbonded} \qquad (10.1)$$

where subscripts refer to stretching, bending, torsion, out-of-plane bending and nonbonded terms, respectively. The above expressions are defined as follows

$$V_{stretch} = \Sigma_{bonds} K_{stretch} \left(r_{ij} - r_{ij}^0 \right)^2 \qquad (10.2)$$

$$V_{bend} = \Sigma_{angles} K_{bend} \left(\theta_{ijk} - \theta_{ijk}^0 \right)^2 \qquad (10.3)$$

$$V_{torsion} = \Sigma_{dihedrals} K_{torsion} (1 + \cos n\varphi) \qquad (10.4)$$

$$V_{outofplane} = \Sigma_{impropers} K_{outofplane} \chi_{ijk;l}^2 \qquad (10.5)$$

$$V_{nonbonded} = \Sigma_{non-bonded\ pairs} \left[A_{nonbonded} \left(C_{12} r_{ij}^{-12} - C_6 r_{ij}^{-6} \right) + q_i q_j / D r_{ij} \right] \qquad (10.6)$$

where r_{ij}, r_{ij}^0, θ_{ijk} and θ_{ijk}^0 are the actual and equilibrium bond lengths and bond angles, respectively. φ is the dihedral angle, $\chi_{ijk;l}$ is the angle between the bond jl

and the plane ijk, where j is the central atom. $K_{stretch}$, K_{bend}, $K_{torsion}$ and $K_{out\ of\ plane}$ refer to respective force constants; n is an integer number. All these, as well as $A_{nonbonded}$, C_6 and C_{12} are adjustable parameters. q_i is the net charge on atom i, D is the so-called dielectric parameter. For review and comparison of various force fields see [10–13].

In order to reduce computational efforts transferable groups of atoms (e.g. alkyl) may be considered as a single interacting unit. Empirical parameters in Eqs. (10.2)–(10.6) are fitted to experimental or calculated data. In order to allow flexibility and better transferability between various molecules, different parameters are used for atoms with different degree of hybridization, in carbonyl or peptide groups or in aromatic systems. Thus, the total number of adjustable parameters may considerably increase. As the number of atom pairs increases quadratically, in order to reduce computation time used for the $V_{nonbonded}$ term most programs use a cutoff value. It has to be noticed that a given set of parameters works only for a given force field, and parameters cannot be transferred between different force fields.

Basically two philosophies of parameterisation can be followed. Class I force fields like AMBER [14] or GROMOS [15], applied to proteins, nucleic acids and carbohydrates are based on experimental data and work with a simpler energy expression. Class II force fields, like the Merck Molecular Force Field [16] include higher order and cross terms. These are calibrated to quantum mechanically calculated energies and gradients, which increases their potential. Lots of freely or commercially available software packages make use of different force fields for biopolymers. Most popular are AMBER [14], GROMACS [17], INSIGHT II [18], SYBYL [19] and CHARMM [20]. These are licensed packages with various components.

Once we have a force field for the calculation of energies related to various structural arrangements of a protein molecule, MD simulation can be used for the analysis of motions that tracks the time-dependent positions of all atoms (for reviews see Refs. [21–23]). This requires the exploration of high-dimensional potential surfaces, where both location of the minima and paths connecting those are relevant to function. Time spans related to various types of motions in proteins are displayed in Table 10.1.

When doing MD simulations, Newton's equations of motion are solved in small integration steps. The potential energy is derived from a chosen force-field and is used for assigning the instantaneous velocities to the atoms of the system in study.

Table 10.1 Time spans and associated motions in proteins

Time span (s)	Associated motions
10^{-15}–10^{-12}	Bond stretching, angle bending
10^{-12}–10^{-9}	Side-chain movements, local rearrangements, loop motions, collective motions
10^{-9}–10^{-6}	Domain movement, secondary structure conversion
10^{-6}–1	Folding, unfolding

For N interacting atoms, the local forces can be calculated by Newtonian mechanics with

$$F_i = -\frac{\partial V}{\partial r_i} \tag{10.7}$$

and

$$F_i = m_i \frac{\partial^2 r_i}{\partial t^2} \tag{10.8}$$

where $V(r_1, r_2, \ldots, r_N)$ is the potential function of the force field. Calculated velocities are taken to be constant for the duration of the time-step (Δt) and the movement of the atoms during that time creates a new conformation. To stabilize the algorithm, velocities and new positions are calculated at times that are shifted by $\Delta t/2$ with respect to one another. A straightforward example is the so-called leap-frog algorithm, where

$$v_i\left(t + \frac{\Delta t}{2}\right) = v_i\left(t - \frac{\Delta t}{2}\right) + \frac{F_i(t)}{m}\Delta t \tag{10.9}$$

with

$$r_i(t + \Delta t) = r_i(t) + v_i\left(t + \frac{\Delta t}{2}\right)\Delta t \tag{10.10}$$

The time step Δt should be adjusted to the fastest motion to be taken into account, while the length of the simulation must refer to the timescale of the motions to be modelled.

Because the atoms are moving, kinetic energy, temperature and pressure can be assigned to the system, its entropy, free energy and average geometric descriptors can be calculated. This creates a direct connection between macroscopic observables and microscopic events. Realistic description of proteins requires inclusion of solvent effects, thus the molecules are immersed in a dielectric medium, most often and preferably, in a box containing explicit water molecules and salt ions. Periodic boundary conditions are applied, which means that we look at the simulation box as if it were surrounded by an infinitely extended series of copies of itself, thus dissolving the boundary of the local system and the artefacts due to edge-effects. This arrangement allows for keeping the particle number constant (any atom leaving the local box will reappear at the opposing side) and for keeping and summing the long-range component of the electrostatic interactions which does not diminish within the usually chosen box sizes.

The length of the simulation should be suited to the most time-consuming process under study; however, the issue of convergence should also be addressed. MD simulations are considered converged, or equilibrated, when all properties of the system become independent of the chosen time of sampling, thus the length of the trajectory. This is especially difficult in case of thermodynamic properties relying on the determination of the entropy, since this quantity continues to increase with simulation time until all local minima are sufficiently sampled, with an estimated convergence time that might exceed 1 ms in case of most protein molecules. However, conformational descriptors and free energy differences are decidedly less sensitive to running time and thus can be estimated much better [24]. Besides lengthening the simulation, converged results can also be reached by increasing sampling efficiency. One such method is replica exchange molecular dynamics (REMD). During a REMD simulation, a set of simulations are carried out in parallel, which describe the same system in very similar states at a well-chosen set of different temperatures. After short intervals, exchange of conformations between the different simulations (the replicas) is attempted, the success of which will depend on whether the given structure is sampled in the replica run sufficiently as compared with a scaled random probability. This algorithm provides a possibility of crossing high-energy barriers, by boosting the probability of arrangements that are only typical at high temperatures in the consideration of lower temperature ensembles as well. The appearance of such high-energy conformers might lead to rearrangement into local minima already mapped at lower temperatures but might also lead to crossing into under-sampled territories [25, 26].

Beyond the description of stable, folded states of proteins, MD trajectories can be used to understand folding and unfolding processes and kinetic behaviour using the ideas of ensemble thermodynamics. For example, so-called Markov state models (MSM) can be derived by clustering the obtained conformers into microstates and estimating the "history-free" crossing probabilities between these states [27, 28].

Even without the application of MD simulation, topology-encoded dynamic nature of proteins can be uncovered by application of elastic network models, such as the anisotropic network model or the Gaussian network model. It has long been known that proteins of similar fold exhibit similar large-scale motions. Normal mode analysis of equilibrium structures allows for fast and effective investigation of both local movements, such as side-chain fluctuations, and of protein collective motions (resulting from the participation of entire domains or substructures) leading to functionally significant rearrangements [29, 30].

Theoretical considerations of protein dynamics have always relied on experimental results for starting or reference structures. Protein crystallography is the overwhelming source of such information reporting stable conformations of these macromolecules with high accuracy. However, a new algorithm, CONTACT, allows for uncovering dynamically connected regions of these structures also, thereby boosting the crystal-frozen states with mechanistic insight. High resolution X-ray structures can be used to identify possible alternative conformations of the amino acid resides at low levels of electron density, and among these contact-networks can be drawn. These networks encompass those sites of heterogeneity which are able to perturb one-another and thus outline the coordinated motions of catalytic function [31].

10.3 NMR Spectroscopy

While the three-dimensional structure of biological macromolecules provides a wealth of information about their architecture, organisation and reactive sites it is evident that just the knowledge of their static structure is not enough to precisely describe their mode of action. Thus, the internal mobility and the different states of proteins, as well as the timescale of these motions during the folding process should also be scrutinised in order to understand the kinetics and thermodynamics of the functional proteins.

Among others, the pioneering work of Nobel laureates Richard R. Ernst and Kurt Wüthrich (Nobel Prize in Chemistry, 1991 and 2002) made NMR a suitable technique to study large macromolecules. Strategy used today for 3D-structure determination of biomolecules by NMR pursuits the following route: (i) preparation of samples by chemical synthesis and biotechnology including DNA and expression techniques, (ii) careful purification of the sample, (iii) spectra acquisition and data processing, (iv) NMR resonance assignment, (v) evaluation of spectral information to decipher spatial restrains, (vi) MD based structure calculations, completed by (vii) iterative refinement and structure validation. As samples from natural sources contain only protons as NMR active nuclei, this method brings in rigorous limitations on size and complexity. However, biotechnology made isotope labelling possible and affordable and thus today it is a routine technique. Selective, residue specific and non-selective ^{13}C and/or ^{15}N labelling schemes are in use as reviewed exhaustively in [32–34]. A handful of conceptually different NMR experiments are available for biomolecules targeting (i) resonance assignment, e.g. correlation spectroscopy (COSY), total correlation spectroscopy (TOCSY), (ii) restrain determination, e.g. nuclear Overhauser effect (NOE), indirect dipole-dipole or J-coupling, residual dipolar coupling (RDC), etc., (iii) determination of aggregation properties, e.g. diffusion-ordered spectroscopy (DOSY), (iv) monitoring internal dynamics, e.g. spin-lattice relaxation rate constants (R_1), spin-spin relaxation rate constants (R_2), heteronuclear NOE, etc. Although, resonance assignment requires prior knowledge of the primary sequence, amino acid composition and sequential order within the protein also reinsures sequential information. Based on the true nature of chemical shifts an assignment table of a folded protein contains not only ^1H, ^{13}C and ^{15}N residue specific resonances, but also data on the secondary structure of the polypeptide chain. The two intimately linked steps of homonuclear resonance assignment are (i) spin system identification, accompanied by (ii) their sequential vicinity determination. The latter information called "sequential assignment" is retrieved from NOE-type NMR experiments and determines the relative order of the spin systems, while the former one, "spin system identification" is the gathering of mutually J-coupled NMR resonances [35]. The latter step relies on experiments transferring coherences via scalar coupling(s), while the former one utilizes magnetization transfer between spatially close spins: 2D- and 3D-NOESY and rotating frame NOE spectroscopy (ROESY) experiments are in use. Sequential assignment of ^{13}C, ^{15}N labelled samples requires the application of more complex NMR pulse sequences; however more information is retrieved and thus this technique is more reliable and more straightforward for automation. Both simple and more sophisticated multidimensional (three or even higher dimensional) heteronuclear experiments are routinely used to solve the latter task. Triple-resonance nD-experiments ($n = 3$, 4, 5, 7) are more and more common, not necessarily lengthening the acquisition time but making the assignment procedure quicker and more robust [36, 37]. It is possible to reduce measurement time, from 1 to 2 weeks to a few days by using NMR experiment schemes where signal-less spectral regions are not (extensively) sampled [38]. These and other "fast" NMR techniques completed by transverse relaxation optimized spectroscopy (TROSY) based pulse sequences make the application of NMR quicker and more efficient, providing a wealth of information on bimolecular structure, interaction and dynamics.

Although protein structure, mobility and its folding can be experimentally assessed by different biophysical techniques, the best choice to study all these properties is NMR spectroscopy. This is a versatile technique which can explore the wide range of time-scales that a protein explores throughout its lifetime in a comprehensive, site-specific and non-invasive manner. Moreover, the perception that NMR studies are limited to small proteins (<20 kDa) is increasingly obsolete. The use of high-field instrumentation combined with cryogenically cooled probe heads and special isotope labelling schemes expanded the size limitations of NMR, reaching up to the 900 kDa [39]. For larger proteins (>12 kDa) it is often inevitable to measure heteronuclear NMR spectra to complete the resonance assignment for structure calculations. But beside resonance assignment and structure determination heteronuclear spectra provide a handful of other useful information about the studied system, e.g. protein-ligand binding constants can be determined, thermal denaturation can be followed, fast and slow timescale protein motion can be studied both for backbone and for side-chain spins.

Solution-state NMR spin relaxation methods for characterizing conformational dynamics and kinetic processes have been extensively reviewed [40–45]. Dynamic processes can be studied using an array of NMR experiments: nuclear spin relaxation measurements are sensitive to ps-ns time-scales, relaxation dispersion measurements are sensitive to μs-ms time-scales, and magnetization exchange spectroscopy is sensitive to ms-sec time-scales. Below we briefly discuss these methods and then we present a case study as well as results from the literature, where MD simulations and NMR spectroscopy were applied to characterize the dynamics of the Trp-cage miniprotein.

10.3.1 Chemical Exchange at Different Time-Scales

In NMR spectroscopy chemical exchange refers to a dynamic process when a NMR sensitive nucleus interconverts between states with different chemical environment in a time-dependent manner. As a first approximation usually a two-state model is assumed. This accounts for a wide variety of dynamic processes, such as the exchange between folded and unfolded states, free and ligand-bound states, monomer and dimer forms or *cis* and *trans* peptide bonds, etc. The exchange process between A and B states can be described by their populations, exchange rate, k_{ex} and chemical shift difference in Hz, $\Delta\nu = |\nu_A - \nu_B|$. In the slow exchange regime ($k_{ex} \ll |\Delta\nu|$) two peaks are observed at the chemical shifts of the individual states; in the fast exchange regime ($k_{ex} \gg |\Delta \nu|$) only one peak is observed reflecting population-weighted average chemical shift and line width. In the intermediate exchange regime ($k_{ex} \approx |\Delta \nu|$) only one peak is observed at the average chemical shift but its line width is substantially "exchange broadened" which can sometime render the signal undetectable.

10.3.2 Nuclear Spin Relaxation

Heteronuclear spin relaxation rate experiments provide a powerful tool for the sequence-specific description of ps-ns time-scale protein dynamics ($k_{ex} \gg |\Delta \nu|$). Motions in this time window include bond vibration and libration, side chain rotamer interconversion, random coil and loop motions as well as backbone torsion angle rotation which processes can affect enzyme catalysis, ligand affinity, allostery or conformational entropy [40].

Nuclear spin relaxation is the process by which non-equilibrium magnetization returns back to its equilibrium state. The rate at which it occurs is governed by conformational motions that cause the energies of nuclear interactions to fluctuate on a ns-ps time-scale. Such oscillations are the consequence of motions of the protein relative to the permanent magnetic field (overall motion) or the motions of magnetic nuclei relative to each other (internal motion). Considering that there are many different types of non-equilibrium magnetizations, various relaxation measurements can be performed. In practice, protein backbone relaxation experiments typically measure the ^{15}N spin-lattice or longitudinal relaxation (R_1), the in-phase ^{15}N spin-spin or transverse relaxation (R_2), and the steady-state heteronuclear $^1H^N$–^{15}N NOE.

The two dominant mechanisms that influence the nuclear spin relaxation are the dipole–dipole interaction (DD) and the chemical shift anisotropy (CSA). The dipole–dipole interaction emerges between a pair of magnetic spins which experience the local magnetic field of each other. The interaction depends on the distance and relative orientation of the spins; thus if their position relative to each other fluctuates with frequencies close to the transition frequency then relaxation occurs. In ^1H–^{15}N spin pairs the ^{15}N relaxation is dominated by the DD interaction of the attached proton over the DD field of all other surrounding nuclei. Chemical shift arises due to the nuclear shielding of the molecule's electron cloud. The anisotropic distribution of the electronic charge in the molecule causes anisotropic nuclear shielding and thus chemical shift anisotropy. The local electronic field varies as the molecule reorients due to molecular motion. Again, if the reorientation occurs on the ps-ns time-scale then CSA is a source of relaxation.

In order to quantitatively relate the relaxation rates to the ps-ns protein motion one should consider the time-dependent rotational correlation function, $C(t)$ or equivalently its Fourier transformed form, the spectral density function, $J(\omega)$, of the oscillating bond vector [46]. The spectral density function quantifies the amplitude of motion at a ω frequency. The relaxation rates can be expressed as the linear combination of $J(\omega)$, evaluated at five critical frequencies: $J(0)$, $J(\omega_N)$, $J(\omega_H)$, $J(\omega_H - \omega_N)$, and $J(\omega_H + \omega_N)$, where ω_H and ω_N are the ^1H and ^{15}N Larmor frequencies, respectively. Note, that other frequencies do not contribute to the measured relaxation rates. The expressions describing the relaxation rates as a function of spectral densities have the following form:

$$R_1 = d^2[J(\omega_H - \omega_N) + 3J(\omega_N) + 6J(\omega_H + \omega_N)] + c^2 J(\omega_N) \quad (10.11)$$

$$R_2 = \left(d^2/2\right)\left[4J(0) + 3J(\omega_N) + 6J(\omega_H) + 6J(\omega_H + \omega_N)\right] + c^2\left[4J(0) + J(\omega_N)\right] \quad (10.12)$$

$$\text{NOE} = 1 + d^2(\gamma_H/\gamma_N)[6J(\omega_H + \omega_N) - 6J(\omega_H - \omega_N)]/R_1 \quad (10.13)$$

with $d = \left(\mu_0 h \gamma_N \gamma_H / 16\pi^2\right)\langle 1/r_{NH}^3 \rangle$ and $c = \omega_N \Delta\sigma/\sqrt{3}$. μ_0 is the vacuum permeability, h is the Planck's constant, γ_N and γ_H are the giromagnetic ratio of ^{15}N and ^1H, respectively, r_{NH} is the length of the N–H bond vector, $\Delta\sigma$ is the chemical shift anisotropy of ^{15}N.

To extract information on dynamics from relaxation rates, we have several options. If the set of relaxation rates measured for each bond vector exceeds the number of unique spectral density samplings, we can directly determine the intrinsic dynamics quantities, and no assumptions are needed. This method is called the spectral density mapping approach [47–50]. If the relaxation rate is dominated by a single spectral density, complete mapping becomes possible via dispersion studies on the magnetic field [51]. In general, data sets are too sparse for direct mapping. In such cases we may follow the most common approach, which is to find an analytical model form for $J(\omega)$, which contains parameters fitted to reproduce experimental relaxation rates. The most popular approach to describe fast-time scale protein motion is the model-free (MF) formalism described by Giovanni Lipari and Attila Szabó in 1982 [52].

The MF approach uses only two simple assumptions: it assumes that (i) the local and global motions are independent and thus separable $C(t) = C_{global}(t) C_{local}(t)$, and (ii) the correlation functions can be approximated with a single exponential decay which gives rise to Lorentzian spectral density functions:

$$J(\omega) = 2/5\left\{S^2\tau_c/\left[1 + (\omega\tau_c)^2\right] + \left(1 - S^2\right)\tau/\left[1 + (\omega\tau)^2\right]\right\} \quad (10.14)$$

where $\tau^{-1} = \tau_c^{-1} + \tau_i^{-1}$. The order parameter, S^2 indicates the relative contribution to relaxation from the overall molecular motion and from the additional local motions and reports on the spatial restriction: for completely restricted motion $S^2 = 1$ and for a completely unrestricted motion $S^2 = 0$. This parameter is the most valuable for solving biological problems because it is related to the intrinsic flexibility of the amide bond vectors and the change of spatial restriction (upon mutation, ligand binding, denaturation etc.) is associated with the change of the conformational entropy which influences both function (e.g. ligand binding affinity or catalysis) and stability.

In the last decades the MF approach became a standard tool for evaluating protein backbone motion where the amide nitrogen and proton can be treated as an isolated two-spin system [40, 46, 53]. Although the method has some limitations, e.g. it cannot take into account the diversity of internal motions, it is now clear that it can serve as a good approximation for describing fast internal backbone amide motion.

10.3.3 Relaxation Dispersion Measurements

Exchange processes occurring on the μs-ms time-scale make the chemical shift a fluctuating quantity, with consequences on the line shape and relaxation parameters. Dynamic processes in this time window include side-chain reorientation, loop motion, secondary structure rearrangement which are often coupled to enzyme catalysis, ligand-binding, folding or allostery. The two main techniques for quantifying the kinetics and populations of the exchanging sites in the intermediate exchange regime are the Carr-Purcell-Meiboom-Gill relaxation dispersion (CPMG RD) [54, 55] and the rotating frame ($R_{1\rho}$) relaxation dispersion measurements [56–58].

CPMG relaxation dispersion measurements are well-suited for the investigation of exchange processes in the intermediate regime (0.3–10 ms), where $k_{ex} \approx \Delta\nu$. This type of exchange results in an apparent increase in R_2 relaxation rate: $R_{2obs} = R_2 + R_{ex}$. The principle of CPMG RD is to refocus exchange broadening by applying a series of pulse elements to transverse magnetization during a special relaxation delay. In general, the spin-echo can refocus a set of magnetization vectors if each individual vector exhibits the same average chemical shift during the first and second τ period. However, if exchange causes a spin to experience a different chemical shift during one τ period and it does not return symmetrically in the other τ period, its magnetization will not be refocused. This results in incomplete refocusing among the ensemble of molecules and therefore leads to signal broadening. The degree of refocusing achieved by the spin-echo element depends on the difference between the average shifts in the first and second τ periods. Importantly, as the duration of the τ period is reduced compared to the exchange time, there will be less signal broadening because the probability of exchange during τ is reduced. The CPMG RD experiment quantitatively explores the relationship between signal broadening and the duration of the spin-echo delay τ.

$R_{1\rho}$ relaxation dispersion can be used to study exchange events in the intermediate-fast regime (20–100 μs) which includes side chain reorientation, loop motion, secondary structure changes and hinged domain movements. Such motions may affect processes including ligand binding and release, as well as folding and unfolding events. By analogy to CPMG RD, where the R_2 relaxation is reduced in the laboratory frame by a series of spin-echo pulses in the range of $\nu_{CPMG} \approx 25$–1,200 Hz, during the $R_{1\rho}$ RD measurement the $R_{1\rho}$ relaxation is attenuated in the rotating frame along an effective field using spin-lock pulses with $\omega_{eff} \approx 1$–50 kHz. Thus the two methods use the same principle with different experimental implementation to study ~ms (CPMG) or ~μs ($R_{1\rho}$) exchange events.

In both experiments transverse relaxation rates are measured as a function of spin-lock frequency which yields the dispersion curve. During the fits, usually a two-site exchange is assumed with a high-population ground state (A) and a low population excited state (B). The shape and profile of the dispersion curves are governed by the population of the two states, the rate of exchange and the difference in chemical shifts between the states ($\Delta v = v_A - v_B$). Thus in optimal cases, fitting can provide both kinetic and thermodynamic information about the exchange process as well as structural information about the minor state. In practice, the structural evaluation is done by comparing the site-specific $|\Delta v|$ values to those observed between the ground state structure and a candidate structure (e.g. ligand-bound, denatured, covalently modified etc.).

10.3.4 Exchange Spectroscopy

Exchange spectroscopy (EXSY) is used to study conformational dynamics occurring on the ms to sec timescale, i.e. when the exchange process is much slower than the difference in the chemical shift of the states ($k_{ex} \ll |\Delta v|$) [59]. In this case, separate peaks are observed and therefore direct information can be gathered about the structural differences of the sites and about their populations (thermodynamics). Both homo- and heteronuclear EXSY variants exist; the former one is performed as a simple NOESY or ROESY experiment while the latter, often called as ZZ-exchange spectroscopy, is measured as a standard correlation experiment with a delay block (T) inserted between the blocks where the spin coherences become labelled with the chemical shifts of the correlated nuclei. The exchange between the A and B states during T delay will manifest in cross-peaks at the $v_A - v_B$ and $v_B - v_A$ resonances. The peak intensities of the auto-peaks (AA and BB) and cross-peaks (AB and BA) are affected by the forward and reverse rate constants, as well as by the longitudinal relaxation rates.

10.4 The Case Study of the Trp-Cage Miniprotein, Tc5b

The Trp-cage miniproteins constitute a group of designed small proteins all related to the first member of this family, Tc5b [60, 61]. However, their physicochemical properties are closely related to that of the single-domain globular proteins than to that of the unstructured polypeptides. Globular proteins have secondary structural elements organized into a well-defined tertiary fold; they display a cooperative profile with a pronounced transition between folded and unfolded states. In the last decade several Trp-cage variants have been designed and analysed by NMR, electronic circular dichroism (ECD) and other spectroscopic methods in order to explore the source of their structural stability and to investigate the folding

Fig. 10.1 Representative structure of Tc5b determined in water. Only the side-chain carbon atoms of the hydrophobic core forming residues are displayed: Tyr3 (*orange*), Ile4 (*cyan*), Trp6 (*coral*), Leu7 (*orchid*), Pro12 (*red*), Arg16 (*magenta*), Pro18 (*gold*) and Pro19 (*blue*); backbone main-chain is displayed for all other residues

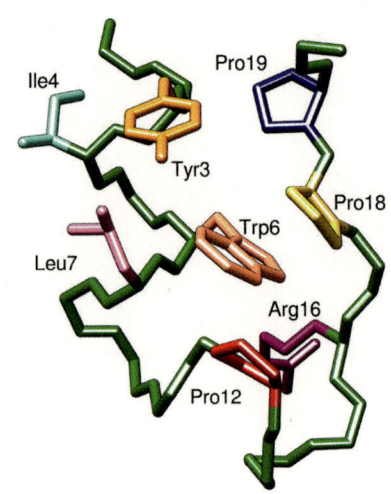

process [62, 63]. The stable and well-defined structure of the Trp-cage is the consequence of a series of truncations and mutations carried out on the naturally occurring 39-residue peptide, Exendin-4 (Ex4). This molecule partially displays the Trp-cage moiety at its C-terminus in the presence of fluoroalcohol or added lipid micelles [64]. Ex4 is a peptide drug in use to treat type II diabetes. By using the optimised Trp-cage structure it is possible to improve the physico-chemical properties of the antidiabetic drug. For a representative structure see Fig. 10.1.

High resolution structural information of small but folded protein models (e.g. Trp-cage) could provide valuable insights into the folding mechanism of globular proteins of larger size. Investigation of the folded (**F**), unfolded (**U**) and on(off)-pathway intermediate structures (**I**) of miniproteins as well as deciphering the fold stabilizing contacts associated with the driving force in between states (e.g. **F ↔ I**, **I ↔ U**) of folding in aqueous media is of high importance. H-bonds, hydrophobic contacts, π-π stacking, salt-bridges, etc. make tertiary structure of the polypeptide chain stable in water. The 20 amino acid long polypeptide chain of the Trp-cage miniprotein consists of common secondary structural elements (α-, 3_{10}- and PPII helices) all grouped spontaneously around the indole ring of Trp6 (Fig. 10.1). Changing the molecular environment (temperature, pH, solvent, etc.), modifying the primary sequence, breaking the existing or forming new structure stabilizing interactions can perturb the 3D-structure of a Trp-cage.

10.4.1 Folding Pathways

In spite of its simplicity, the exact folding pathway and the driving force behind the folding process of Trp-cage is still under debate. To date, fluorescence correlation spectroscopy [65], UV resonance Raman spectroscopy [66], photochemically induced dynamic nuclear polarization spectroscopy [67] and IR T-jump spectroscopy [68] have been used to experimentally characterize the molecular details of the folding mechanism. Most studies conclude that a molten globule-like intermediate state evolves, which enhances the folding efficiency and accelerates the velocity of the spontaneous process in water [68, 69]. In the intermediate state the main tertiary contacts are already formed, while the secondary structures are still disjointedly hydrated. Neuweiler et al. reported that the hydrophobic collapse is assisted by non-native tertiary contacts between the N-terminal (residues N1-G11) and the proline-rich C-terminal (residues P12-S20) segments [65]. Such a molten-globule like intermediate state is responsible for the efficient folding. Furthermore, they also pointed out the significance of I4 side-chain in stabilizing the intermediate fold. In another study, Mok et al. specified the residues that take part in the hydrophobic core formation. They found that in the intermediate state the side-chains of the I4, L7, P12 and R16 residues are closer to the indole ring than they are in the native state and thus, the hydrophobic collapse in H_2O must precede the α-helical secondary structure formation [67]. However, Ahmed et al. measured residual helical structure in the denatured state suggesting the early formation of the helical segment which process is consistent with a diffusion-collision model [66]. This assumption was also confirmed by Culik et al. using IR T-jump experiments [68]. An analysis of different Tc5b mutants implied that only the α-helix is formed when the folding reaches the transition state and neither the D9-R16 salt-bridge nor the 3_{10}-helix are present in the folding intermediate.

The existence of a cornerstone intermediate state has also been demonstrated with computational simulations [70–77]. Chowdhury et al. gave a qualitative description of the folding events of Trp-cage and identified four possible intermediate states [75]. They found that already in the early stage of the simulation a native-like backbone topology is spontaneously formed and stabilized by the clustering of key hydrophobic residues (I4, L7, P17, P18 and P19) around the Trp6 side-chain. Thereafter the α-helix is formed, starting from its N-terminus and the event is completed with the formation of the salt-bridge between Asp9 and Arg16 accompanied by the rearrangement of Trp6 within the preformed hydrophobic core. This very last event is the rate-limiting step of Trp-cage formation [72].

Juraszek and Bolhuis used a more sophisticated replica exchange molecular dynamic (REMD) method performed in explicit water to analyse the same process [73, 74]. They revealed that the Trp-cage could follow two major alternative routes between the native and the unfolded states. The dominant route occurs trough a single loop state, while the alternative path goes through two consecutive intermediate states. The first, highly populated route is consistent with the NC model, as the tertiary contacts precede the formation of the secondary elements. An alternative route resembles the DC model that occurs in the opposite order, by first forming the helix and thereafter the tertiary contacts. These results are consistent with the simulations of Chowdhury et al. [75] and were later affirmed by similar REMD simulations by Zheng et al. [76]. The slow-folding rate of the secondary pathway is attributed to the loss of side-chain rotational freedom, due to the early core collapse, which impedes the helix formation [77, 78]. A low-temperature kinetic intermediate, stabilized by a salt bridge between residues Asp9 and Arg16, was located by the simulation.

Nikiforovich et al. [79] concluded on the basis of simulations based on a novel molecular dynamics method that the probable folding pathway starts with folding of the α-helical fragment 4–9, followed by the formation of the final three-dimensional structure of fragments 4–12 and 4–18. The structures of Trp-cage obtained by this study by independent energy calculations are in excellent agreement with experimental data obtained by NMR spectroscopy. It is also possible to follow folding kinetics by molecular dynamics simulations [80].

Molecular dynamics simulations indicate that water plays a role in folding. Paschek et al. [81, 82]. identified states with buried internal water molecules, which interconnect parts of the molecule by hydrogen bonds. The loss of hydrogen bonds formed by these buried water molecules in the folded state is likely to destabilize it at elevated temperatures. Dependence of folding on viscosity can also be studied by simulations [83].

Day et al. [84] have calculated the free energy function of Trp-cage with remarkably good agreement with the experimental folding transition temperature, free energy, and specific heat changes. However, changes in enthalpy and entropy are significantly different than the experimental values. The folding is very fast, as also found by simulations in excellent agreement with experiment [85]. It was concluded from simulation studies that substituting a key glycine in Tc5b with D-Gln dramatically stabilizes the fold without altering the protein backbone [86].

The comprehensive analysis of the above experimental and MD results implies that both the number and the explicit atomic resolution structure of the intermediate state(s) are still ambiguous. We have experimental evidence, obtained by NMR spectroscopy, indicating how the temperature induced unfolding of Tc5b occurs. Our NMR studies show that for such a small and fast-folding system like Tc5b conventional temperature dependent ^{1}H–^{15}N and ^{1}H–^{13}C Heteronuclear single quantum coherence (HSQC) spectra are appropriate to monitor the global rearrangements related to the unfolding process. We assert that the unfolding of these miniproteins is not a simple two-state process, rather different intermediate states evolve under different conditions, i.e. at neutral or acidic pH values.

We studied the temperature-induced unfolding of Tc5b by monitoring resonance changes of the ^{1}H–^{15}N and ^{1}H–^{13}C HSQC spectra at various temperatures (cf. Fig. 10.2) [72]. Some of these changes are not linear, rather curved, indicating that unfolding is a complex process in which at least one fast-exchanging intermediate state is formed.

In order to characterise the thermodynamics of the unfolding process, we fitted a three-state model ($\mathbf{F} \rightleftharpoons \mathbf{I} \rightleftharpoons \mathbf{U}$) iteratively to the observed chemical shifts. In the first step the mole fractions of $\mathbf{F}$, $\mathbf{I}$ and $\mathbf{U}$ were set by using initial parameters, like transition temperature, T_m^{F-I} and T_m^{F-U}, enthalpy of transition at T_m, $\Delta H^{F-I}(T_m^{F-I})$ and $\Delta H^{I-U}(T_m^{I-U})$, as well as heat capacity change upon transition, ΔC_p^{F-I} and ΔC_p^{I-U}. With these fractions the chemical shifts of the pure states were deconvoluted from the observed weighted sum of chemical shifts. In the second step, chemical shifts calculated for the first one were used and the thermodynamic

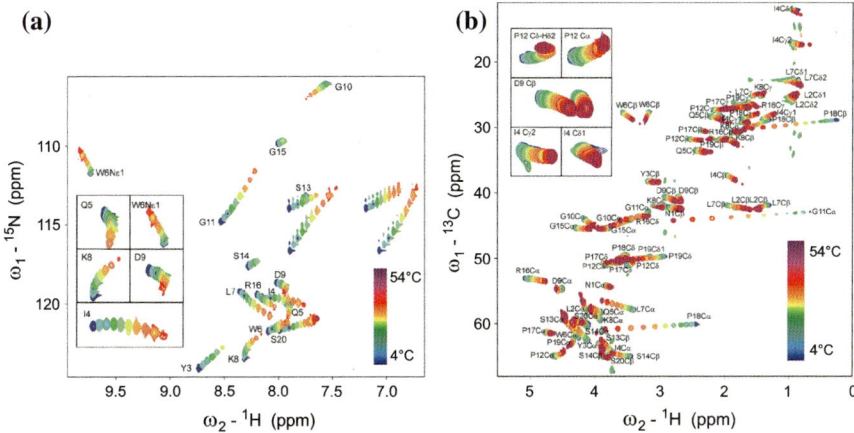

Fig. 10.2 **a** ^{1}H–^{15}N HSQC chemical shift changes versus increase of temperature for ^{13}C ^{15}N Tc5b at neutral pH. The colour of resonances changes gradually from blue (4 °C) to purple (54 °C). *Insets* indicate expanded views of the Ile4NH, Gln5NH, Trp6Nε1, Lys8NH and Asp9NH resonances. **b** The same as in (**a**) for ^{1}H–^{13}C HSQC chemical shift changes upon temperature increase. *Insets* display expanded views of the aromatic region and the resonances of Pro12Cδ–Hδ2; Pro12Cδ–Hα; Asp9Cβ–Hβ1, Hβ2; Ile4Cγ2–Hγ2# and Ile4Cδ1–Hδ1#

parameters were simultaneously fitted to the same system of equations. The procedure was repeated until convergence was reached. Thus, we obtained the following values: $\Delta H^{F-I}(T_m^{F-I}) = 36$ kJ/mol, $\Delta H^{I-U}(T_m^{I-U}) = 44$ kJ/mol, $T_m^{F-I} = 24$ °C, $T_m^{I-U} = 35$ °C, $\Delta C_p^{F-I} = 0.0$ kJ/mol, $\Delta C_p^{I-U} = 0.3$ kJ/mol. Chemical shifts of the pure states were also calculated by this procedure.

Secondary structures of the different states were assessed by analysing chemical shifts with the program TALOS+. This indicated that both the **F** and **I** states are well-folded. There is a difference between only at the G11–G15 segment, where the helix propensity is increased in **I** as related to **F**. This may be due to a 3_{10}- to $\propto$-helix backbone rearrangement. It was found that **U** is not a true random-coil structure, since it retains some residual turns and helices at its N-terminal region.

In order to follow the $3_{10} \rightleftharpoons \propto$-helix rearrangement of the Trp-cage, MD simulations were carried out using the CHARMM27 (with CMAP correction) [87], AMBER ff99SB-ILDN [88] and the OPLS-AA [89] force fields as incorporated in the GROMACS [90] and Desmond MD packages, respectively. Energy minimized initial structures were equilibrated at the target temperatures in three 200 ps steps by subsequent removal of restraints on the protein atoms and a 200 ps NVT step to stabilize pressure, followed by a 200 ns NPT MD simulation at temperatures between 250 and 400 K.

Various force-fields were chosen due to their range of helix formation propensities. The Tc5b system proved to be a sensitive test-case. At 300 K temperature, where all three conformations of the G10–G15 turn region should exist (3_{10}-helix (folded), α-helix (intermediate) and coil (unfolded)) with the domination of the 3_{10} helix, we found the following distributions: 2.2/83.2/14.6 % using CHARMM27,

98.7/0.8/0.5 % using AMBER ff99SB-ILDN and 24.0/0.0/76.0 % using OPLS-AA for the 3_{10}-helix/α-helix/coil content of the equilibrium ensembles, respectively. It has been shown that both CHARMM27 and AMBER ff99SB-ILDN are more successful for small proteins than OPLS-AA, but CHARMM27 overestimates, while AMBER ff99SB-ILDN underestimates the helical content [91]. Raising the temperature from 250 to 400 K did not reverse trends, the helical content of G10-G15 was reduced to 42.4 % with CHARMM27, and increased to only 3.2 % with AMBER ff99SB-ILDN. The unfolded state became the most populated at higher temperatures. We also tried to increase the ionic strength of the solvent by increasing the concentration of NaCl from 0.1 to 1 M, but this had only a minor effect on the helical content.

Thus we selected the mid-structure of the most populated cluster of the equilibrium ensemble obtained using the AMBER ff99SB-ILDN force-field to be our final model of the folded state, and the intermediate conformer is represented by that of the trajectory obtained with CHARMM27 (see Fig. 10.3).

The $3_{10} \rightleftharpoons \alpha$-helix transition is easy. The 3_{10}-helix conformation of the turn region is stabilized by two main-chain H-bonds, both of the i + 3 $\rightarrow$ i type: the Ser13NH $\rightarrow$ Gly10CO and Ser14NH $\rightarrow$ Gly11CO H-bonds. A small tilt of the Gly10 carbonyl oxygen results in reshuffling of the H-bond pattern, with the formation of Ser14NH $\rightarrow$ Gly10CO and Gly15NH $\rightarrow$ Gly11CO hydrogen bonds instead of the previous, carrying the characteristic i + 4 $\rightarrow$ i H-bond motif of α-helices (Fig. 10.4).

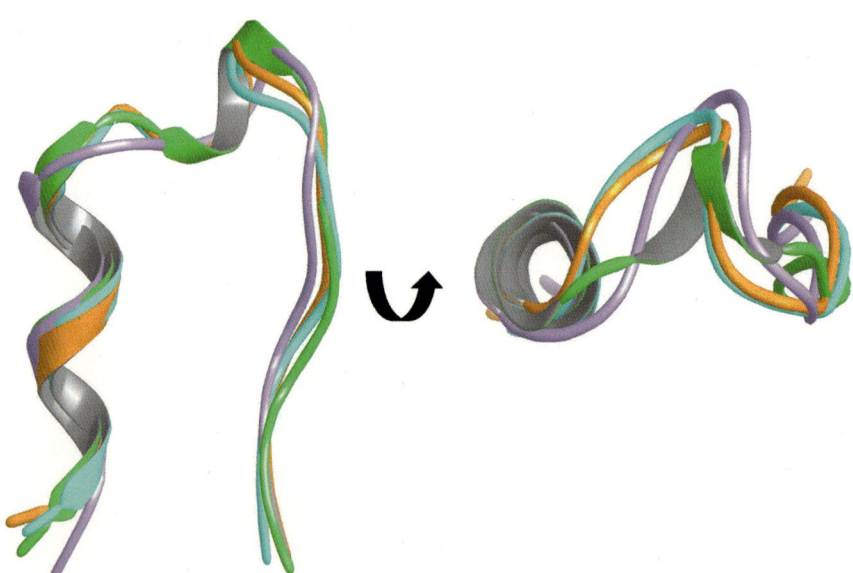

Fig. 10.3 *Side* and *top views* of the MD derived structures of Tc5b as compared with the NMR derived structure. *Orange* NMR structure, *cyan* the mid-structure of the most populated cluster of the equilibrium ensembles of the 300 K MD simulations using AMBER ff99SB-ILDN, *green* CHARMM27, *lilac* OPLS-AA

Fig. 10.4 MD derived structures of the folded (Gly11–Gly15 segment in 3_{10} helix conformation) and intermediate states (Gly11–Gly15 segment in α-helical conformation), shown in *cyan* and *green*, respectively

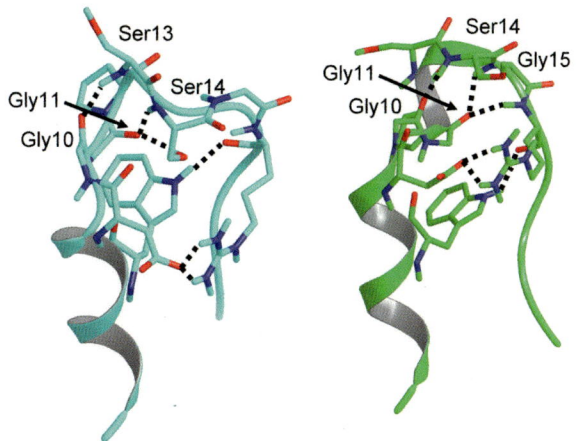

The transformation seems to be energetically feasible also. The Ramachandran map of both conformers indicates a stable arrangement, with all amino-acid main-chain conformations falling into the preferred regions—while the greatest change between the two involves the shift of the Gly11 residue from one of the preferred regions to another (Fig. 10.5).

NMR spectroscopic studies indicate that the rearrangement results in an altered conformation of Pro12. The latter residue gets closer to Trp6, the distance between the ring centroids decreases by 0.8 Å, in agreement with the tighter core seen for **I** in the in vitro studies. The protein core remains intact in over 85 % of the helical structures, with either the Arg16 or the Pro17 carbonyl oxygen atoms within H-bonding distance to Trp6Nε1, similarly as in case of the Asp9-Arg16 salt bridge (Fig. 10.6).

The energy landscape of the Tc5b folding funnel was also investigated with the introduction of acidic conditions. Acidification reduces the stability of **F** by decreasing the electrostatic attraction between the Asp9 and Arg16 side chains and thus facilitates the formation of an alternative conformer. Using ^{13}C, ^{15}N double labelling and other 3D NMR techniques, many of the minor resonances could by assigned unambiguously, too. Several major resonances have one (Leu2, Tyr3, Ile4, Gln5, Trp6, Leu7, Lys8, Asp9 and Ser20), or two (Gly10, Gly11, (Pro12), Ser13, Ser14, Gly15 and Arg16) additional sets of resonances, which indicates the presence of two minor conformers. ZZ-exchange NMR-experiments confirmed that the **F** state is not only in a fast exchange with both **I** and **U**, but it is also in a slow exchange with the highly mobile minor forms **U′** and **U″**. The low intensity of the minor signals did not allow to derive exact rate constants for the **F** ⇌ **U′** and **F** ⇌ **U″** transitions. However, we could identify the source of the slow exchange, which is a *cis-trans* isomerization of the Gly11-Pro12 amide bond.

The temperature dependent experiments were repeated under acidic conditions. We have found that the fast and slow exchanging intermediate states (**I** and **U′/U″**) coexist. Some of the HSQC lines of the major signals remain curved

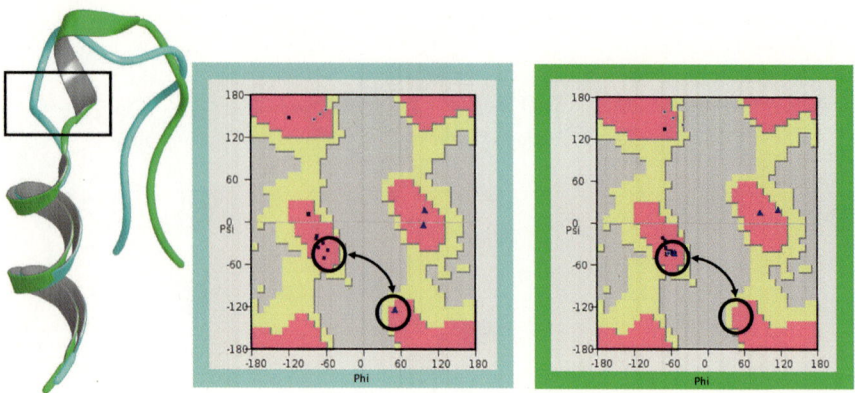

Fig. 10.5 Superimposition of the proposed folded and intermediate structures (*cyan* and *green*) and their corresponding Ramachandran maps. *Blue squares* and *triangles* (indicating glycine residues) show location of each residue of the two conformers and the *black arrow* indicates the effect of the 3_{10} *left right arrow* α-helix transition: re-location of Gly11 (framed in *black* on the superimposed structures)

Fig. 10.6 Superimposed structures of the folded (*blue*) and intermediates (*orange*) states as obtained by NMR spectroscopy

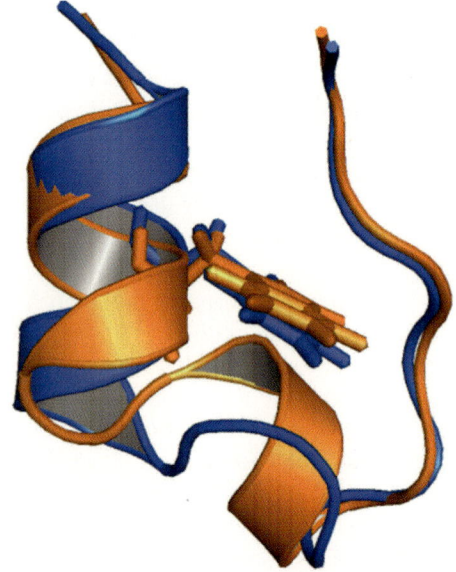

while the corresponding minor sets of resonances show linear shifts. The curved nature of the HSQC lines indicates that the fast-exchanging **I** state is yet populated, although to a smaller extent than at a neutral pH. The minor (**U′** and **U″**) and the major (**F** and the hidden **I**) forms are in a fast-exchange with the unfolded (**U**) state, since all of these resonances converge to the very same pure **U** state with increasing temperatures.

Fig. 10.7 Schematic
representation of the Trp-
cage folding funnel. At
neutral pH the folded state
is in a fast exchange with
an intermediate state; these
states interconvert by a 3_{10}
α-helix rearrangement. At
acidic pH the folded and
intermediate states are in
slow exchange with two
alternative states. These two
latter states interconvert by
a Gly11-Pro12 peptide bond
isomerization

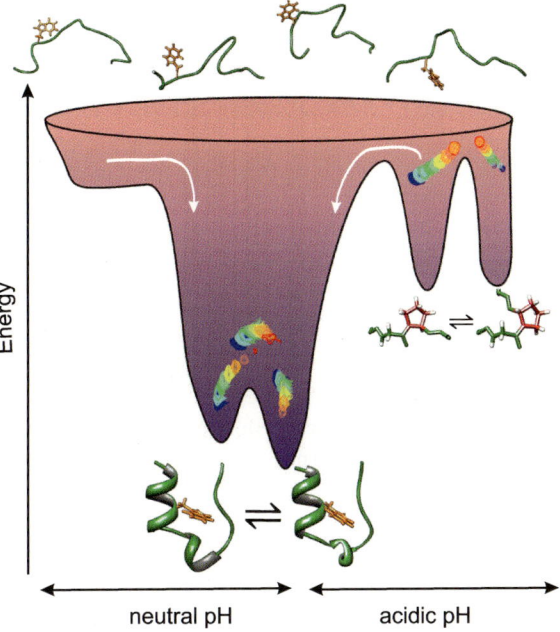

In summary, we were able to demonstrate that during the folding route of
Trp-cage two intermediate states evolve under native or close to native conditions.
(i) A fast-exchanging intermediate state **I**, with a native-like 3D-structure and
(ii) two slow-exchanging intermediate states, **U′** and **U″** with a considerable random coil feature. The population of **I** is higher at neutral pH, and lower under
destabilising conditions, like acidic pH or salt-bridge deletion, which supports that
the **I** state is on the transition pathway (**F** ⇌ **I** ⇌ **U**). In contrast, the population of
the **U′** and **U″** states increases under fold destabilising conditions, like acidification and increase of the temperature. This implies that both **U′** and **U″** are off-pathway intermediates. Under acidic conditions the off-pathway slow-exchanging (**U′**
and **U″**) and the on-pathway fast-exchanging (**I**) intermediates coexist. This veri-
fies a folding scenario which is more complex than previously assumed for such a
small protein. Our conclusions are summarised in Fig. 10.7.

According to the NMR and ECD analysis state **I** has the following characteristics: (i) a well-formed α-helix at the N-terminus (Leu2-Asp9) and at the Gly11-
Gly15 region; (ii) a compact tertiary structure with a tight core which resembles
to the native state; (iii) a trans Xaa-Pro peptide bonds for all four proline residues;
(iv) highest population at 28 °C; and (v) fast-exchange with both **F** and **U** states.
On the other hand, the slow-exchanging intermediates (**U′** and **U″**) have significantly different characteristics. (i) A transient secondary structure or tertiary interactions are absent; (ii) chemical shifts are close to the random coil reference values;
(iii) both *cis* and *trans* Gly11-Pro12 peptide bonds are of a comparable population;
(iv) there is a slow-exchange with **F** and a fast-exchange with **U** states. In vitro

dynamic studies shed light on the first step of the thermal unfolding by providing information about the temperature induced changes of backbone mobility. With the increase of temperature the amplitude of backbone motion decreases, while the timescale of internal motion increases in the Gly11–Gly15 region implying a local structural rearrangement: 3_{10}- to α-helix transformation. MD trajectory analysis supports the basic assumptions for this transition: the Ser13NH → Gly10CO and Ser14NH → Gly11CO H-bonds break and new Ser14NH → Gly10CO and Gly15NH → Gly11CO H-bonds are formed. This finding regarding the structural reshuffling is in agreement with the recently published crystal structure of a stabilized Trp-cage. Scian et al. found that in the crystal structure the Gly10-Arg16 loop, including the 3_{10}-helix, presents the greatest structural versatility [92].

Structural elements with such elevated mobility are prone to induce structural reshuffling within a molecule. Our results indicate that during the folding of Tc5b the α-helix formation precedes the hydrophobic collapse since the **U** state has a nascent tendency to form helices at the N-terminal segment (L2–K8). This observation is consistent with UV-Resonance Raman experiments which reported a broad α-helix melt for Tc5b [66]. The α-helix formation restricts the conformational space so that the folded state can be achieved by fewer structural transitions and it also accelerates the folding process. The hydrophobic collapse is fast and generates a compact intermediate structure with high similarity to the folded structure. This intermediate state differs from the folded state mainly in the structure of the G^{11}–G^{15} region implying that the 3_{10}-helix is the most unstable part of Tc5b and it rearranges easily to other conformations. These findings are consistent with the observations of recent experimental [68] and computational studies [73, 74, 93] stating that the unfolding of Tc5b begins at the 3_{10}-helix; although none of these studies observed the 3_{10} to α-helix rearrangement of Gly11-Gly15 before the complete unfolding. Molecular dynamics simulations also failed to generate the slow-exchanging and completely unstructured alternative conformer of Tc5b with a *cis* Gly11–Pro12 peptide bond that appears under acidic pH [94].

Recently the denatured state of Tc5b was analysed by NMR relaxation studies. It was found that the molecule has structural features different from those obtained by thermal denaturation [95]. This alternative **U** state, called here as **UU**, presents both *cis* (20 %) and *trans* (80 %) isomers of P12, as well as native and non-native residue contact forming largely mobile and thus unfolded backbones. Both the temperature (**U**) and the urea induced unfolded states (**UU**) have a high degree of mobility and random-coil features. There are clear differences between these sates: (i) the temperature induced unfolded state has an inherent tendency to form turns and short helices in the Leu2–Lys8 segment while no trace of this is seen in the urea induced **UU** state; (ii) the non-native contact between residues 4 and 6 of the **UU** state is not detected in **U**; (iii) higher than average R_2/R_1 ratios of residues 4, 5, 7 and 11 are present in the **UU** state, while absent in the **U** state. These observed differences emphasize that there is no single unfolded state for any protein; the residual structure and behaviour of the unfolded states depend fundamentally on the way how the unfolding was induced.

Although Tc5b consists of only twenty residues and represents the smallest and simplest protein model ever investigated we found that nothing is simple either for this model. Our analysis revealed that a dense, well-balanced non-covalent, weak interaction network stabilizes the structure, H-bonds, salt-bridges, dispersion forces act together to maintain the three-dimensional fold. A small perturbation to any of these interaction has a dramatic effect on the stability. Similarly, the folding pathway is rather complex with structured and unstructured intermediates which exchange with the major, folded state at different time-scales and possess substantially different dynamic properties. All the states coexist in equilibrium at the same time thus their thermodynamic description involves multiple parameters. Therefore, the question arises if the structure-stabilizing interactions, the folding process and the time-scale of atomic motions of the simplest model system are so complicated then what shall we expect for a real protein consisting of hundreds of amino acid residues? Much should be learned about the factors that influence the interactions defining the structure and internal dynamics and govern the folding processes. Hence, Tc5b is an excellent test case both for future experimental and computational studies.

10.5 The Case Study of Podocin

In the following we present an example for the successful application of MD techniques to a complex problem, where the impairing the flexibility of a small folded protein leads to severe consequences. A specific kidney disease, nephrotic syndrome type 2, is due to the pathogenicity of an allele encoding the R229Q mutant of podocin, a small protein that localizes in the podocytes of the kidney, anchoring members of its filtration system [96]. Malfunction of mutant podocin claims the life of patients during their young ages. It is caused by the Arg229Gln mutation of the protein, but only under peculiar circumstances. Both parents carrying the polymorphism associated with the illness (causing a single mutation) will not necessarily have an unhealthy child, only the polymorphism in association with any of a certain set of other mutations will be detrimental. Based on these observations it was proposed that the Arg229Gln mutation must impair the ability of podocin to dimerize: in pair with a wild-type monomer the resultant dimer will be functional, while pairing the Arg229Gln variant with certain "difficult cases", distorted dimers will emerge.

In order to study this problem, homology models of the wild-type and mutant podocin monomers were subjected to a 40 ns MD simulation in our laboratory. The last 10 ns of the trajectory was averaged and clustered. Podocin monomers contain a globular head domain and a long, helical tail. The mutated Arg229 residue can be found in the head-domain, turning inside, stabilized by at least two H-bonds formed with negatively charged amino acids of the head-domain (Glu233, Glu237 and Asp244). In the Arg229Gln mutant monomer, where Gln229 is too short to allow these contacts, Glu233 and Glu237 flip toward the tail domain and form H-bonds with positively charged amino acids of it (Fig. 10.8).

While the domains themselves are unperturbed by the Arg229Gln switch, these added interactions between the head and tail domains restrict their inter-domain hinge-like movement, greatly affecting the dynamic nature of the monomer (Fig. 10.9).

(a)

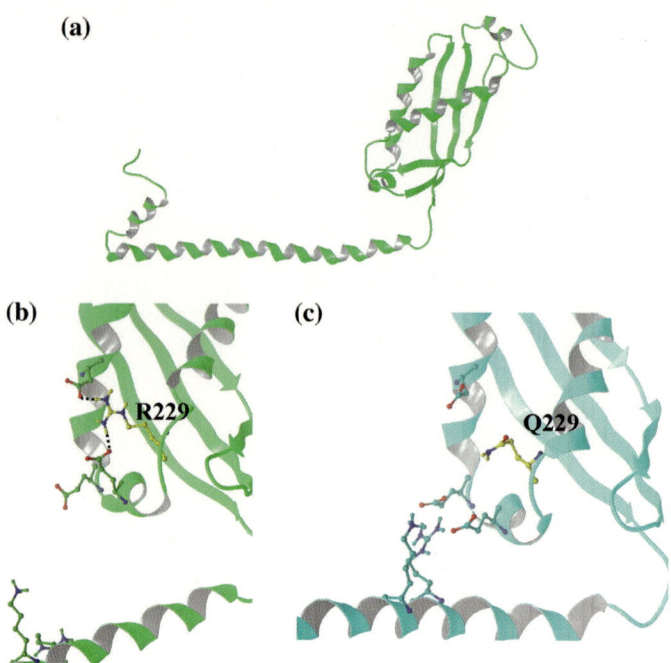

Fig. 10.8 **a** Structure of the native podocin monomer obtained as an average of the most populated clusters of the last 10 ns of our simulations. **b** Interactions of the Arg229 residue of the wild-type protein with negatively charged groups of the head domain. **c** Interactions of the Gln229 residue of the polymorph variant, a new link is formed between the head and tail domains

Fig. 10.9 Difference in the flexibility of the wild type protein (*green*) and the polymorph (*cyan*) monomers. All clusters of more than 0.1 % population of the last 10 ns of our simulations are shown, with their head domains superimposed

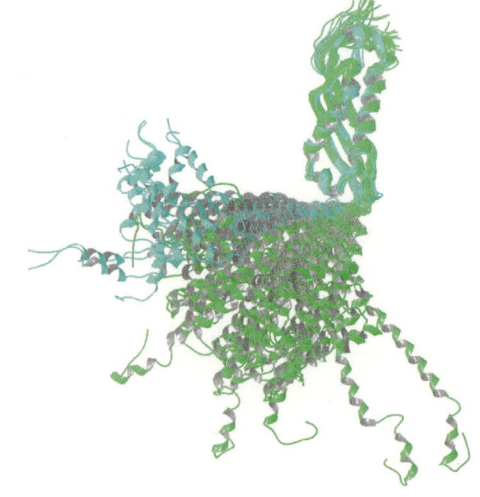

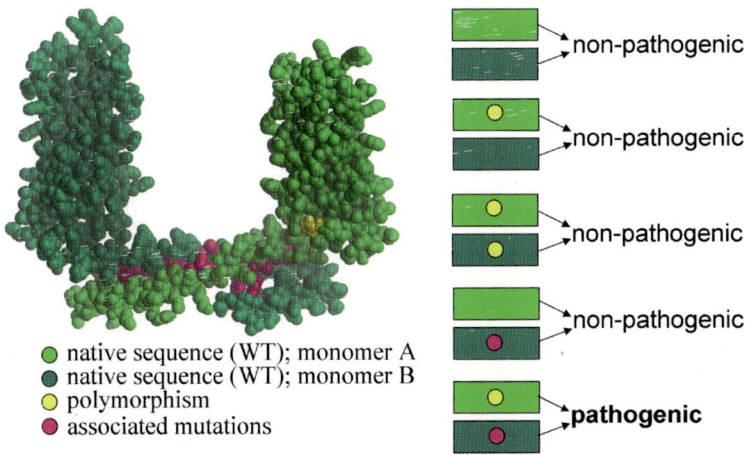

Fig. 10.10 Wild-type podocin dimer formation. Residue 229 is shown in *yellow*, mutation sites that lead to pathogenic dimer formation when faced with the polymorph monomer are shown in *magenta*. *WT* wild type

All the associated mutations that lead to unhealthy dimer formation when paired with the Arg229Gln polymorph are found in the helical tail region of podocin, therefore it was suggested that this must be the primary site of dimerization. A coiled-coil type dimer was created and the next set of calculations (100 ns MD simulations) involved dimer pairs, both non-pathogenic and pathogenic combinations (Fig. 10.10).

Interestingly, it was found that the loss of pliability, as a result of the Arg229Gln switch, leads to formation of distorted dimers in the case of the Arg229Gln polymorph pairing up with monomers carrying a mutation on the helical tail region of podocin. On the other hand, it seems that the flexibility of the wild-type monomer allows for correction of small misfits in the dimerization process, thus, paired-up with the same tail-mutated variants, the resulting dimers will not differ greatly from the wild-type dimer structure (Fig. 10.11). It was, therefore shown that an inheritance pattern that defies the laws of Mendel, could be explained by the impaired dimer forming capacity of the Arg229Gln polymorph, which is the result of the change of its overall flexibility, caused by a single mutation. MD simulation proved to be an ideal method to follow such changes.

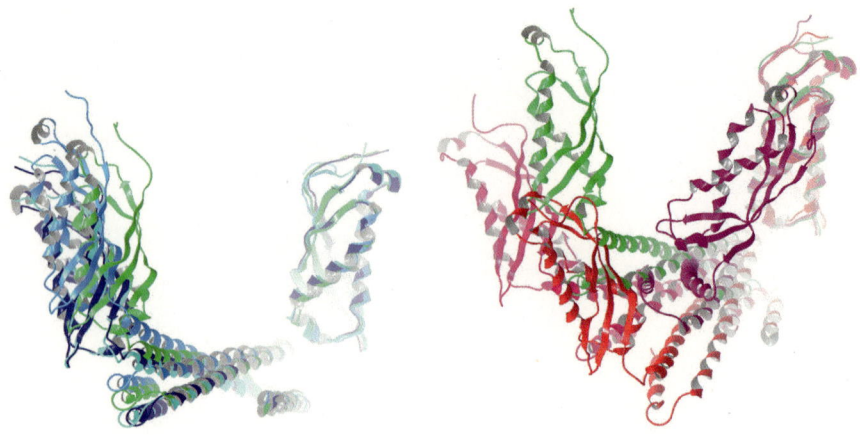

Fig. 10.11 Structure of non-pathogenic and pathogenic dimers. Average structures of the last 30 ns of the trajectory are shown. (*Left*) non-pathogenic dimers (coloured by indifferent shades of *blue*: WT/R229Q, R229Q/R229Q, A284V/WT) superimposed on the WT/WT dimer (shown in *green*). (*Right*) pathogenic dimers (coloured indifferent shades of red: A284V/A284V, A284V/R229Q, A297V/R229Q) superimposed on the WT/WT dimer (shown in *green*). *WT* wild type

References

1. Bu Z, Callaway DJ (2011) Adv Protein Chem Struct Biol 83:163–221
2. Zhao Q (2013) Rev Theor Sci 1:83–101
3. Baker CM, Best RB (2013) Insights into the binding of intrinsically disordered proteins from molecular dynamics simulation. WIREs Comput Mol Sci. doi:10.1002/wcms.1167
4. Namanja AT, Wang XJ, Xu B, Mercedes-Camacho AY, Wilson BD, Wilson KA, Etzkorn FA, Peng JW (2010) J Am Chem Soc 132:5607–5609
5. Boehr DD, Nussinov R, Wright PE (2009) Nat Chem Biol 5:789–796
6. Smock RG, Gierasch LM (2009) Science 324:198–203
7. Kamerzell TJ, Middaugh CR (2008) Sending signals dynamically. J Pharm Sci 97:3494–3517
8. Csermely P, Palotai R, Nussinov R (2010) Induced fit, conformational selection and independent dynamic segments: an extended view of binding events. Trends Biochem Sci 35:539–546
9. Khersonsky O, Tawfik DS (2010) Annu Rev Biochem 79:471–505
10. Khoruzhii O, Butin O, Illarionov A, Leontyev I, Olevanov M, Ozrin V, Pereyaslavets L, Fain B, Chapter 5 of this book
11. Wang W, Donini O, Reyes CM, Kollman PA (2001) Annu Rev Biophys 30:211–243
12. Mackerell AD, Feig M, Brooks CL III (2004) J Comput Chem 25:1400–1415
13. Hu Z, Jiang J (2010) J Comput Chem 31:371–380
14. Case DA, Cheatham TE III, Darden T, Gohlke H, Luo R, Merz KM Jr, Onufriev A, Simmerling C, Wang B, Woods RJ (2005) J Comput Chem 26:1668–1688
15. http://www.gromacs.org/Documentation/Terminology/Force_Fields/GROMOS. Accessed 21 March 2014
16. Halgren TA (1996) J Comp Chem 17:490–519
17. Van Der Spoel D, Lindahl E, Hess B, Groenhof G, Mark AE, Berendsen HJ (2005) J Comput Chem 26:1701–1718
18. http://accelrys.com/products/discovery-studio/. Accessed 21 March 2014

19. https://www.certara.com/products/molmod/sybyl-x. Accessed 21 March 2014
20. http://www.charmm.org/. Accessed 21 March 2014
21. Adcock SA, McCammon JA (2006) Chem Rev 106:1589–1615
22. Allison JR, Hertig S, Missimer JH, Smith LJ, Steinmetz MO, Dolenc J (2012) J Chem Theory Comput 8:3430–3444
23. Dror RO, Dirks RM, Grossman JP, Xu H, Shaw DE (2012) Annu Rev Biophys 41:429–452
24. Genheden S, Ryde U (2012) Phys Chem Chem Phys 14:8662–8677
25. Zhou R (2007) Methods Mol Biol 350:205–223
26. Sugita Y, Okamoto Y (1999) Chem Phys Lett 314:141–151
27. Weber JK, Jack RL, Pande VS (2013) J Am Chem Soc 135:5501
28. McGibbon R, Pande VS (2013) J Chem Theory Comput 9:2900–2906
29. Tirion MM (1996) Phys Rev Lett 77:1905–1908
30. Cui Q, Bahar I (2006) Normal mode analysis: theory and applications to biological and chemical systems. Chapman and Hall/CRC, Boca Raton
31. van den Bedem H, Bhabha G, Yang K, Wright PE, Fraser JS (2013) Nat Methods 10:896–902
32. Staunton D, Schlinkert R, Zanetti G, Coelbrook SA, Campbell ID (2006) Magn Reson Chem 44:S2–S9
33. Kainosho M, Torizawa T, Iwashita Y, Terauchi T, Ono AM (2006) Nature 440:52–57
34. Gáspári Z, Pál G, Perczel A (2008) BioEssays 30:772–780
35. Wüthrich K (1986) NMR of proteins and nucleic acids. Wiley
36. Moseley HN, Monleon D, Montelione GT (2001) Methods Enzymol 339:91–108
37. Jung YS, Zweckstetter M (2004) J Biomol NMR 30:11–23
38. Hiller S, Wider G, Wüthrich K (2008) J Biomol NMR 42:179–195
39. Fernandez C, Wider G (2003) Curr Opin Struct Biol 13:570–580
40. Jarymowycz VA, Stone MJ (2006) Chem Rev 106:1624–1671
41. Peng JW (2012) Phys Chem Lett 3:1039–1051
42. Sapienza PJ, Lee AL (2010) Curr Opin Pharmacol 10:723–730
43. Palmer AG (2001) Annu Rev Biophys Biomol Struct 30:129–155
44. Igumenova TI, Frederick KK, Wand AJ (2006) Chem Rev 106:1672–1699
45. Mittermaier AK, Kay LE (2009) Trends Biochem Sci 34:601–611
46. Fischer MWF, Majumdar A, Zuiderweg ERP (1998) Prog Nucl Magn Reson Spectrosc 33:207–272
47. Peng JW, Wagner GJ (1992) J Magn Reson 98:308–322
48. Peng JW, Wagner GJ (1992) Biochemistry 31:8571–8586
49. Farrow NA, Zhang O, Forman-Kay JD, Kay LE (1995) Biochemistry 34:868–878
50. Ishima R, Nagayama K (1995) J Magn Reson Ser B 108:73–76
51. Redfield AG (2012) J Biomol NMR 52:159–177
52. Lipari G, Szabo A (1982) J Am Chem Soc 104(4546–4559):4559–4570
53. Palmer AG (1997) Curr Opin Struct Biol 7:732–737
54. Carr HY, Purcell EM (1954) Phys Rev 94:630–638
55. Meiboom S, Gill D (1958) Rev Sci Instrum 29:688–691
56. Peng JW, Thanabal V, Wagner G (1991) J Magn Reson 94:82–100
57. Mulder FAA, de Graaf RA, Kaptein R, Boelens R (1998) J Magn Reson 131:351–357
58. Palmer AG III, Massi F (2006) Chem Rev 106:1700–1719
59. Jeener J, Meier BH, Bachmann P, Ernst RR (1979) J Chem Phys 71:4546–4553
60. Neidigh JW, Fesinmeyer RM, Andersen NH (2002) Nat Struct Biol 9:425–430
61. Barua B, Lin JC, Williams VD, Kummler P, Neidigh JW, Andersen NH (2008) Protein Eng Des Sel 21:171–185
62. Williams DV, Barua B, Andersen NH (2008) Org Biomol Chem 6:4287–4289
63. Williams DV, Byrne A, Stewart J, Andersen NH (2011) Biochemistry 50:1143–1152
64. Neidigh JW, Fesinmeyer RM, Prickett KS, Andersen NH (2001) Biochemistry 40:13188–13200
65. Neuweiler H, Doose S, Sauer M (2005) Proc Natl Acad Sci USA 102:16650–16655
66. Ahmed Z, Beta IA, Mikhonin AV, Asher SA (2005) J Am Chem Soc 127:10943–10950

67. Mok KH, Kuhn LT, Goez M, Day IJ, Lin JC, Andersen NH, Hore PJ (2007) Nature 447:106–109
68. Culik RM, Serrano AL, Bunagan MR, Gai F (2011) Angew Chem Int Ed Engl 50:10884–10887
69. Qiu L, Pabit SA, Roitberg AE, Hagen SJ (2002) J Am Chem Soc 124:12952–12953
70. Mok KH, Kuhn LT, Goez M, Day IJ, Lin JC, Andersen NH, Hore PJ (2007) Nature 447:106–109
71. Neuweiler H, Doose S, Sauer M (2005) Proc Natl Acad Sci USA 102:16650–16655
72. Rovó P, Stráner P, Láng A, Bartha I, Huszár K, Nyitray L, Perczel A (2013) Chem Eur J 19:2628–2640
73. Juraszek J, Bolhuis PG (2008) Biophys J 95:4246–4257
74. Juraszek J, Bolhuis PG (2006) Proc Natl Acad Sci USA 103:15859–15864
75. Chowdhury S, Lee MC, Xiong G, Duan Y (2003) J Mol Biol 327:711–717
76. Zheng W, Gallicchio E, Deng N, Andrec M, Levy RM (2011) J Phys Chem B 115:1512–1523
77. Zhou R (2003) Proc Natl Acad Sci USA 100:13280–13285
78. Linhananta J, Boer I, MacKay J (2005) J Chem Phys 122:114901
79. Nikiforovich GV, Andersen NH, Fesinmeyer RM, Frieden C (2003) Proteins: Struct Funct Bioinf 52:292–302
80. Snow CD, Zagrovic B, Pande VS (2002) J Am Chem Soc 124:14548–14549
81. Paschek D, Nymeyer H, Garcia AE (2007) J Struct Biol 157:524–533
82. Paschek D, Hempel S, García AE (2008) Proc Natl Acad Sci USA 105:17754–17759
83. Zagrovic B, Pande V (2003) J Comput Chem 24:1432–1436
84. Day R, Paschek D, Garcia AE (2010) Proteins: Struct Funct Bioinf 78:1889–1899
85. Dyer RB (2007) Curr Opin Struct Biol 17:38–47
86. Rodriguez-Granillo A, Annavarapu S, Zhang L, Koder RV, Nanda V (2011) J Am Chem Soc 133:18750–18759
87. Mackerell AD Jr, Feig M, Brooks CL III (2004) J Comput Chem 25:1400–1415
88. Lindorff-Larsen K, Piana S, Palmo K, Maragakis P, Klepeis JL, Dror RO, Shaw DE (2010) Proteins 78:1950–1958
89. Kaminski GA, Friesner RA, Tirado-Rives J, Jorgensen WL (2001) J Phys Chem B 105:6474–6487
90. Hess B, Kutzner C, van der Spoel D, Lindahl E (2008) J Chem Theory Comput 4:435–447
91. Lindorff-Larsen K, Maragakis P, Piana S, Eastwood MP, Dror RO, Shaw DE (2012) PLoS ONE 7(2):e32131
92. Scian M, Lin JC, Trong IL, Makhatadze GI, Stenkamp RE, Andersen NH (2012) Proc Natl Acad Sci USA 109:12521–12525
93. Marinelli F, Pietrucci F, Laio A, Piana S (2009) PLoS Comput Biol. doi:10.1371/journal.pcbi.1000452
94. Jimenez-Cruz CA, Makhatadze GI, Garcia AE (2011) Phys Chem Chem Phys 13:17056–17063
95. Rogne P, Ozdowy P, Richter C, Saxena K, Schwalbe H, Kuhn LT (2012) PLoS ONE. doi:10.1371/journal.pone.0041301
96. Tory K, Menyhárd DK, Woerner S, Nevo F, Gribouval O, Kerti A, Stráner P, Arrondel C, Cong EH, Tulassay T, Mollet G, Perczel A, Antignac C (2014) Nat Genet 46:299–304

Chapter 11
Protein Ligand Docking in Drug Discovery

N.F. Brás, N.M.F.S.A. Cerqueira, S.F. Sousa, P.A. Fernandes
and M.J. Ramos

11.1 Introduction

In the past decades, the number of protein structures publicly available in the Research Collaboratory for Structural Bioinformatics (RCSB) database has grown from two structures in 1972 to approximately 99,000 protein structures in April 2014, with thousands being added each year. The development in this field has been partially sponsored by the enormous advances in genomics, X-ray crystallography and NMR spectroscopy, which have paved the way for a large number of new potential therapeutic targets. Simultaneously, a demand for powerful and reliable technologies that can identify high quality lead drug candidates for those potential targets is growing.

At the end of the XXth century, the main approaches to increase lead discovery were experimentally based. These methods included sophisticated combinatorial chemistry and high-throughput assays that provided automated screening of hundreds of thousands or even millions of potential compounds. However, despite the impressive technological advances observed in this field, these methodologies failed to provide many new drugs. In addition, the process was very costly, time demanding and in the end there was no guarantee that new lead compounds would be obtained [1]. All these factors declined the interest in these methodologies for drug discovery and, at the same time, advanced the search for new methodologies.

This scenario led to the development of several computational methodologies and algorithms that could help researchers to understand and predict molecular

Authors N.F. Brás, N.M.F.S.A. Cerqueira and S.F. Sousa contributed equally.

N.F. Brás · N.M.F.S.A. Cerqueira · S.F. Sousa · P.A. Fernandes · M.J. Ramos (✉)
REQUIMTE, Departamento de Química e Bioquímica, Faculdade de Ciências,
Universidade do Porto, Rua do Campo Alegre s/n, 4169-007 Porto, Portugal
e-mail: mjramos@fc.up.pt

© Springer International Publishing Switzerland 2014
G. Náray-Szabó (ed.), *Protein Modelling*, DOI 10.1007/978-3-319-09976-7_11

recognition from a structural and energetic point of view. Molecular docking is perhaps the most popular method in structure-based drug design that is employed to this purpose [2].

Molecular Docking can be defined as a computational method that allows predicting the preferred position, orientation and conformation (i.e. the pose) of one molecule (*ligand*) in relation to a second one (often much larger and called protein or *receptor*), when the binding between the two forms a stable complex. The preferred orientation of the molecule in relation to the protein can then be used to predict the strength of association or the binding affinity between both intervenients.

Currently, there is a relatively large and ever increasing number of molecular docking programs, such as DOCK [3], AutoDock [4, 5], FlexX [6, 7], FlexE [8], GEMDOCK [9], MEDock [10], MolDock [11], Tribe-PSO [12], SODOCK [13], Surflex [14], GOLD [15–17], ICM [18, 19], Glide [20], Cdocker [21], LigandFit [22], MCDock [23], RDock [24], ZDock [25], M-ZDOCK [26], and MSDOCK [27], among others. Generally speaking, these programs have similar implementations and only require the knowledge of the tridimensional structure of the protein (Fig. 11.1). The structure of most of the ligands can be created with any molecular modeling software or obtained from chemical databases.

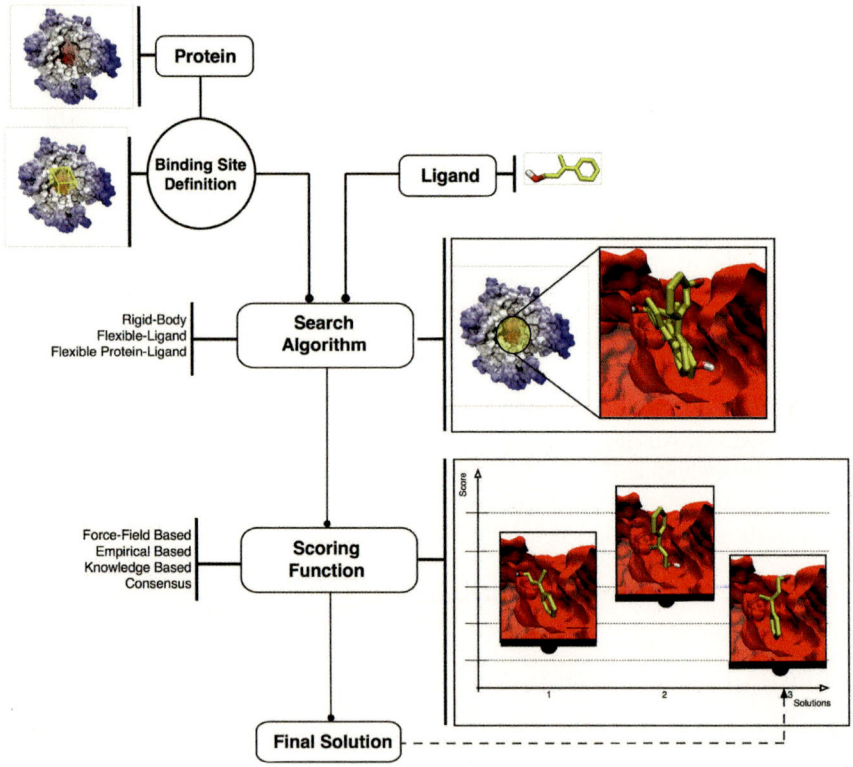

Fig. 11.1 General protocol of molecular docking programs

In the majority of these programs, the first step of a docking study involves the definition of a binding region, i.e. an area in the protein where the ligand may be bound. This can be done in two ways: if the location of the binding site is known, the programs generally allow the user to restrict the binding area to a small portion of the protein. Otherwise, if nothing is known about the location of the binding site, then a blind docking can be chosen instead. In this case, the entire surface of the target is scanned to explore putative binding pockets [28, 29]. The docking prediction is much less reliable in the latter case, and this strategy should be used only as a last resort.

The next step involves the search for the preferred position (and conformation) of the ligand in relation to the protein. To this end, two types of algorithms are used: the search algorithm and the scoring function. The search algorithm generates a number of possible poses (ligand, and eventually protein, conformations and mutual orientations) that fit the ligand into the binding pocket of the protein [30, 31]. The scoring function generates a score for each pose and then ranks the different poses that are generated by the search algorithm. This value should ideally represent the thermodynamics of interaction of the protein-ligand system (the binding free energy) in order to correctly distinguish the true binding modes from all the others explored [32].

At the end of this process, the best-scored solution(s) should correspond to true binding poses and should be very close to the one that is observed experimentally, if such information exists.

The majority of molecular docking programs are developed to be fast, since they are supposed to be applied to large databases of millions of compounds when they are used for drug discovery. To this end, several assumptions and simplifications are included in the search algorithms and scoring functions. In order to minimize their impact on the final results, several different strategies are adopted to maintain them fast and simultaneously accurate.

In this review, different types of search algorithms and scoring functions available in the most popular molecular docking programs are analysed and discussed. The current status in this research field is examined also and the current challenges and future directions discussed.

11.2 Search Algorithms

The goal of a search algorithm is to generate an ensemble of low-energy protein-ligand poses, with the correct one among them. As both molecules are flexible, it is fundamental to explore enough degrees of freedom of the system.

Generating poses for a real biological protein-ligand system requires searching over a space of $N*M + 6$ dimensions (N and M are the protein and ligand conformational degrees of freedom, respectively, and 6 corresponds to the rotation and translation components of the spatial arrangement of one unit onto the other). This high dimensional space is computationally untreatable [33]. A simple and small protein-ligand system with 50 rotatable bonds, having a minimum at every 60° in each dihedral angle, will exhibit 6^{50} conformations ($\approx 10^{39}$ conformations), which is a number trillions of times larger than the number of stars in the observable universe, even excluding clashes. To overcome this issue, the docking algorithms integrate different approximations to efficiently sample the pose space. The number of available search algorithms is continuously increasing, to optimize the speed, reliability, coverage, and accuracy in sampling the relevant pose space. The speed is particularly crucial for drug discovery studies with virtual screening tools, in which millions of different ligands are docked. Currently, the docking algorithms can be categorized into rigid-body, flexible-ligand, and flexible-protein methods.

Rigid-body algorithms are the simplest and fastest ones because they only sample the rotational/translational space (6 degrees of freedom) and consider essentially the geometrical complementarities between the protein and the ligand. This approach was widely applied in the earlier protein-ligand docking studies or in the initial stages of virtual screening studies. These tools frequently use a hierarchical sophistication docking protocol in a later stage, in which a less demanding and less precise rigid docking method is firstly performed, followed by more time-consuming and accurate procedures (flexible-ligand and flexible-protein docking) to refine and optimize the pose of the ligands.

Flexible-ligand docking algorithms emerged afterwards. These algorithms consider the protein as a rigid body and implement partial or full flexibility on the ligand. Hence, they explore the 6 translational and rotational degrees of freedom of the complex and the conformational degrees of freedom of the ligand, which makes these approaches more computationally demanding.

Proteins are dynamic molecules and their binding regions can acquire many different conformations or have significant structural changes upon ligand binding. Hence, these small rearrangements may influence negatively the accuracy of the docking results and the flexibility of the protein should not be neglected. To overcome this challenging problem, the flexible-protein search algorithms have emerged. These take into account the partial flexibility of the protein, in addition to the ligand flexibility and the 6 translational and rotational degrees of freedom of the complex. These powerful tools require, however, a much larger computational effort.

Table 11.1 summarizes the main docking algorithms widely used nowadays, examples of programs where they are implemented and some relevant and recent applications. However, it is important to note that the boundaries between different categories are not rigid and, in fact, some approaches could easily fall in more than one category. A detailed description will be subsequently provided for each category.

11.2.1 Rigid-Body Search Algorithms

Rigid-body are the most basic and fastest algorithms to sample the conformational space because they do not take into account the conformational flexibility of neither ligand nor protein, thus sampling only the rotational/translational space (6

Table 11.1 Main categories of search docking algorithms

Docking search algorithm category		Search method	Examples of programs	Applications
Rigid-body docking		Fast fourier transform/shape matching	ZDOCK, FTDOCK, SYSDOC, EUDOC, DOCK, MSDOCK	[27, 34]
Flexible-ligand docking	Systematic	Conformational	DOCK	[30, 35, 36]
		Fragmentation	LUDI, DOCK, Surflex, FlexX, ADAM, FLOG, Hammerhead, eHiTS	[37, 38]
		Database/geometric distance	FLOG	[39]
	Random or stochastic	Monte carlo	AutoDock, MCDOCK, ICM, QXP, Prodock, DockVision, LigandFit, Glide	[20, 22, 23, 38, 40]
		Genetic algorithms	AutoDock, GOLD, DIVALI, DARWIN, DOCK	[41–49]
		Differential evolutionary	GEMDOCK, MolDock, SADock, AutoDock	[9, 50]
		Tabu search	PRO_LEADS, PSI-DOCK	[51, 52]
		Particle swarm optimization	AutoDock, SODOCK, PSO@AUTODOCK, FIPSDock	[13, 53]
	Molecular simulations	Molecular dynamics and energy minimization[a]	AMBER[b], GROMACS[b], CHARMM[b], Prodock, ICM, QXP, DARWIN, DOCK, Hammerhead, ADAM	[54, 55]
Flexible-protein docking		Molecular dynamics, monte carlo	AMBER[b], GROMACS[b], CHARMM[b]	[56–59]
		Simulated annealing	AutoDock	[60]
		Rotamer libraries	MADAMM, GOLD, AutoDock, FlexX (FlexE extension), Dolina, SCARE	[56, 61–64]
		Protein-ensemble grids	FlexX (FlexE extension)	[65]
		Soft-receptor modelling	DOCK	[52, 66, 67]
		Collective degrees of freedom	ICM (IFREDA extension)	[133, 134]

More Complete

Time Consuming →

[a]These methods are only used to generate (MD) and refine (minimization) ligand poses

[b]These programs are used to generate different poses and they are not docking programs

degrees of freedom) and considering essentially the geometrical complementarities between both molecules. This simple approach was widely applied in the earlier protein-ligand docking studies [30] and, currently, in protein–protein docking protocols [68, 69] or in the initial stages of virtual screening studies.

The main rigid-body algorithms are based on the Fast Fourier Transform (FFT) [25, 70–72] and Shape Matching (SM) [73, 74] approaches. The FFT-based algorithms systematically sample the relative position of protein and ligand with an orthogonal grid, using correlation-type scoring functions to calculate the degree of overlap between pairs of grids in different relative orientations [25, 71, 72, 75, 76]. The SM algorithms [30] consider the geometrical overlap between the protein and ligand by doing several alignments of both molecules. They identify possible binding sites of a protein, which are then compared, to generate a small number of trial poses for grid-scoring [77]. ZDOCK [25] and FTDOCK [70] docking programs combine both SM and FFT search algorithms. Furthermore, SM algorithms are also widely used in flexible docking algorithms as a part of their search strategies, such as in SYSDOC [78], EUDOC [79], DOCK [3, 30], MSDOCK [27], LigandFit [22] and Glide [20].

Rigid-body docking has severe limitations. The 3D structures of biological systems (such as protein-ligand complexes) are very dynamic and can acquire many different conformations that influence the accuracy of the docking results [80]. Furthermore, due to their lack of sensibility, rigid-body docking methods are not adequate to find the pose of different ligands with a common substitution pattern or to discriminate new scaffolds with similar size. These limitations come from an absence of sampling of the ligand conformational space. These reinforce the progressive replacement of rigid-body docking methods by methods with full or partial flexibility at the ligand and, sometimes, at the protein as well.

11.2.2 Flexible-Ligand Search Algorithms

The most common flexible-ligand docking algorithms consider the protein as a rigid body and full or partial flexibility of the ligand, analysing its conformational space [81]. Since these challenging docking approaches explore the 6 translational and rotational degrees of freedom of the complex and the conformational degrees of freedom of the ligand, they became more computationally demanding [82–84], implying several approximations to allow their application in a proficient way. These algorithms are divided in three main classes: systematic methods, random or stochastic methods and molecular simulations methods. In all cases the algorithms need to explore the orientations, and translational and rotational degrees of freedom. The first two have been addressed in the rigid-body section above, and here we will emphasize the methods that confer flexibility to the ligand by exploring its conformational space.

11.2.2.1 Systematic Search Docking Algorithms

Systematic search algorithms try to explore most (ideally all) conformational degrees of freedom of the ligand, and they are divided into three categories: (i) conformational search methods, (ii) fragmentation methods, and (iii) database methods.

Conformational search methods systematically explore the conformational space of all rotatable bonds in the ligand by 360°, using a fixed increment to generate all possible conformations. The higher number of combinations due to a large number of rotatable bonds limits the application of these methods. In general, drug-like compounds have 10 (or lower) rotatable bonds, and if the conformational space of each rotatable bond uses a 15° increment, there are 24^{10} possibilities to explore during the docking procedure, which is computationally impossible nowadays. To overcome this issue in general, several constraints and restraints on the ligand bonds are applied to reduce the dimensionality of the problem. This systematic conformational approach is implemented in the DOCK [3] program.

Fragmentation search methods act by a "place-and-join" approach, in which the ligand is split into several pieces and the various fragments are successively docked into the binding site, and covalently linked to recreate the global ligand. Alternatively, the ligand can be divided into a rigid core fragment that is firstly docked, and the remaining flexible parts are subsequently added in an "incremental construction" or "anchor and grow procedure" approaches. The fragment-based incremental method is one of the most applied, and is employed in various flexible-ligand docking programs, such as LUDI [85], FlexX [6], DOCK [3], ADAM [86], Hammerhead [87], Surflex [14, 88], eHiTS [89], and FLOG [90].

Database search methods use known libraries of pre-generated conformations (conformational ensembles) to account for the ligands' flexibility. These search methods make use of intra- and intermolecular distances. The assembling of a small set of constrained distances enables the calculation of several structures or conformations [91]. The most popular docking software that applies this algorithm is FLOG [90].

11.2.2.2 Random or Stochastic Algorithms

Random search algorithms sample the ligand conformational space by doing stochastic modifications in its conformation, which can be accepted or rejected based on a predefined probability function [4, 92]. There are six main types of docking methods that use random algorithms: (i) Monte Carlo [23], (ii) Genetic Algorithms [4, 15], (iii) Tabu Search [52], (iv) Particle Swarm Optimization, (v) Differential Evolutionary Algorithms and (vi) Evolutionary Gaussians Algorithms.

Monte Carlo (MC) methods take into account a Boltzmann probability function as the acceptance criterium for a newly generated ligand pose. MC algorithms dock the ligand inside the protein binding site through many random translations, rotations and conformations, decreasing the probability of being trapped in local minima. These methods enhance the sampling in regions that are energetically favourable to create more reliable complexes. The simple energy minimization functions used in the MC methods do not require any sort of derivative information [93], and are also very efficient in stepping energy barriers, which allows a good sampling of the conformational space. Prodock [55], ICM [18], MCDOCK [23], DockVision [94], and QXP [95] are some examples of docking programs that have an MC-based algorithm. The LigandFit [22] and Glide (grid-based ligand docking with energetics) [20] programs combine the MC sampling with shape matching methods to use grids to explore the poses of a ligand in the binding site of the target.

Genetic algorithms (GA), popularized by John Holland in the 1970s [96], are a global searching strategy that belongs to the evolutionary programming methods with the purpose of finding solutions for search problems and, in the molecular docking case, trying to find the pose closest to the global energy minimum for a given protein conformation. GA methods are heuristic algorithms that emerged from genetics and the theory of biological evolution. They start from an initial population of several different ligand poses (*n* chromosomes) generated randomly, and each ligand pose is characterized by a set of state variables (defined as genes) that describes its translation and orientation in relation to the protein, and its conformation. In GA methods, the full set of the ligand state variables is defined as the genotype, whereas the ligand atomic coordinates are the phenotype. Various genetic operators, such as mutations, crossovers, and migrations are applied to the population to sample the pose space, until a population that optimizes a predefined fitness function is obtained. Programs GOLD [15, 17], AutoDock [4], DIVALI [97], and DARWIN [98] use or include a GA, or a GA-like algorithm. In particular, AutoDock uses a hybrid Lamarckian-GA method (LGA) [99], for which the GA plays a global search and a subsequent energy minimization [100] refines and improves the searching efficiency [101]. Since the LGA development in 1990s, it has suffered several improvements to optimize virtual screening speed and accuracy [102, 103]. GA algorithm is also used by the DOCK [3] program that is able to dock either the whole ligand inside the binding site or dock the ligand using fragmentation search methods.

Differential Evolutionary (DE) algorithms are based in a heuristic and population-based methodology derived from GA methods [104]. GEMDOCK [9], MolDock [11] and SADock [50] programs possess a DE algorithm.

Tabu Search (TS) algorithms [52] were developed by Glover and have been used to solve a large variety of hard optimization problems [105]. They correspond to a meta-heuristic method characterized by an iterative procedure that moves from one pose to another and imposes several restrictions to prevent revisiting previously considered poses. Earlier visited poses are stored in a "tabu list" and the root-mean-square deviation (RMSd) of a new conformation, in relation to the previous ones, is calculated and used as criterion to accept or reject the new conformation relatively to the previous ones. PRO_LEADS [52, 106] is the most popular docking program that uses a TS algorithm. The Tabu-Enhanced Genetic Algorithm is a search tool derived from the TS method that is implemented in the PSI-DOCK [51] program to explore in the first step the potential binding poses of a ligand, and the predicted binding poses are then optimized by a GA method.

Particle Swarm Optimization (PSO) is a heuristic and evolutionary optimization algorithm [107] that was inspired by social behaviour of organisms, such as the flocking of birds or fish school. It is simpler and converges faster than standard GA methods. In PSO methods, the population of ligand poses is called a "swarm" and the individual poses are called "particles", which are determined by three types of parameters: translation, orientation, and torsions. Each ligand pose moves within the search space and retains in its memory the best position (pose with the lowest energy) that has been encountered. Recently, a few PSO-based search algorithms, such as SODOCK [13] and PSO@AutoDock [108] have been implemented within the framework of the docking package AutoDock to improve the docking performance. The Fully Informed Particle Swarm (FIPS) [109] is a search algorithm derived from a variant of PSO that exploits a population of individuals to detect promising regions in the search space.

11.2.2.3 Molecular Simulation Algorithms

These methods are also applied in protein-ligand docking studies and there are two main types: (i) Molecular Dynamics simulations and (ii) Energy minimization.

Molecular dynamics (MD) simulations of the ligand are based on the calculation of the solutions to Newton's equations of motion. They are versatile and widely used in many computational studies [110], whereas the application of these approaches to search for ligand conformations show some limitations due to their difficulties in sampling the configurational space within a feasible simulation time, due to the lack of ergodicity. They can have problems in navigating a rugged hypersurface of a biological ligand and crossing high-energy rotational barriers. Strategies generally employed to overcome these limits are the use of high temperatures in some parts of the MD simulation, or starting from different ligand configurations [111]. The advantage that they have is that they can include explicit solvation, and explore essentially low-energy conformations (even though around the initial one).

Energy minimization methods are a complementary search tool to refine the ligand poses and not an actual search technique. They include gradient methods (e.g. steepest descend), conjugate-gradient methods (e.g. Fletcher-Reeves), second derivative methods (e.g. Newton-Raphson), and least-squares methods (e.g. Marquardt). Some of the docking algorithms previously described also use energy minimization methods, which are present in the Prodock [55], ICM [18], QXP [95], DARWIN [98], DOCK 4.0 [3], ADAM [86], and Hammerhead [87] docking programs. By definition, energy minimization will look to the relative minimum closest from the initial configuration. Therefore, they do not "search poses", they only refine poses obtained by other methods.

11.2.3 Flexible Protein Search Algorithms

The flexibility of the protein is one of the major challenges in the search for the correct pose, and consequently the number of degrees of freedom that are considered greatly influence the searching success [81]. Several protein-ligand docking studies have shown that the application of the flexible-ligand docking algorithms only give successful results when the protein is rather rigid and its 3D structure is representative of the protein conformation in the docked complex [112, 113]. Basically, this corresponds to the cases where the lock-and-key model of molecular recognition applies. However, many proteins display significant structural changes upon ligand binding, such as the local rearrangement of side chains or loops at and near the binding site, without affecting most of the protein backbone conformation or the protein overall folding. Others proteins display much more extensive rearrangements, including backbone movements, but these are less common (although not rare). In terms of molecular recognition this corresponds to the induced-fit paradigm. This is particularly important for enzymes because they could acquire different conformations to recognize their substrates and also for the transition states' stabilization along catalysis. Hence, these small movements could have an adverse effect on docking results and the flexibility of the protein cannot be neglected.

An example of this issue is shown in Fig. 11.2. In this case, the molecular docking software that treats the protein as a rigid body did not give the correct

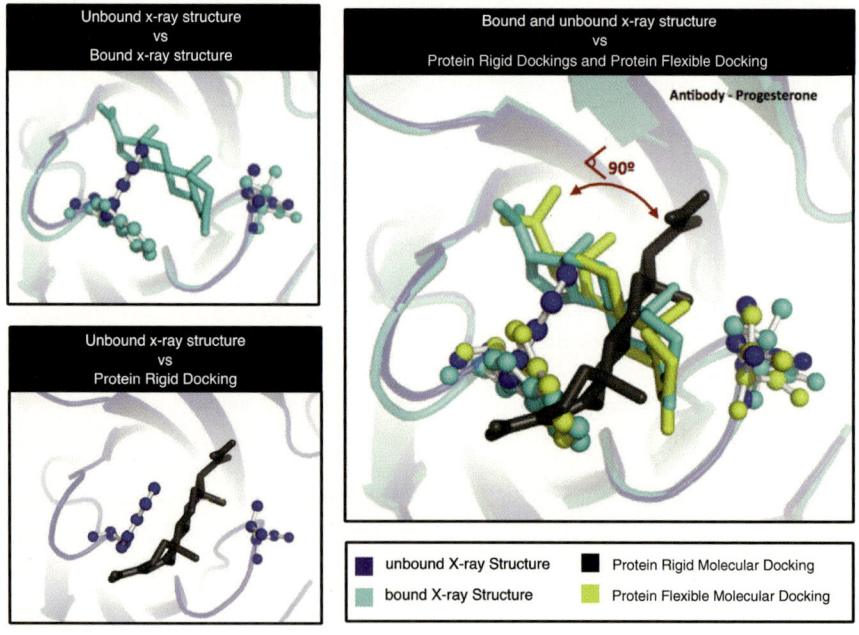

Fig. 11.2 Illustrative example of the importance of including protein flexibility in a molecular docking protocol [114]

binding pose of the ligand, as it is observed in the co-crystallized X-ray structure. The correct binding pose of the ligand was only obtained when flexibility was introduced in two residues of the binding site. It is has nothing to do with the performance of the ligand search algorithm or scoring function of the molecular docking software, but rather with the conformation of two residues in the protein binding site that preclude the correct binding of the ligand in the unbound X-ray structure. This illustrates how important the inclusion of protein flexibility in a protein-ligand docking program can be.

To solve this challenging docking problem, some specialized search algorithms and computational strategies were developed to accurately account for the partial flexibility of the protein, in addition to the ligand flexibility, and nowadays, several docking programs offer these treatments [56, 115–119]. Approaches addressing the protein flexibility can be classified as (i) MD and MC methods [4, 94, 120], (ii) simulated annealing, (iii) rotamer libraries [63, 121–125], (iv) protein ensemble grids [65, 83, 126], (v) soft-receptor modeling [112, 127–129], and (vi) collective degrees of freedom.

11.2.3.1 MD Simulations and MC Methods

MD simulations and MC methods were successfully applied in a wide range of protein-ligand studies that consider the flexibility of the protein [130]. These methods generate different configurations for the system, and their main advantage is that they are very accurate and can model explicitly all the degrees of freedom of the protein-ligand system and may also include the solvent if necessary. However, the high-dimensionality of the search space involved in these simulations tools, makes an ergodic exploration of the protein conformations unfeasible, due to the higher computational time required (several days of computation) [67]. With these methods, ergodicity cannot be attained even within the nanosecond time scale, which prevents the complete (possibly the relevant) sampling of the configurational space.

To reduce the computational cost and simplify the molecular description of the system, more realistic approaches that take into account only partial flexibility of the protein (e.g. binding site and surrounding residues) were developed [56].

11.2.3.2 Simulated Annealing

Simulated Annealing (SA) [60] methods carry every docking pose into a simulation with high temperature, which allows for transitions over energy barriers separating energetic valleys. Subsequently, the temperature is gradually decreased along regular intervals of time in each simulation cycle. SA methods consider flexibility in different thermodynamic states during an interval of time. However, the annealing cycle must be repeated many times, transforming this approach in a very computationally expensive alternative. To prevent trapping in local minima,

SA-based methods combine this procedure with other search algorithms, such as MC, GA and LGA [4], to explore a wider range of possible conformations with high accuracy. The popular docking software AutoDock uses a Monte Carlo Simulated Annealing protocol that specifically combines both SA and MC algorithms, in which random alterations in ligand pose were performed inside protein binding pocket during each SA temperature cycle.

11.2.3.3 Rotamer Library Based Methods

Rotamer library based methods are the ones most used and they represent the protein conformational space as a set of experimentally observed and preferred rotameric states for each residue side chain [55, 56, 58, 66]. The conformational changes in the protein binding site induced by ligand binding are considered by combinatorial rearrangement of side chains lining the binding site. Hence, the side-chains of the binding site residues containing H-bond donors or acceptors, and the side-chains whose original conformers may bump on the ligand pose, are rearranged to account for possible induced-fit in several steps. The main handicaps of this approach are: the requirement of previous knowledge in which side-chains should be flexible; the reduced number of key residues possible to consider (typically less than 10) in a feasible docking process; and the absence of any real change in the backbone of the protein [131].

The algorithm implemented in the Dolina [64] software uses a pool of low-energy ligand poses to obtain a valid pose, and then a combinatorial scan of energetically favourable side-chain rotamers in the ligand vicinity is performed, to ensure optimal interaction with the ligand pose and with other side-chains. Several residue rearrangements are grouped in sterically independent families and clusters of side-chain conformers are employed to achieve a good accuracy in the generated poses. The induced fit docking algorithm included in the SCARE (SCan Alanines and REfine) [63] protocol only requires one initial protein pocket conformation and identifies most of the correct ligand positions as the lowest score. It systematically scans pairs of neighbouring side chains, replaces them by alanine residues, and then docks the ligand to each 'gapped' pocket site. Subsequently, all docked positions are scored, refined with original side chains and flexible backbone, and finally re-scored.

11.2.3.4 Ensemble of Protein Conformations

An ensemble of protein conformations, obtained from X-ray crystallography, NMR or MD/MC simulations, can be used as another strategy to include protein flexibility [127, 131]. However, these search methods have two disadvantages: how the initial protein conformations are generated and how they are combined among themselves. One of the most popular ensemble docking methods is FlexE [8], an extension of the docking tool FlexX. The great difference between this search method and MD and MC techniques is the fact that this algorithm superimposes the set of conformations available for a given protein and merges similar parts of the structures. Dissimilar substructures are treated as independent alternatives and, then, the algorithm selects the combination of substructures that best complement a given ligand with respect to the scoring function.

11.2.3.5 Soft-Receptor Modelling Approach

The soft-receptor modelling approach combines the information derived from several different experimental and computational protein conformations to generate one energy weighted average grid, which is subsequently used to dock the ligands [112, 127]. This protein flexibility docking technique is the less computationally demanding approach, due to the use of a single energy-grid as a target for docking, which is almost equivalent to using a single structure, in terms of computational efficiency. However, it cannot manage large scale motion and another disadvantage is the fact that mutual exclusive binding regions can be simultaneously considered, leading to an enlargement of the binding pocket of the protein, which may wrongly influence the docking results. The soft docking models are also used to improve convergence during energy minimization and to avoid becoming trapped in local minima during the search.

11.2.3.6 Collective Degrees of Freedom

Collective Degrees of Freedom are global protein motions that result from a simultaneous change of all or part of the native degrees of freedom of the protein. This tool allows for the introduction of large-scale protein flexibility, including the backbone, loops and domains. Collective degrees of freedom can be determined using different methods, such as normal mode analysis. This approach takes into account protein flexibility by exploring the low frequency normal modes. The main advantage of this method is that protein flexibility is not limited to a specific small region of the protein. However, as the degrees of freedom searched are collective modes of motion that try to account for most of the variance observed during protein motion, this may result in an increased difficulty to get the "true solution". The soft modes approach has been applied for protein-ligand [132] and protein-protein docking problems [133], and examples of these approaches are IFREDA (ICM-flexible receptor docking algorithm implemented in the ICM program) [134], and elastic network normal modes [135]. The IFREDA algorithm considers both side-chain rearrangements and essential backbone movements and, even in some cases, large loop movements.

In general, the success for flexible protein-ligand searching passes through a combination of different algorithmic approaches, such as the computationally cheaper ensemble docking and the more demanding induced fit approach [136]. Furthermore, taken into account the fast evolution of searching algorithms, the approaches currently used to simulate partial protein flexibility will probably be replaced gradually by new tactics that allow for full protein movements. However, in general, the choice of a search algorithm to apply in a molecular docking study greatly depends on:

(i) the problem one wants to address: e.g. for small ligands with a small number of rotatable bonds we could use a systematic search algorithm; however, to dock a larger ligand or a larger number of ligands, a random algorithm such as GA should be used.

(ii) the biological background: for instance, if we know the size and location of the binding region, and the number of flexible residues present in the binding pocket of a protein, this could justify the use of the rotamer-libraries flexible-protein search algorithm, which is more systematic and covers better the conformational space of a set of residues at the expense of neglecting all the others.

(iii) the available computational power: e.g. rigid-body or flexible-ligand algorithms should be used when we have reduced computational resources (or many lists of compounds to dock), whilst the opposite situation could allow for a more demanding algorithm, such as flexible-protein and flexible-ligand docking algorithms.

11.3 Scoring Functions

The main goal of a scoring function is to calculate an energy that estimates the binding affinity between the protein and the ligand.

The binding affinity of a complex can be expressed by the binding free energy (ΔG_{Bind}) and can be estimated by Eq. 11.1 (see also Fig. 11.3).

$$\Delta G_{Bind} = G_{Complex} - G_{Protein} - G_{Ligand} \qquad (11.1)$$

where $G_{Complex}$ is the free energy of the complex formed by the protein and the ligand, $G_{Protein}$ is the free energy of the protein and G_{Ligand} is the free energy of the ligand. Generally, the binding affinity between a protein and a ligand is experimentally determined, given by the dissociation constant for the complex, e.g. the inhibition constant K_i, for most inhibitors. ΔG_{Bind} is related to the K_i by the following equation:

$$\Delta G_{Bind} = +RT \ln K_i \qquad (11.2)$$

with R as the perfect gas constant.

There are a wide variety of different methods capable of predicting computationally the binding free energy of a protein-ligand complex (Fig. 11.3), and generally they differ significantly in accuracy and speed. Very accurate binding free energies can be obtained for instance with free energy perturbation (FEP) or Thermodynamic Integration (TI) methods. However, these methods are very time-consuming and laborious and not many complexes can be analysed in a short period of time. To analyse the binding free energy of hundreds or thousands of

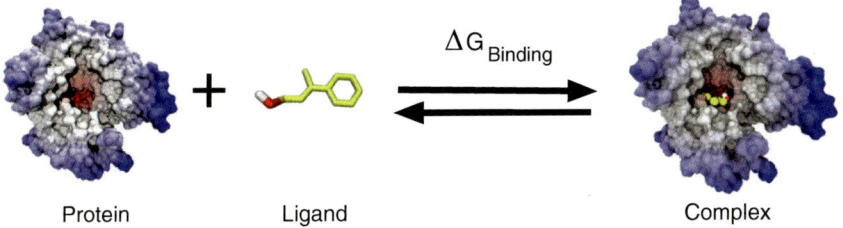

Fig. 11.3 Schematic representation of the free energy of binding ΔG_{Bind}

protein-ligand complexes as generated by virtual screening campaigns, then "scoring functions" are used instead.

A "scoring function" relies on several assumptions and simplifications to estimate protein-ligand binding. These methods are very fast since the complexity and computational cost required for the calculation of protein ligand binding is dramatically reduced [137]. However, the accuracy of the final results can be compromised, as a number of physical phenomena that determine molecular recognition are not included in the calculation or are modelled by predefined parameters that are obtained from experimental observations or quantum chemical calculations. The development of "scoring functions" is thus not an easy task, as it can have a major impact on the quality of molecular docking results [111].

Generally speaking, the accuracy of a scoring function can be evaluated taking into account its capability to follow the following criteria: (i) it must be capable of estimating the interaction between the receptor and the ligand and this value should be proportional to the free energy of binding; (ii) the poses of a given ligand must be ranked correctly, and the best-scored ones, if existent, should be close to what is observed experimentally; (iii) if multiple ligands are docked, it must be possible to discriminate between molecules that bind the target and molecules that do not, and the ones that bind should be ranked accurately; (iv) a scoring function must be sufficiently fast to be applied in a docking algorithm [138].

Currently, the number of scoring functions available for predicting protein-ligand interactions is large and increasing. Many algorithms share common methodologies, some with novel extensions, and the diversity in both their complexity and computational speed provides a plethora of techniques to tackle modern structure-based drug design problems. Roughly speaking, the scoring functions can be grouped into four main categories: force field scoring functions, empirical scoring functions, knowledge-based potentials, and consensus scoring. Each of these categories will be briefly described in the following sections.

11.3.1 Force Field Based Scoring Function

Force-field-based scoring functions have been used for more than 2 decades and apply classical molecular mechanics energy functions to compute the binding affinity between the protein and the ligand. These scoring functions are mainly based on the non-bonded terms of the molecular mechanics force field. This means, therefore, that they do not estimate the free energy of binding but rather the interaction energy between the protein and the ligand. In Eq. 11.3, it is represented a simple force-field scoring function (from the DOCK software [139]) with energy parameters taken from the AMBER force field. In this case, the binding energy is approximated to an interaction energy through a combination of non-bounded terms, i.e. the van der Waals and electrostatic energy terms. The van der Waals term is estimated using a Lennard-Jones dispersion/repulsion term, and the electrostatic term is computed by a Coulombic formulation with a distance-dependent dielectric function that reduces the contribution from charge–charge interactions [140, 141].

$$Score = \sum_i \sum_j \left(\frac{A_{ij}}{r_{ij}^{12}} - \frac{B_{ij}}{r_{ij}^6} + \frac{q_i q_j}{\varepsilon(r_{ij}) r_{ij}} \right) \tag{11.3}$$

where r_{ij} stands for the distance between protein atom i and ligand atom j, A_{ij} and B_{ij} are the Lenard-Jones parameters, and q_i and q_j are the atomic charges. Here, the effect of solvent is implicitly considered by introducing a simple distance-dependent dielectric constant $\varepsilon(r_{ij})$ in the Coulombic term.

Nowadays, the available force-field scoring functions have additional terms in their formulations, since the forces that govern the protein-ligand binding do not have an exclusively non-bonded nature. Indeed the intermolecular forces, such as ionic bonds and hydrogen bonds play a major role. Furthermore, when the ligand binds to a protein it changes its chemical conformation (three-dimensional shape) and therefore this effect must also be taken into account while estimating protein ligand interaction (energy of the ligand).

Some examples of force field based scoring functions are displayed in Table 11.2. All the scoring functions are almost identical among them, and only differ in a few terms, in the force field that is used to calculate their parameters and on how the interaction energy is calculated. For instance, the D-Score scoring functions is based on the TRIPOS force field [142], while DOCK [139] and AutoDock [143, 144] are both based on the Amber force field [145]. D-Score only calculates the interaction energy between the protein and the ligand, whereas AutoDock and GoldScore also take into account the energy of the ligand in order to approximate the calculation to the binding free energy. The terms used to calculate these energies are also different between the scoring functions. For example AutoDock has a hydrogen-bonding term (although with different functional forms) in an attempt to increase the potential of specific molecular recognition, while such term is not taken into account in the D-Score and Gold Score scoring function.

Despite the good results and the clear physical meaning of the force field scoring functions, they have several drawbacks. The force fields that are used in their core structure were primarily formulated to model enthalpy gas-phase contributions to structure and energetics, and do not include implicit solvation and entropic terms that are important to assess the interaction energy between the protein and the ligand. This is commonly corrected in the scoring functions through the inclusion of additional terms. For instance, the desolvation energies of the ligand and

Table 11.2 Examples of some force-field based scoring functions

Scoring function	Formula	
	Protein ligand interaction	Energy of the ligand
D-score [146]	$\Delta E_{int} = E_{vdW} + E_{electrostatic}$ $$= \sum_i^{prot} \sum_j^{lig} \left[\left(\frac{A_{ij}}{d_{ij}^{12}} - \frac{B_{ij}}{d_{ij}^6} \right) + 332.0 \frac{q_i q_j}{\varepsilon(d_{ij})d_{ij}} \right]$$	
AutoDock v2.4 [143]	$\Delta E_{int} = E_{vdW} + E_{H-bond} + E_{electrostatic}$ $$= \sum_i^{prot} \sum_j^{lig} \left[\left(\frac{A_{ij}}{d_{ij}^{12}} - \frac{B_{ij}}{d_{ij}^6} \right) + E(t) \times \left(\frac{C_{ij}}{d_{ij}^{12}} - \frac{D_{ij}}{d_{ij}^{10}} \right) \\ + 332.0 \frac{q_i q_j}{\varepsilon(d_{ij})d_{ij}} \right]$$	$\Delta E = E_{vdW} + E_{H-bond} + E_{electrostatic}$ $$= \sum_{i,j}^{lig} \left[\left(\frac{A_{ij}}{d_{ij}^{12}} - \frac{B_{ij}}{d_{ij}^6} \right) + E(t) \times \left(\frac{C_{ij}}{d_{ij}^{12}} - \frac{D_{ij}}{d_{ij}^{10}} \right) \\ + 332.0 \frac{q_i q_j}{4(d_{ij})d_{ij}} \right]$$
GoldScore [147]	$\Delta E_{int} = E_{vdW} + E_{electrostatic}$ $$= \sum_{prot} \sum_{lig} \left[\left(\frac{A_{ij}}{d_{ij}^8} - \frac{B_{ij}}{d_{ij}^4} \right) + 332.0 \frac{q_i q_j}{\varepsilon(d_{ij})d_{ij}} \right]$$	$\Delta E = E_{vdW} + E_{electrostatic}$ $$= \sum_{i,j}^{lig} \left[\left(\frac{A_{ij}}{d_{ij}^{12}} - \frac{B_{ij}}{d_{ij}^6} \right) + 332.0 \frac{q_i q_j}{\varepsilon(d_{ij})d_{ij}} \right] + (E_{HBond})_{optional}$$

of the protein are sometimes taken into account using implicit solvation methods such as GBSA or PBSA. Entropic corrections are included through a torsional entropy term that estimates the conformational entropy lost upon binding.

Force-field based scoring is further complicated by the fact that it generally requires the introduction of cut-off distances for the treatment of non-bonded interactions, which are more or less arbitrarily chosen and complicate the accurate treatment of long-range effects involved in the binding process.

11.3.2 Empirical Based Scoring Function

The functional form of empirical scoring functions is often simpler than force-field based scoring functions, although many of the individual contributing terms have counterparts in the force-field molecular mechanics terms. Generally, the empirical scoring functions decompose the overall binding free energy into several energetic terms as it is displayed in Eq. 11.4.

$$Score = \sum_i W_i \times \Delta G_i \tag{11.4}$$

where ΔG_i represents different energy terms such as vdW energy, electrostatic, hydrogen bond, desolvation, entropy, etc., that are calculated by a somewhat intuitive algorithm. The corresponding coefficients W_i are derived from a regression analysis on a set of protein–ligand complexes with known binding affinities. For this reason, empirical scoring functions are also referred to as regression-based methods.

Currently, several empirical scoring functions are available in diverse molecular docking programs, such as FlexX [6], F-Score [148], the Piecewise Linear Potential (PLP) [149], ChemScore [150, 151], Glide SP/XP [152], SCORE [153], Fresno [154] and X-SCORE [155]. Some examples are displayed in Table 11.3.

Several of these programs contain a modified form of the scoring function that was initially developed for the molecular docking software called LUDI. The general equation of the original implementation of this scoring function is displayed in Eq. 11.5 [156]:

$$\begin{aligned} \Delta G = {} & \Delta G_0 + \Delta G_{rot} NR \\ & + \Delta G_{hb} \sum_{h-bonds} f(\Delta R) f(\Delta \alpha) \\ & + \Delta G_{ion} \sum_{ionic} f(\Delta R) f(\Delta \alpha) + \Delta G_{lipo} A_{lipo} \end{aligned} \tag{11.5}$$

where ΔG coefficients are unknown and are determined by multi-linear regression in order to fit the experimental measured binding affinities.

The first terms are a fixed ground term and a term taking into account the loss of translational and rotational entropy during ligand binding by hindrance of rotatable bonds (ΔG_{rot} corresponds to the energy that is lost per rotatable bond and NR

Table 11.3 Formula of some empirical scoring functions

Name	Scoring function formula
F-score [148]	$\Delta G_{bind} = \Delta G_0 + \Delta G_{rotor} N_{rotor}$ $+ \Delta G_{H-bond} \sum_{H-bond} f(\Delta R, \Delta \alpha) + \Delta G_{ionic} \sum_{ionic} f(\Delta R, \Delta \alpha)$ $+ \Delta G_{aromatic} \sum_{aromatic} f(\Delta R, \Delta \alpha) + \Delta G_{lipo} \sum_{lipo} f^*(\Delta R)$ where the ΔG coefficients are unknown and will be obtained by multiple linear regression, f is a penalty function for deviations from ideal geometry for each kind of interaction, and f^* is a function penalizing for lipophilic interactions deviating from an ideal separation distance. This scoring function is very similar to the initial implementation of the LUDI scoring function (Eq. 11.5), but has two additional terms: one that takes into account the interaction between aromatic groups, and a second one that calculates the lipophilic interactions through a sum of pairwise atom-atom contacts
ChemScore [150]	$\Delta G_{bind} = \Delta G_0 + \Delta G_{H-bond} \sum_{il} g_1(\Delta R) g_2(\Delta \alpha) + \Delta G_{metal} \sum_{aM} f(r_{aM})$ $+ \Delta G_{lipo} \sum_{il} f(r_{IL}) + \Delta G_{rot} H_{rot}$ where, the ΔG coefficients are unknown and will be obtained by multiple linear regression. The hydrogen bond term, $\sum_{il} g_1(\Delta R) g_2(\Delta \alpha)$, is calculated for all complementary possibilities of hydrogen bonds between ligand atoms i, and protein atoms j. The metal term, $\sum_{aM} f(r_{aM})$, is calculated for all acceptor and acceptor/donor atoms, a, in the ligand and any metal atoms, M, in the protein. The lipophilic term, $\sum_{il} f(r_{IL})$, is calculated for all lipophilic ligand atoms i, and all lipophilic protein atoms l. The final term, H_{rot}, identifies frozen rotatable bonds
Glide XP [152]	$Glide_{XP} Score = E_{coul} + E_{vdW} + E_{bind} + E_{penalty}$ $E_{bind} = E_{hyd\_enclosure} + E_{hyd\_n\_motif} + E_{hb\_cc\_motif} + E_{PI} + E_{hb\_pair} + E_{probic\_pair}$ $E_{penalty} = E_{desolv} + E_{ligand\_strain}$ where $E_{hyd\_enclosure}$ assigns scores to lipophilic ligand atoms, $E_{hyd\_n\_motif}$ is an improved model of protein-ligand hydrogen bonding, $E_{hyd\_n\_motif}$ is a term that identifies neutral–neutral hydrogen-bond motifs that are found in many if not most pharmaceutical targets, $E_{hb\_cc\_motif}$ identifies special charged–charged hydrogen-bond motifs, E_{PI} is a term that accounts for pi stacking and pi-cation interactions, E_{hb_pair} and E_{probic_pair} are hydrogen bond and lipophilic pair terms, respectively. The E_{desolv} accounts for the desolvation effects and E_{ligand_strain} is a function that is used to penalize poses with close internal contacts
Fresno [154]	$\Delta G_{bind} = K + \alpha_{hbond} + \beta_{lipo} + \gamma_{rot} + \delta_{bp} + \varepsilon_{desolv}$ where constant K as well as regression coefficients $\alpha, \beta, \gamma, \delta, \varepsilon$ are unknown and will be optimized for each protein-ligand series by multiple linear regression. The H-bond term (HB) estimates the favorable contribution from hydrogen bonds between the ligand and the protein. The lipophilic term (LIPO) estimates the favorable contribution to binding given by the contacts of lipophilic atoms of the ligand with lipophilic atoms of the protein. The rotational term (ROT) estimates the loss of entropy due to the freezing of rotatable bonds of the ligand upon binding. The buried-polar term (BP) is used to describe the unfavorable interactions arising from the contact of polar atoms with lipophilic atoms between the ligand and the protein. The desolvation term (DESOLV) accounts for the desolvation effects during binding and it is obtained by solving the linear form of the Poisson-Boltzmann equation using a finite-difference method

represents the number of rotatable bonds that the ligand has). ΔG_{hb} and ΔG_{ion} give the binding energy for each optimal hydrogen bond and salt bridge, respectively. $f(\Delta R, \Delta \alpha)$ is a scaling function penalizing deviations from the ideal interaction geometry in terms of distance (ΔR) and angle $(\Delta \alpha)$. The ΔG_{lipo} term represents the contribution from lipophilic interactions. The lipophilic contribution is assumed to be proportional to the lipophilic contact surface, A_{lipo}, between the receptor and the fragment.

In spite of the similarities between the terms present in the empirical scoring functions, presented in Table 11.3, to that displayed in Eq. 11.5, we can notice that most of them have completely different implementations for the computation of each term. For example, in the early LUDI formulation, there are two independent terms that estimate the hydrogen-bond and ionic salt bridges interaction, whereas in ChemScore, the last term is absent. The LUDI function calculates hydrophobic contributions (ΔG_{lipo}) on the basis of the representation of a molecular surface area, whereas ChemScore evaluates contacts between hydrophobic atom pairs. In the case of F-Score, an additional term was added to account for aromatic an interaction that for instance was not present in the original implementation of the LUDI scoring function. The empirical scoring functions also include non-enthalpic contributions, such as the so-called rotor term, which approximates entropy penalties on binding from a weighted sum of the number of rotatable bonds in ligands. ChemScore implements ligand rotational entropy in a more complicated form than LUDI, which describes the molecular environment surrounding each rotatable bond. Fresno also differs from the LUDI algorithm mainly by the explicit treatment of ligand desolvation and of unfavorable protein–ligand contacts [154].

Currently, the empirical scoring functions have been applied with great success to predict the pose and scoring of several molecules. The main advantage of this method, when compared with the force-field based scoring functions, comes from the fast and easy computation of the terms that compose them. In addition, with the rapid increase in the number of protein-ligand complexes with known 3D structures and affinities, we are now closer of developing a relatively general scoring function.

Based on the success of the empirical scoring functions, several force-field scoring functions were changed in order to include some empirical parameters into their formulations. This is for instance the case of AutoDock (after version 4 [157]), in which various terms in the molecular mechanics energy function have been re-scaled by new coefficients (W), including the new term that estimates the desolvation free energy of the ligand (Eq 11.6).

$$+W_{elec} \sum_{i,j} \frac{q_i q_j}{e(r_{ij}) r_{ij}} + W_{sol}\left(S_i V_j + S_j V_i\right) e^{\left(\frac{-r_{ij}^2}{2\sigma^2}\right)} \tag{11.6}$$

The main drawback of the empirical scoring functions is their dependence on the experimental data set used in the parameterization process (not very versatile and transferable). This means that there is no guarantee whether these scoring functions are able to predict the binding affinity of ligands that are structurally very different from those that are used in the training set. In addition, the terms of different fitted scoring functions cannot be easily recombined into a new scoring function, since they are a result of different weighting factors.

11.3.3 Knowledge-Based Scoring Functions

Knowledge-based scoring functions are purely statistical methods, designed to reproduce experimental structures rather than to reproduce binding affinities (such as force field and empirical based scoring functions). These scoring functions use simple statistical potentials that estimate the frequency of occurrence or non-occurrence (i.e. negative data) of different atom–atom pair contacts and other typical interactions that are obtained from the structural information embedded in experimentally determined atomic structures. In this process, it is assumed that if an interatomic distance occurs more often than some average value, it should represent a favourable contact, and vice versa. Additionally, the observed distribution of distances between pairs of different atom types must reflect their interaction energies [158].

The typical formula of a knowledge-based scoring function is displayed in Eq. 11.7.

$$Score = \sum_{i,j}^{N} u_{ij}(r) \tag{11.7}$$

where i and j stand for a protein atom type and a ligand atom type, respectively, r is the atom pair distance, N is the number of all possible atom pairs in the system, and u_{ij} corresponds to the pairwise potentials between atom types i and j. This value is directly obtained from the occurrence frequency of atom pairs in a pre-defined database, using the inverse formulation of the Boltzmann law [159] according to Eq. 11.8.

$$u(r) = -k_B T \ln \left(\frac{\rho(r)}{\rho^*(r)} \right) \tag{11.8}$$

where k_B is the Boltzmann constant, and T is the absolute temperature of the system, $\rho(r)$ is the number density of the protein–ligand atom pair at distance r, and $\rho^*(r)$ is the atom pair density in a "reference" state where interatomic interactions are zero.

The various knowledge-based scoring functions differ between each other in the sets of protein-ligand complexes used to obtain these potentials, the form of the energy function, the definition of protein and ligand atom types, distance cut-off, and several additional parameters. Muegges's Potential of Mean Force (PMF) [160–162], DrugScore [163, 164] and SMall Molecule Growth (SMoG) [165] are the most popular examples of knowledge-based scoring functions, whose formulas can be found in Table 11.4. Other knowledge-based scoring functions are also available such as BLEEP [166, 167] and M-score [168].

Table 11.4 Selected knowledge-based scoring functions [158]

Name	Scoring function formula
PMF [161]	$$PMF = \sum_{kl} A_{ij}(r) = -k_B T \ln \left[f_{Vol\_corr}^{j}(r) \frac{\rho_{seg}^{ij}(r)}{\rho_{bulk}^{ij}} \right]$$ where k_B is the Boltzmann constant, $f_{Vol\_corr}^{j}(r)$ is a ligand volume correction factor and $\frac{\rho_{seg}^{ij}(r)}{\rho_{bulk}^{ij}}$ indicates a radial distribution function for a protein atom i and a ligand atom j
Drug Score [163]	$$\Delta W = \gamma \sum_{i} \sum_{j} \Delta W_{ij}(r) + (1-\gamma) \left[\sum_{i} \Delta W_{i}(SAS, SAS_0) + \sum_{j} \Delta W_{j}(SAS, SAS_0) \right]$$ where SAS corresponds to the surface accessible area terms, W_{ij} is a distance dependent pairwise potential and γ is an adjustable weight factor
SMoG [165]	$$F = \sum_{p} \sum_{l} - \ln \left(\frac{P_{(\sigma_p, \sigma_l)}}{p^{ref}} \right) \Delta(p, l)$$ where p denotes a protein atom of type σ_p, l, denotes a ligand atom of type σ_l, $\Delta(p,l)$ is the characteristic function of the contact (1 if atoms p and l are in contact and 0 otherwise), $P_{(\sigma_p, \sigma_l)}$ denotes the measure of frequency of the contacts between atom types σ_p and σ_l in the training database and p^{ref} is the probability of those contacts in the hypothetical reference state
AutoDock Vina [169]	$$\Delta G_{bind} = \Delta G_{gauss} + \Delta G_{repulsion} + \Delta G_{hbond} + \Delta G_{hydrophobic} + \Delta G_{tors}$$ where ΔG_{gauss} is an attractive term for dispersion, $\Delta G_{repulsion}$ is a function that is used to penalize close internal contacts, ΔG_{hbond} is a function that estimates favourable contribution from the hydrogen bonds between the ligand and the protein, $\Delta G_{hydrophobic}$ is a Ramp function that treats hydrophobic contributions and ΔG_{tors} is a term proportional to the number of rotable bonds

Compared to force field and empirical scoring functions, the knowledge-based scoring functions offer a good balance between accuracy and speed. The major attraction of many knowledge-based scoring functions is their computational simplicity. The deduction of the potentials required for the scoring function only requires the knowledge of a set of protein–ligand complex structures, and such knowledge is relatively rich and still increasing due to the contributions from structural biologists [170]. Due to the pairwise characteristic, the knowledge-based scoring functions are also very fast, similarly to what happens with the empirical scoring functions. Other advantages of the knowledge-based methods are their ability to capture implicitly binding effects that are difficult to model explicitly, such as sulphur-aromatic or cation-π interactions, which are generally badly handled explicitly. Furthermore, these scoring functions can be easily applied to systems that were not used for the setup of the scoring function, which turn them attractive to apply for very large compound databases.

The major disadvantage of these methods is that their parameterization is limited by the sets of protein–ligand complex structures that are known, which can lead to sub-optimal differentiation of atom types.

11.3.4 Consensus Scoring Functions

Consensus scoring functions combine the information obtained from different scores to improve the probability of finding the correct solution [171].

The main idea behind the consensus scoring is that the scoring functions perform very well for the purpose of pose prediction, but cannot predict the binding affinity of protein ligand complexes in statistically rigorous terms since the functional forms used to describe the chemistry and physics of ligand binding are incomplete. This is particularly important in virtual screening campaigns where these functions often fail in the comparison of the binding affinity of different ligands and thus fail to distinguish between inactive and active compounds. This often causes the presence of many false positives among the top scored solutions of a single ranking list, a factor that can compromise the efficiency of a virtual screening campaign.

Despite its short existence, the consensus scores have now become a common method. Several consensus-scoring functions are already available, such as X-score [155], GFscore [172], DS LigandScore by Accelrys, and Model-Composer, MOE by Chemical Computing Group, among others. In the majority of cases, they are employed as a post-processing step after docking runs. Many studies have suggested that employing consensus-scoring approaches can improve the performance by compensating for the deficiencies of the scoring functions [173–175]. In addition they tend to reduce the number of false positives that are identified by individual scoring functions, leading to a significant enhancement in hit-rates [176, 177].

11.3.5 Typical Problems in Molecular Docking

Over the past decades docking has become increasingly widespread in drug design and development efforts. In spite of the progress that has accompanied this evolution, several features in docking still remain important challenges for the users, often limiting its range of applicability. Here we highlight four important challenges in docking: (1) validation of the docking protocol; (2) covalent docking; (3) presence of structural water molecules; (4) treatment of entropy.

11.3.5.1 Validation of Molecular Docking Protocol

The validation of the docking protocol is an important requirement for successful docking. A variety of features are normally involved in a standard protein-ligand docking campaign. Many of these features are dependent on choices that are made by the user. Examples include the specific search algorithm and scoring function to be used, but also a variety of parameters, including the box size and position, grid size, number of structures, number of energy evaluations, specific protonation state of an active site residue, choice of flexible residues, etc. Such choices can vary significantly from target to target, and often with the specific type of ligands. Even the best combination of a scoring function and search algorithm can fail dramatically with a poor choice of docking conditions. While the user's experience can ease significantly this process of selection, only a careful validation of the docking protocol can ensure a reasonable level of accuracy.

Strategies to validate the docking protocol can vary significantly, in terms of the level of detail and sophistication. Ideally, independent experimental validation would be the preferred choice. In practice, however, other alternatives have to be used. The most basic approach consists in starting from a crystallographic structure of a protein-ligand complex and trying to re-dock the ligand to its correct position in the original experimental PDB structure. The different parameters are tested and optimized, until a good pose prediction is achieved, measured in terms of RMSD.

Logically, as the amino acid residues along the binding pocket in the target protein are already prearranged for that specific ligand and for its pose, re-docking the ligand provides only a rough assessment of the quality of the docking protocol in reproducing the pose of that specific ligand in that specific structure target. It corresponds to a lock-and-key recognition in a protein that eventually may use the induced-fit recognition model.

A better and more general validation approach would involve the use of an independent PDB structure of the target to dock the ligand, typically from another complexed ligand or substrate, or even from a structure of the free target. The final pose is then compared with the real pose of the ligand in its native PDB structure. In this case, docking becomes much more challenging, but also approaches significantly more the conditions of a real docking campaign. Logically, if this validation process can be applied to several different experimental ligands, the conclusions become more general and the ability in correctly docking novel ligands with success improves significantly. In addition, this type of analysis provides a glimpse on how the different side chains at the binding pocket are oriented for different ligands in different structures, allowing an understanding of the model of protein recognition and flexibility for that specific protein target.

11.3.5.2 Covalent Docking

Most docking methods focus exclusively on non-covalent interactions, including van der Waals, electrostatic interactions and hydrogen bonding. The main stream of rational drug design relies on non-covalent interactions as the mechanism of the functionality of the drugs [178, 179], which in itself justifies the emphasis precisely on non-covalent interactions. However, some specific drugs, called covalent drugs, employ as part of their binding mechanism the formation of a covalent interaction with the target.

The existence of such bond(s) confers to the covalent drugs a higher affinity towards their targets [180, 181]. For this reason, covalent drugs are often able to exhibit a very strong potency allied with relatively small molecular size, features that make them very attractive pharmaceutically [182]. In fact, 3 of the 10 top-selling drugs in U.S. in 2009 were reported to be covalent drugs [180].

In spite of the popularity of covalent drugs, their discovery so far has relied almost exclusively in serendipity [180]. Covalent docking could enable the large-scale application of the huge potential that has been traditionally associated with conventional docking, also to this class of molecules. Most docking programs do not allow the treatment of covalent ligands. Some mainstream docking alternatives, like Autodock, GOLD and FlexX, already offer a "covalent docking" option that can be used. However, they are in general of limited applicability [178]. The search for improved alternatives in this area continues.

Recently Oyuang et al. [178] have reported a new docking package specifically designed to allow reliable covalent docking. The program, termed CovalentDock includes an automatic procedure that recognizes and prepares all covalently bondable chemical groups, together with a specific energy term that integrates the energy contribution from the covalent interaction with the scoring function. The program is compatible with common scoring functions used in docking, and has been shown to significantly improve pose prediction of covalent ligands [178]. A web server for automated covalent docking using this approach was recently implemented [183].

However, it is important to take into account that covalent docking has still fundamental limitations that ideally would require an explicit treatment of the electronic structure for being properly addressed. Research into the development of new algorithms incorporating covalent docking (including new potentials for metal-ligand binding) continues.

11.3.5.3 Structural Waters

Currently, the importance of solvation in the binding ability of drugs is well known [184, 185]. Many scoring functions used in protein-ligand docking already include, at least partially, solvation. However, more than solvation, it is the presence of structural water molecules that remains a hard challenge in present day docking. In fact, when analysing ligand-binding pockets in protein crystallographic structures, a common feature is the presence of interacting water molecules or in close vicinity of the ligand. An analysis of a representative set of 392 high-resolution protein-ligand complexes from the Protein Data Bank revealed an average of 4.6 ligand-bound water molecules per structure, 76 % of which interacting simultaneously with both the ligand and the protein [186]. These water molecules provide indirect interactions between the protein and the ligand through the formation of hydrogen bonds with both partners [187]. Their presence and precise number and positioning can affect significantly both the ligand binding affinity and range of most favoured conformations, aspects that can be essential for accurate protein-ligand docking [188–193].

Such issues cannot be handled simply by a scoring function and require a more explicit inclusion of water molecules. The strategies to adopt from this point onwards differ, depending on whether there is or not some a priori knowledge of the presence of water molecules in the binding pocket [194].

For example, if the docking study targets a protein for which there are several good resolution X-ray structures complexed with different ligands showing the presence of a conserved water molecule at the binding pocket, then the atomistic inclusion of that specific water molecule in the docking process (as part of the target) would be a natural alternative to consider. One way to test the potential importance of that conserved water molecule on docking, would be to test its inclusion in the validation stage with the ligands for which structural information is available. If the tests without the presence of the water molecule fail to predict the correct pose, while the ones that include it provide an accurate prediction of the ligand position, than its presence is shown to be essential. Naturally, this strategy only works if the ligands to be docked are structurally similar or have a common scaffold with those for which a complexed X-ray structure is already available. Furthermore, the reader should be aware that most conserved water molecules, even the tightly bound ones, change at least slightly their position or orientation when varying the ligand, and that such difference might prevent the identification of the correct pose.

Often, however, this problem takes a much more complicated form: the molecule to be docked is a novel entity, very different from those complexed in available X-ray structures, or no target-ligand X-ray structure exists at all. The number of possible relevant water molecules for ligand binding and their position is not known and can vary from ligand to ligand.

If a reasonable guess on the preferred hydration sites can be performed from the available non-complexed structures of the target, it becomes necessary to be able to anticipate which water molecules are more likely to be displaced to allow ligand binding. Methods like WaterScore [195], HINT [196], or Consolv [197] can be used to distinguish between water molecules that should be included in the docking process and those that should be replaced to make room for the ligand, helping to prepare initial structures for docking. In addition, some docking programs have also included approaches that change the position of water molecules (enabling also its addition or removal) during the docking process, normally by employing an energy penalty [190].

In cases where no information is available, more sophisticated approaches need to be adopted. An example is the "Just Add Water Molecules" (JAWS) approach [198]. This method employs a double-decoupling scheme that compares the energetic cost associated to the appearance and disappearance of water molecules on a binding-site grid. The JAWS methodology has been shown to work particularly well for water molecules well-buried in cavities, in which the grid is isolated from the bulk water [198]. Other alternatives include AQUARIUS [199], CS-Map [200], MCSS [201], SuperStar [202] and GRID [203].

Giving the difficulty of the task, alternative techniques such as molecular dynamics simulations can often play an important role in identifying relevant water molecules at the binding pocket and in refining their specific positions and arrangement. Two strategies involving molecular dynamics are normally employed: (1) Pre-Docking MD, in which the molecular dynamics simulations are performed on X-ray structures of relevant protein-ligand complexes properly solvated (typically in a box of waters under periodic boundary conditions), and the residence times of individual water molecules, in the binding pocket, are analysed to identify persistent contacts that could be of importance for docking new ligands; (2) Post-Docking MD, in which the protein-ligand docking complex resulting from docking is subjected to MD simulations in water, and the reorientation of individual water molecules at the binding site, and the movement of new water molecules from the bulk solvent into the binding pocket is taken into account to refine the structure of the complex.

11.3.5.4 Entropy

Entropy can make an important difference on the evaluation of the protein-ligand affinity [204–207]. However, it is often neglected or the subject of drastic simplifications in most computational methodologies that handle protein-ligand complexes [205, 206, 208], including even accurate free energy calculations [209, 210]. The reasons for this choice are not difficult to understand. In fact, the determination of the entropic contribution of the ligand and protein can be computationally very demanding, requiring particularly well minimized structures for a normal mode analysis, or large numbers of conformations for a quasi-harmonic analysis [211–213]. In problems such as protein-ligand docking, for which speed is often a critical issue, such approaches are not normally feasible.

The entropy contribution to the binding free energy emerges essentially from the reduction of the translational and rotational degrees of freedom in the ligand, and from changes in the normal modes of the protein and most of all of the ligand, during binding [205, 206, 214–217]. Incorporating such effects in an efficient scoring function is presently an important challenge in the field of protein ligand docking. While some attempts to incorporate entropy in scoring functions have been reported in the literature (particularly in knowledge-based scoring functions) [207], most approaches involve re-scoring schemes [218–220].

One way to estimate the translational, rotational and torsional entropy in docking, was developed by Ruvinsky and Kozintsev [220]. In their approach, multiple docking experiments are performed and the results are clustered by similarity. A measure of the size of each cluster is then used to estimate the entropic contribution, assuming that large clusters of conformations are indicative of favourable entropic contributions of the local energy landscapes. Naturally, the method assumes that the search algorithm can provide a reasonable exploration of the conformational space associated. Tests with different scoring functions (Autodock, D-Score, LigScore, PLP, LUDI, F-Score, ChemScore, X-Score, PMF, DrugScore, etc.) have shown some small improvement in the docking accuracy [219, 220].

Another approach was suggested by Lee and Seok [218] that proposed the introduction of a probability function to analyse the populations of different binding modes in the context of statistical mechanics. Such approach allows an estimate of the contribution of the state represented by a sampled conformation of the configurational integral, applying the notion of colony energy, proposed by Xiang et al. [221]. Its application in combination with several common scoring functions has resulted in improved accuracy [218]. Furthermore, its low computational cost enables its combination with other pre-existing scoring functions.

In spite of these developments, properly handling entropy in docking remains challenging. In addition to the entropic contributions of the ligand and protein, other key components still require our attention. For example, the entropic contribution of the solvent molecules, particularly those at the binding pocket is normally forgotten. The very limited movement of buried waters in the protein-ligand complex and the huge variation arising from dessolvation contribute to making this an important term.

11.3.6 Future Developments and Perspectives

A widely spread concept is that the major weakness of today's docking programs lies not on the sampling methods but on the scoring functions, particularly in those cases in which the protein rearrangement has been shown to be limited to a small and predictable number of side-chains. As a matter of fact, considerable efforts continue to be devoted to the development of computational methods for describing protein–ligand interactions.

In spite of the large number of protein-ligand docking alternatives, we are still far from a perfect docking algorithm. In terms of search algorithms, efficiently accounting for protein flexibility remains a challenging task. In terms of scoring functions, consensus scoring is emerging as the best alternative, compensating for the particular deficiencies of the individual scoring functions.

Features such as the presence of structural water molecules and the treatment of entropy, still pose considerable problems for protein-ligand docking. Covalent docking remains a challenging issue.

New developments continue to be reported every year. This fact, together with the increasing number of programs available and the different way in how they deal with the diverse challenges posed by protein-ligand docking, demonstrate the richness of the field and show that the future of docking is promising.

Acknowledgments This work has been funded by FEDER/COMPETE and Fundação para a Ciência e a Tecnologia through projects EXCL/QEQ-COM/0394/2012, EXPL/QEQ-COM/ 1125/2013 and PEst-C/EQB/LA0006/2011.

References

1. Sousa SF, Cerqueira NM, Fernandes PA, Ramos MJ (2010) Virtual screening in drug design and development. Comb Chem High Throughput Screening 13:442–453
2. Cerqueira NM, Sousa SF, Fernandes PA, Ramos MJ (2009) Virtual screening of compound libraries. Methods Mol Biol 572:57–70
3. Ewing TJA, Makino S, Skillman AG, Kuntz ID (2001) DOCK 4.0: search strategies for automated molecular docking of flexible molecule databases. J Comput Aided Mol Des 15:411–428
4. Morris GM, Goodsell DS, Halliday RS, Huey R, Hart WE, Belew RK, Olson AJ (1998) Automated docking using a Lamarckian genetic algorithm and an empirical binding free energy function. J Comput Chem 19:1639–1662
5. Goodsell DS, Morris GM, Olson AJ (1996) Automated docking of flexible ligands: applications of AutoDock. J Mol Recognit 9:1–5
6. Rarey M, Kramer B, Lengauer T, Klebe G (1996) A fast flexible docking method using an incremental construction algorithm. J Mol Biol 261:470–489
7. Kramer B, Rarey M, Lengauer T (1999) Evaluation of the FLEXX incremental construction algorithm for protein-ligand docking. Proteins-Struct Funct Genet 37:228–241
8. Claussen H, Buning C, Rarey M, Lengauer T (2001) FlexE: efficient molecular docking considering protein structure variations. J Mol Biol 308:377–395
9. Yang JM, Chen CC (2004) GEMDOCK: a generic evolutionary method for molecular docking. Proteins-Struct Funct Bioinform 55:288–304
10. Chang DTH, Oyang YJ, Lin JH (2005) MEDock: a web server for efficient prediction of ligand binding sites based on a novel optimization algorithm. Nucleic Acids Res 33:W233–W238
11. Thomsen R, Christensen MH (2006) MolDock: a new technique for high-accuracy molecular docking. J Med Chem 49:3315–3321
12. Chen K, Li TH, Cao TC (2006) Tribe-PSO: a novel global optimization algorithm and its application in molecular docking. Chemometr Intell Lab Syst 82:248–259
13. Chen H-M, Liu B-F, Huang H-L, Hwang S-F, Ho S-Y (2007) SODOCK: swarm optimization for highly flexible protein-ligand docking. J Comput Chem 28:612–623
14. Jain AN (2003) Surflex: fully automatic flexible molecular docking using a molecular similarity-based search engine. J Med Chem 46:499–511
15. Jones G, Willett P, Glen RC, Leach AR, Taylor R (1997) Development and validation of a genetic algorithm for flexible docking. J Mol Biol 267:727–748
16. Verdonk ML, Cole JC, Hartshorn MJ, Murray CW, Taylor RD (2003) Improved protein-ligand docking using GOLD. Proteins-Struct Funct Genet 52:609–623
17. Jones G, Willett P, Glen RC (1995) Molecular recognition of receptor-sites using a genetic algorithm with a description of desolvation. J Mol Biol 245:43–53
18. Abagyan R, Totrov M, Kuznetsov D (1994) ICM—a new method for protein modeling and design—applications to docking and structure prediction from the distorted native conformation. J Comput Chem 15:488–506
19. Totrov M, Abagyan R (1997) Flexible protein-ligand docking by global energy optimization in internal coordinates. Proteins Suppl 1:215–220
20. Friesner RA, Banks JL, Murphy RB, Halgren TA, Klicic JJ, Mainz DT, Repasky MP, Knoll EH, Shelley M, Perry JK, Shaw DE, Francis P, Shenkin PS (2004) Glide: a new approach for rapid, accurate docking and scoring. 1. docking accuracy. J Med Chem 47:1739–1749
21. Wu GS, Robertson DH, Brooks CL, Vieth M (2003) Detailed analysis of grid-based molecular docking: a case study of CDOCKER—a CHARMm-based MD docking algorithm. J Comput Chem 24:1549–1562
22. Venkatachalam CM, Jiang X, Oldfield T, Waldman M (2003) LigandFit: a novel method for the shape-directed rapid docking of ligands to protein active sites. J Mol Graph Model 21:289–307

23. Liu M, Wang SM (1999) MCDOCK: a Monte Carlo simulation approach to the molecular docking problem. J Comput Aided Mol Des 13:435–451
24. Li L, Chen R, Weng ZP (2003) RDOCK: refinement of rigid-body protein docking predictions. Proteins-Struct Funct Genet 53:693–707
25. Chen R, Li L, Weng ZP (2003) ZDOCK: an initial-stage protein-docking algorithm. Proteins-Struct Funct Genet 52:80–87
26. Pierce B, Tong WW, Weng ZP (2005) M-ZDOCK: a grid-based approach for C-n symmetric multimer docking. Bioinformatics 21:1472–1478
27. Sauton N, Lagorce D, Villoutreix BO, Miteva MA (2008) MS-DOCK: accurate multiple conformation generator and rigid docking protocol for multi-step virtual ligand screening, BMC Bioinform 9(1):184
28. Hocker HJ, Cho K-J, Chen C-YK, Rambahal N, Sagineedu SR, Shaari K, Stanslas J, Hancock JF, Gorfe AA (2013) Andrographolide derivatives inhibit guanine nucleotide exchange and abrogate oncogenic Ras function. Proc Natl Acad Sci USA 110:10201–10206
29. Grant BJ, Lukman S, Hocker HJ, Sayyah J, Brown JH, McCammon JA, Gorfe AA (2011) Novel allosteric sites on Ras for lead generation. Plos One 6(10):e25711
30. Kuntz ID, Blaney JM, Oatley SJ, Langridge R, Ferrin TE (1982) A geometric approach to macromolecule-ligand interactions. J Mol Biol 161:269–288
31. Morris GM, Goodsell DS, Huey R, Olson AJ (1996) Distributed automated docking of flexible ligands to proteins: parallel applications of AutoDock 2.4. J Comput Aided Mol Des 10:293–304
32. Sousa SF, Fernandes PA, Ramos MJ (2006) Protein-ligand docking: current status and future challenges. Proteins-Struct Funct Bioinform 65:15–26
33. Moitessier N, Englebienne P, Lee D, Lawandi J, Corbeil CR (2008) Towards the development of universal, fast and highly accurate docking/scoring methods: a long way to go. Br J Pharmacol 153:S7–S26
34. Yi H, Qiu S, Cao Z, Wu Y, Li W (2008) Molecular basis of inhibitory peptide maurotoxin recognizing Kv1.2 channel explored by ZDOCK and molecular dynamic simulations. Proteins-Struct Funct Bioinform 70:844–854
35. Imberty A, Hardman KD, Carver JP, Perez S (1991) Molecular modelling of protein-carbohydrate interactions. Docking of monosaccharides in the binding site of concanavalin A. Glycobiology 1:631–642
36. Leach AR, Kuntz ID (1992) Conformational-analysis of flexible ligands in macromolecular receptor-sites. J Comput Chem 13:730–748
37. Harkcom WT, Bevan DR (2007) Molecular docking of inhibitors into monoamine oxidase B. Biochem Biophys Res Commun 360:401–406
38. Rajamani R, Good AC (2007) Ranking poses in structure-based lead discovery and optimization: current trends in scoring function development. Curr Opin Drug Discov Devel 10:308–315
39. Lorber DM, Shoichet BK (2005) Hierarchical docking of databases of multiple ligand conformations. Curr Top Med Chem 5:739–749
40. Farag NA, Mohamed SR, Soliman GAH (2008) Design, synthesis, and docking studies of novel benzopyrone derivatives as H(1)-antihistaminic agents. Bioorg Med Chem 16:9009–9017
41. Sato H, Shewchuk LM, Tang J (2006) Prediction of multiple binding modes of the CDK2 inhibitors, anilinopyrazoles, using the automated docking programs GOLD, FlexX, and LigandFit: an evaluation of performance. J Chem Inf Model 46:2552–2562
42. Bras NF, Cerqueira NMFSA, Fernandes PA, Ramos MJ (2008) Carbohydrate-binding modules from family 11: understanding the binding mode of polysaccharides. Int J Quantum Chem 108:2030–2040
43. Bras NF, Fernandes PA, Ramos MJ (2009) Docking and molecular dynamics studies on the stereoselectivity in the enzymatic synthesis of carbohydrates. Theoret Chem Acc 122:283–296

44. Bras NF, Goncalves R, Fernandes PA, Mateus N, Ramos MJ, de Freitas V (2010) Understanding the binding of procyanidins to pancreatic elastase by experimental and computational methods. Biochemistry 49:5097–5108
45. Bras NF, Goncalves R, Mateus N, Fernandes PA, Ramos MJ, Do Freitas V (2010) Inhibition of pancreatic elastase by polyphenolic compounds. J Agric Food Chem 58:10668–10676
46. Moorthy NSHN, Bras NF, Ramos MJ, Fernandes PA (2012) Virtual screening and QSAR study of some pyrrolidine derivatives as alpha-mannosidase inhibitors for binding feature analysis. Bioorg Med Chem 20:6945–6959
47. Francisco CS, Rodrigues LR, Cerqueira NM, Oliveira-Campos AM, Rodrigues LM, Esteves AP (2013) Synthesis of novel psoralen analogues and their in vitro antitumor activity. Bioorg Med Chem 21:5047–5053
48. Gupta S, Rodrigues LM, Esteves AP, Oliveira-Campos AM, Nascimento MS, Nazareth N, Cidade H, Neves MP, Fernandes E, Pinto M, Cerqueira NM, Bras N (2008) Synthesis of N-aryl-5-amino-4-cyanopyrazole derivatives as potent xanthine oxidase inhibitors. Eur J Med Chem 43:771–780
49. Francisco CS, Rodrigues LR, Cerqueira NM, Oliveira-Campos AM, Rodrigues LM (2012) Synthesis of novel benzofurocoumarin analogues and their anti-proliferative effect on human cancer cell lines. Eur J Med Chem 47:370–376
50. Chung HW, Cho SJ, Lee K-R, Lee K-H, IOP (2013) Self-adaptive differential evolution algorithm incorporating local search for protein-ligand docking. In: Ic-Msquare 2012: international conference on mathematical modelling in physical sciences
51. Pei JF, Wang Q, Liu ZM, Li QL, Yang K, Lai LH (2006) PSI-DOCK: towards highly efficient and accurate flexible ligand docking. Proteins-Struct Funct Bioinform 62:934–946
52. Baxter CA, Murray CW, Clark DE, Westhead DR, Eldridge MD (1998) Flexible docking using Tabu search and an empirical estimate of binding affinity. Proteins-Struct Funct Genet 33:367–382
53. Liu Y, Zhao L, Li W, Zhao D, Song M, Yang Y (2013) FIPSDock: a new molecular docking technique driven by fully informed swarm optimization algorithm. J Comput Chem 34:67–75
54. Mustard D, Ritchie DW (2005) Docking essential dynamics eigenstructures. Proteins-Struct Funct Bioinform 60:269–274
55. Trosset JY, Scheraga HA (1999) PRODOCK: software package for protein modeling and docking. J Comput Chem 20:412–427
56. Cerqueira NMFSA, Bras NF, Fernandes PA, Ramos MJ (2009) MADAMM: a multistaged docking with an automated molecular modeling protocol. Proteins-Struct Funct Bioinform 74:192–206
57. Leone V, Marinelli F, Carloni P, Parrinello M (2010) Targeting biomolecular flexibility with metadynamics. Curr Opin Struct Biol 20:148–154
58. Pak YS, Wang SM (2000) Application of a molecular dynamics simulation method with a generalized effective potential to the flexible molecular docking problems. J Phys Chem B 104:354–359
59. Caflisch A, Fischer S, Karplus M (1997) Docking by Monte Carlo minimization with a solvation correction: application to an FKBP-substrate complex. J Comput Chem 18:723–743
60. Goodsell DS, Olson AJ (1990) Automated docking of substrates to proteins by simulated annealing. Proteins-Struct Funct Genet 8:195–202
61. Hartmann C, Antes I, Lengauer T (2009) Docking and scoring with alternative side-chain conformations. Proteins-Struct Funct Bioinform 74:712–726
62. Taylor RD, Jewsbury PJ, Essex JW (2003) FDS: Flexible ligand and receptor docking with a continuum solvent model and soft-core energy function. J Comput Chem 24:1637–1656
63. Bottegoni G, Kufareva I, Totrov M, Abagyan R (2008) A new method for ligand docking to flexible receptors by dual alanine scanning and refinement (SCARE). J Comput Aided Mol Des 22:311–325
64. Smiesko M (2013) DOLINA—docking based on a local induced-fit algorithm: application toward small-molecule binding to nuclear receptors. J Chem Inf Model 53:1415–1423

65. Totrov M, Abagyan R (2008) Flexible ligand docking to multiple receptor conformations: a practical alternative. Curr Opin Struct Biol 18:178–184

66. Schnecke V, Swanson CA, Getzoff ED, Tainer JA, Kuhn LA (1998) Screening a peptidyl database for potential ligands to proteins with side-chain flexibility. Proteins-Struct Funct Genet 33:74–87

67. Apostolakis J, Pluckthun A, Caflisch A (1998) Docking small ligands in flexible binding sites. J Comput Chem 19:21–37

68. Ritchie DW (2008) Recent progress and future directions in protein-protein docking. Curr Protein Pept Sci 9:1–15

69. Hashmi I, Shehu A (2013) HopDock: a probabilistic search algorithm for decoy sampling in protein-protein docking. Proteome Sci 11(Suppl 1):S6

70. Gabb HA, Jackson RM, Sternberg MJE (1997) Modelling protein docking using shape complementarity, electrostatics and biochemical information. J Mol Biol 272:106–120

71. Katchalskikatzir E, Shariv I, Eisenstein M, Friesem AA, Aflalo C, Vakser IA (1992) Molecular-surface recognition—determination of geometric fit between proteins and their ligands by correlation techniques. Proc Natl Acad Sci USA 89:2195–2199

72. Kozakov D, Brenke R, Comeau SR, Vajda S (2006) PIPER: an FFT-based protein docking program with pairwise potentials. Proteins-Struct Funct Bioinform 65:392–406

73. Wolfson HJ, Rigoutsos I (1997) Geometric hashing: an overview. IEEE Comput Sci Eng 4:10–21

74. Schneidman-Duhovny D, Inbar Y, Nussinov R, Wolfson HJ (2005) Geometry-based flexible and symmetric protein docking. Proteins-Struct Funct Bioinform 60:224–231

75. Sternberg MJE, Gabb HA, Jackson RM (1998) Predictive docking of protein-protein and protein-DNA complexes. Curr Opin Struct Biol 8:250–256

76. Smith GR, Sternberg MJE (2002) Prediction of protein-protein interactions by docking methods. Curr Opin Struct Biol 12:28–35

77. Fischer D, Lin SL, Wolfson HL, Nussinov R (1995) A geometry-based suite of molecular docking processes. J Mol Biol 248:459–477

78. Pang YP, Kozikowski AP (1994) Prediction of the binding-site of 1-benzyl-4- (5, 6-dimethoxy-1-indanon-2-yl)methyl piperidine in acetylcholinesterase by docking studies with the SYSDOC program. J Comput Aided Mol Des 8:683–693

79. Perola E, Xu K, Kollmeyer TM, Kaufmann SH, Prendergast FG, Pang YP (2000) Successful virtual screening of a chemical database for farnesyltransferase inhibitor leads. J Med Chem 43:401–408

80. Teague SJ (2003) Implications of protein flexibility for drug discovery. Nat Rev Drug Discov 2:527–541

81. Muegge I, Rarey M (2001) Small molecule docking and scoring. Rev Comput Chem 17(17):1–60

82. Amaro RE, Baron R, McCammon JA (2008) An improved relaxed complex scheme for receptor flexibility in computer-aided drug design. J Comput Aided Mol Des 22:693–705

83. Bolstad ESD, Anderson AC (2009) In pursuit of virtual lead optimization: pruning ensembles of receptor structures for increased efficiency and accuracy during docking. Proteins-Struct Funct Bioinform 75:62–74

84. Cavasotto CN, Singh N (2008) Docking and high throughput docking: successes and the challenge of protein flexibility. Curr Comput Aided Drug Des 4:221–234

85. Bohm HJ (1992) The computer-program LUDI—a new method for the denovo design of enzyme-inhibitors. J Comput Aided Mol Des 6:61–78

86. Mizutani MY, Tomioka N, Itai A (1994) Rational automatic search method for stable docking models of protein and ligand. J Mol Biol 243:310–326

87. Welch W, Ruppert J, Jain AN (1996) Hammerhead: fast, fully automated docking of flexible ligands to protein binding sites. Chem Biol 3:449–462

88. Jain AN (2007) Surflex-Dock 2.1: robust performance from ligand energetic modeling, ring flexibility, and knowledge-based search. J Comput Aided Mol Des 21:281–306

89. Zsoldos Z, Reid D, Simon A, Sadjad SB, Johnson AP (2007) eHiTS: a new fast, exhaustive flexible ligand docking system. J Mol Graph Model 26:198–212

90. Miller MD, Kearsley SK, Underwood DJ, Sheridan RP (1994) FLOG—a system to select quasi-flexible ligands complementary to a receptor of known 3-dimensional structure. J Comput Aided Mol Des 8:153–174
91. More JJ, Wu ZJ (1999) Distance geometry optimization for protein structures. J Global Optim 15:219–234
92. Thomsen R, IEEE (2003) Flexible ligand docking using differential evolution. IEEE
93. Taylor RD, Jewsbury PJ, Essex JW (2002) A review of protein-small molecule docking methods. J Comput Aided Mol Des 16:151–166
94. Hart TN, Read RJ (1992) A multiple-start monte-carlo docking method. Proteins-Struct Funct Genet 13:206–222
95. McMartin C, Bohacek RS (1997) QXP: powerful, rapid computer algorithms for structure-based drug design. J Comput Aided Mol Des 11:333–344
96. Holland HJ (1975) Adaptation in natural and artificial systems: an introductory analysis with applications to biology, control, and artificial intelligence. U Michigan Press
97. Clark KP, Ajay J (1995) Flexible ligand docking without parameter adjustment across 4 ligand-receptor complexes, J Comput Chem 16:1210–1226
98. Taylor JS, Burnett RM (2000) DARWIN: a program for docking flexible molecules. Proteins-Struct Funct Genet 41:173–191
99. Morris GM, Huey R, Lindstrom W, Sanner MF, Belew RK, Goodsell DS, Olson AJ (2009) AutoDock4 and AutoDockTools4: automated docking with selective receptor flexibility. J Comput Chem 30:2785–2791
100. Solis FJ, Wets RJB (1981) Minimization by random search techniques. Math Oper Res 6:19–30
101. Brooijmans N, Kuntz ID (2003) Molecular recognition and docking algorithms. Annu Rev Biophys Biomol Struct 32:335–373
102. Jiang X, Kumar K, Hu X, Wallqvist A, Reifman J (2008) DOVIS 2.0: an efficient and easy to use parallel virtual screening tool based on AutoDock 4.0. Chem Cent J 2(1):1–7
103. Zhang S, Kumar K, Jiang X, Wallqvist A, Reifman J (2008) DOVIS: an implementation for high-throughput virtual screening using AutoDock. BMC Bioinform 9(1):126
104. Storn R, Price K (1997) Differential evolution—a simple and efficient heuristic for global optimization over continuous spaces. J Global Optim 11:341–359
105. Glover F (1986) Future paths for integer programming and links to artificial-intelligence. Comput Oper Res 13:533–549
106. Murray CW, Baxter CA, Frenkel AD (1999) The sensitivity of the results of molecular docking to induced fit effects: application to thrombin, thermolysin and neuraminidase. J Comput Aided Mol Des 13:547–562
107. Eberhart RC, Shi YH (2004) Special issue on particle swarm optimization. IEEE Trans Evol Comput 8:201–203
108. Namasivayam V, Guenther R (2007) PSO@AUTODOCK: a fast flexible molecular docking program based on swarm intelligence. Chem Biol Drug Des 70:475–484
109. Mendes R, Kennedy J, Neves J (2004) The fully informed particle swarm: simpler, maybe better. IEEE Trans Evol Comput 8:204–210
110. Norberg J, Nilsson L (2003) Advances in biomolecular simulations: methodology and recent applications. Q Rev Biophys 36:257–306
111. Kitchen DB, Decornez H, Furr JR, Bajorath J (2004) Docking and scoring in virtual screening for drug discovery: methods and applications. Nat Rev Drug Discov 3:935–949
112. Osterberg F, Morris GM, Sanner MF, Olson AJ, Goodsell DS (2002) Automated docking to multiple target structures: incorporation of protein mobility and structural water heterogeneity in AutoDock. Proteins-Struct Funct Genet 46:34–40
113. Erickson JA, Jalaie M, Robertson DH, Lewis RA, Vieth M (2004) Lessons in molecular recognition: the effects of ligand and protein flexibility on molecular docking accuracy. J Med Chem 47:45–55
114. Cerqueira NM, Bras NF, Fernandes PA, Ramos MJ (2009) MADAMM: a multistaged docking with an automated molecular modeling protocol. Proteins-Struct Funct Bioinform 74:192–206

115. Biesiada J, Porollo A, Velayutham P, Kouril M, Meller J (2011) Survey of public domain software for docking simulations and virtual screening. Hum Genomics 5:497–505
116. B-Rao C, Subramanian J, Sharma SD (2009) Managing protein flexibility in docking and its applications. Drug Discov Today 14:394–400
117. Lill MA (2011) Efficient incorporation of protein flexibility and dynamics into molecular docking simulations. Biochemistry 50:6157–6169
118. Sinko W, Lindert S, McCammon JA (2013) Accounting for receptor flexibility and enhanced sampling methods in computer-aided drug design. Chem Biol Drug Des 81:41–49
119. Durrant JD, McCammon JA (2010) Computer-aided drug-discovery techniques that account for receptor flexibility. Curr Opin Pharmacol 10:770–774
120. Oshiro CM, Kuntz ID, Dixon JS (1995) Flexible ligand docking using a genetic algorithm. J Comput Aided Mol Des 9:113–130
121. Desmet J, Demaeyer M, Hazes B, Lasters I (1992) The dead-end elimination theorem and its use in protein side-chain positioning. Nature 356:539–542
122. Leach AR (1994) Ligand docking to proteins with discrete side-chain flexibility. J Mol Biol 235:345–356
123. Koska J, Spassov VZ, Maynard AJ, Yan L, Austin N, Flook PK, Venkatachalam CM (2008) Fully automated molecular mechanics based induced fit protein-ligand docking method. J Chem Inf Model 48:1965–1973
124. Meiler J, Baker D (2006) ROSETTALIGAND: protein-small molecule docking with full side-chain flexibility. Proteins-Struct Funct Bioinform 65:538–548
125. Kokh DB, Wenzel W (2008) Flexible side chain models improve enrichment rates in in silico screening. J Med Chem 51:5919–5931
126. Korb O, Olsson TSG, Bowden SJ, Hall RJ, Verdonk ML, Liebeschuetz JW, Cole JC (2012) Potential and limitations of ensemble docking. J Chem Inf Model 52:1262–1274
127. Knegtel RMA, Kuntz ID, Oshiro CM (1997) Molecular docking to ensembles of protein structures. J Mol Biol 266:424–440
128. Jiang F, Kim SH (1991) Soft docking—matching of molecular-surface cubes. J Mol Biol 219:79–102
129. Ferrari AM, Wei BQQ, Costantino L, Shoichet BK (2004) Soft docking and multiple receptor conformations in virtual screening. J Med Chem 47:5076–5084
130. Bras NF, Fernandes PA, Ramos MJ (2014) Molecular dynamics studies on both bound and unbound renin protease. J Biomol Struct Dyn 32:351–363
131. Carlson HA (2002) Protein flexibility is an important component of structure-based drug discovery. Curr Pharm Des 8:1571–1578
132. May A, Zacharias M (2008) Protein-ligand docking accounting for receptor side chain and global flexibility in normal modes: evaluation on kinase inhibitor cross docking. J Med Chem 51:3499–3506
133. May A, Zacharias M (2008) Energy minimization in low-frequency normal modes to efficiently allow for global flexibility during systematic protein-protein docking. Proteins-Struct Funct Bioinform 70:794–809
134. Cavasotto CN, Abagyan RA (2004) Protein flexibility in ligand docking and virtual screening to protein kinases. J Mol Biol 337:209–225
135. Rueda M, Bottegoni G, Abagyan R (2009) Consistent improvement of cross-docking results using binding site ensembles generated with elastic network normal modes. J Chem Inf Model 49:716–725
136. Cavasotto CN, Kovacs JA, Abagyan RA (2005) Representing receptor flexibility in ligand docking through relevant normal modes. J Am Chem Soc 127:9632–9640
137. Yuriev E, Agostino M, Ramsland PA (2011) Challenges and advances in computational docking: 2009 in review. J Mol Recognit 24:149–164
138. Yuriev E, Ramsland PA (2013) Latest developments in molecular docking: 2010–2011 in review. J Mol Recognit 26:215–239
139. Makino S, Kuntz ID (1997) Automated flexible ligand docking method and its application for database search. J Comput Chem 18:1812–1825

140. Kitchen DB, Decornez H, Furr JR, Bajorath J (2004) Docking and scoring in virtual screening for drug discovery: methods and applications. Nat Rev Drug Discov 3:935–949

141. Bissantz C, Folkers G, Rognan D (2000) Protein-based virtual screening of chemical databases. 1. Evaluation of different docking/scoring combinations. J Med Chem 43:4759–4767

142. Kramer B, Rarey M, Lengauer T (1999) Evaluation of the FLEXX incremental construction algorithm for protein-ligand docking. Proteins-Struct Funct Bioinform 37:228–241

143. Morris GM, Goodsell DS, Huey R, Olson AJ (1996) Distributed automated docking of flexible ligands to proteins: parallel applications of AutoDock 2.4. J Comput Aided Mol Des 10:293–304

144. Morris GM, Huey R, Olson AJ (2008) Using AutoDock for ligand-receptor docking. Current protocols in bioinformatics/editoral board, Andreas D. Baxevanis ... [et al.] (Chapter 8, Unit 8 14)

145. Weiner SJ, Kollman PA, Nguyen DT, Case DA (1986) An all atom force-field for simulations of proteins and nucleic-acids. J Comput Chem 7:230–252

146. Meng EC, Shoichet BK, Kuntz ID (1992) Automated docking with grid-based energy evaluation. J Comput Chem 13:505–524

147. Jones G, Willett P, Glen RC, Leach AR, Taylor R (1997) Development and validation of a genetic algorithm for flexible docking. J Mol Biol 267:727–748

148. Rarey M, Kramer B, Lengauer T, Klebe G (1996) A fast flexible docking method using an incremental construction algorithm. J Mol Biol 261:470–489

149. Gehlhaar DK, Verkhivker GM, Rejto PA, Sherman CJ, Fogel DB, Fogel LJ, Freer ST (1995) Molecular recognition of the inhibitor Ag-1343 by Hiv-1 protease—conformationally flexible docking by evolutionary programming. Chem Biol 2:317–324

150. Eldridge MD, Murray CW, Auton TR, Paolini GV, Mee RP (1997) Empirical scoring functions: I. The development of a fast empirical scoring function to estimate the binding affinity of ligands in receptor complexes. J Comput Aided Mol Des 11:425–445

151. Murray CW, Auton TR, Eldridge MD (1998) Empirical scoring functions. II. The testing of an empirical scoring function for the prediction of ligand-receptor binding affinities and the use of Bayesian regression to improve the quality of the model. J Comput Aided Mol Des 12:503–519

152. Friesner RA, Murphy RB, Repasky MP, Frye LL, Greenwood JR, Halgren TA, Sanschagrin PC, Mainz DT (2006) Extra precision glide: docking and scoring incorporating a model of hydrophobic enclosure for protein-ligand complexes. J Med Chem 49:6177–6196

153. Wang RX, Liu L, Lai LH, Tang YQ (1998) SCORE: a new empirical method for estimating the binding affinity of a protein-ligand complex. J Mol Model 4:379–394

154. Rognan D, Lauemoller SL, Holm A, Buus S, Tschinke V (1999) Predicting binding affinities of protein ligands from three-dimensional models: application to peptide binding to class I major histocompatibility proteins. J Med Chem 42:4650–4658

155. Wang RX, Lai LH, Wang SM (2002) Further development and validation of empirical scoring functions for structure-based binding affinity prediction. J Comput Aid Mol Des 16:11–26

156. Bohm HJ (1994) The development of a simple empirical scoring function to estimate the binding constant for a protein-ligand complex of known three-dimensional structure. J Comput Aided Mol Des 8:243–256

157. Huey R, Morris GM, Olson AJ, Goodsell DS (2007) A semiempirical free energy force field with charge-based desolvation. J Comput Chem 28:1145–1152

158. Vaque M, Ardrevol A, Blade C, Salvado MJ, Blay M, Fernandez-Larrea J, Arola L, Pujadas G (2008) Protein-ligand docking: a review of recent advances and future perspectives. Curr Pharm Anal 4:1–19

159. Huang SY, Zou X (2006) An iterative knowledge-based scoring function to predict protein-ligand interactions: I. Derivation of interaction potentials. J Comput Chem 27:1866–1875

160. Muegge I (2006) PMF scoring revisited. J Med Chem 49:5895–5902

161. Muegge I, Martin YC (1999) A general and fast scoring function for protein-ligand interactions: a simplified potential approach. J Med Chem 42:791–804

162. Muegge I (2001) Effect of ligand volume correction on PMF scoring. J Comput Chem 22:418–425
163. Gohlke H, Hendlich M, Klebe G (2000) Knowledge-based scoring function to predict protein-ligand interactions. J Mol Biol 295:337–356
164. Velec HFG, Gohlke H, Klebe G (2005) DrugScore(CSD)-knowledge-based scoring function derived from small molecule crystal data with superior recognition rate of near-native ligand poses and better affinity prediction. J Med Chem 48:6296–6303
165. Ishchenko AV, Shakhnovich EI (2002) SMall molecule growth 2001 (SMoG2001): an improved knowledge-based scoring function for protein-ligand interactions. J Med Chem 45:2770–2780
166. Mitchell JBO, Laskowski RA, Alex A, Forster MJ, Thornton JM (1999) BLEEP—potential of mean force describing protein-ligand interactions: II. Calculation of binding energies and comparison with experimental data. J Comput Chem 20:1177–1185
167. Mitchell JBO, Laskowski RA, Alex A, Thornton JM (1999) BLEEP—potential of mean force describing protein-ligand interactions: I. Generating potential. J Comput Chem 20:1165–1176
168. Yang CY, Wang RX, Wang SM (2006) M-score: a knowledge-based potential scoring function accounting for protein atom mobility. J Med Chem 49:5903–5911
169. Trott O, Olson AJ (2010) AutoDock Vina: improving the speed and accuracy of docking with a new scoring function, efficient optimization, and multithreading. J Comput Chem 31:455–461
170. Coupez B, Lewis RA (2006) Docking and scoring–theoretically easy, practically impossible? Curr Med Chem 13:2995–3003
171. Baber JC, William AS, Gao YH, Feher M (2006) The use of consensus scoring in ligand-based virtual screening. J Chem Inf Model 46:277–288
172. Betzi S, Suhre K, Chetrit B, Guerlesquin F, Morelli X (2006) GFscore: a general nonlinear consensus scoring function for high-throughput docking. J Chem Inf Model 46:1704–1712
173. Wang RX, Lu YP, Wang SM (2003) Comparative evaluation of 11 scoring functions for molecular docking. J Med Chem 46:2287–2303
174. Charifson PS, Corkery JJ, Murcko MA, Walters WP (1999) Consensus scoring: a method for obtaining improved hit rates from docking databases of three-dimensional structures into proteins. J Med Chem 42:5100–5109
175. Bissantz C, Folkers G, Rognan D (2000) Protein-based virtual screening of chemical databases. 1. Evaluation of different docking/scoring combinations. J Med Chem 43:4759–4767
176. Halperin I, Ma BY, Wolfson H, Nussinov R (2002) Principles of docking: an overview of search algorithms and a guide to scoring functions. Proteins-Struct Funct Genet 47:409–443
177. Feher M (2006) Consensus scoring for protein-ligand interactions. Drug Discov Today 11:421–428
178. Ouyang X, Zhou S, Su CTT, Ge Z, Li R, Kwoh CK (2013) CovalentDock: automated covalent docking with parameterized covalent linkage energy estimation and molecular geometry constraints. J Comput Chem 34:326–336
179. Schroeder J, Klinger A, Oellien F, Marhoefer RJ, Duszenko M, Selzer PM (2013) Docking-based virtual screening of covalently binding ligands: an orthogonal lead discovery approach. J Med Chem 56:1478–1490
180. Singh J, Petter RC, Baillie TA, Whitty A (2011) The resurgence of covalent drugs. Nat Rev Drug Discov 10:307–317
181. Potashman MH, Duggan ME (2009) Covalent modifiers: an orthogonal approach to drug design. J Med Chem 52:1231–1246
182. Smith AJT, Zhang X, Leach AG, Houk KN (2009) Beyond picomolar affinities: quantitative aspects of noncovalent and covalent binding of drugs to proteins. J Med Chem 52:225–233
183. Ouyang XC, Zhou S, Ge ZM, Li RT, Kwoh CK (2013) CovalentDock Cloud: a web server for automated covalent docking. Nucleic Acids Res 41:W329–W332
184. Yuriev E, Agostino M, Ramsland PA (2011) Challenges and advances in computational docking: 2009 in review. J Mol Recognit 24:149–164

185. Lipinski CA, Lombardo F, Dominy BW, Feeney PJ (1997) Experimental and computational approaches to estimate solubility and permeability in drug discovery and development settings. Adv Drug Deliv Rev 23:3–25
186. Lu Y, Wang R, Yang C-Y, Wang S (2007) Analysis of ligand-bound water molecules in high-resolution crystal structures of protein-ligand complexes. J Chem Inf Model 47:668–675
187. Lemmon G, Meiler J (2013) Towards ligand docking including explicit interface water molecules, Plos One 8(6):e67536
188. Roberts BC, Mancera RL (2008) Ligand-protein docking with water molecules. J Chem Inf Model 48:397–408
189. Huang N, Shoichet BK (2008) Exploiting ordered waters in molecular docking. J Med Chem 51:4862–4865
190. Wong SE, Lightstone FC (2011) Accounting for water molecules in drug design. Expert Opin Drug Discov 6:65–74
191. de Graaf C, Pospisil P, Pos W, Folkers G, Vermeulen NPE (2005) Binding mode prediction of cytochrome P450 and thymidine kinase protein-ligand complexes by consideration of water and rescoring in automated docking. J Med Chem 48:2308–2318
192. Thilagavathi R, Mancera RL (2010) Ligand-protein cross-docking with water molecules. J Chem Inf Model 50:415–421
193. Elokely KM, Doerksen RJ (2013) Docking challenge: protein sampling and molecular docking performance. J Chem Inf Model 53:1934–1945
194. Sousa SF, Ribeiro AJM, Coimbra JTS, Neves RPP, Martins SA, Moorthy NSHN, Fernandes PA, Ramos MJ (2013) Protein-ligand docking in the new millennium—a retrospective of 10 years in the field. Curr Med Chem 20:2296–2314
195. Garcia-Sosa AT, Mancera RL, Dean PM (2003) WaterScore: a novel method for distinguishing between bound and displaceable water molecules in the crystal structure of the binding site of protein-ligand complexes. J Mol Model 9:172–182
196. Amadasi A, Spyrakis F, Cozzini P, Abraham DJ, Kellogg GE, Mozzarelli A (2006) Mapping the energetics of water-protein and water-ligand interactions with the "natural" HINT forcefield: predictive tools for characterizing the roles of water in biomolecules. J Mol Biol 358:289–309
197. Raymer ML, Sanschagrin PC, Punch WF, Venkataraman S, Goodman ED, Kuhn LA (1997) Predicting conserved water-mediated and polar ligand interactions in proteins using a K-nearest-neighbors genetic algorithm. J Mol Biol 265:445–464
198. Michel J, Tirado-Rives J, Jorgensen WL (2009) Prediction of the water content in protein binding sites. J Phys Chem B 113:13337–13346
199. Pitt WR, Goodfellow JM (1991) Modeling of solvent positions around polar groups in proteins. Protein Eng 4:531–537
200. Kortvelyesi T, Dennis S, Silberstein M, Brown L, Vajda S (2003) Algorithms for computational solvent mapping of proteins. Proteins-Struct Funct Genet 51:340–351
201. Miranker A, Karplus M (1991) Functionality maps of binding-sites—a multiple copy simultaneous search method. Proteins-Struct Funct Genet 11:29–34
202. Verdonk ML, Cole JC, Taylor R (1999) SuperStar: a knowledge-based approach for identifying interaction sites in proteins. J Mol Biol 289:1093–1108
203. Goodford PJ (1985) A computational-procedure for determining energetically favorable binding-sites on biologically important macromolecules. J Med Chem 28:849–857
204. Salaniwal S, Manas ES, Alvarez JC, Unwalla RJ (2007) Critical evaluation of methods to incorporate entropy loss upon binding in high-throughput docking. Proteins-Struct Funct Bioinform 66:422–435
205. Murray CW, Verdonk ML (2002) The consequences of translational and rotational entropy lost by small molecules on binding to proteins. J Comput Aided Mol Des 16:741–753
206. Finkelstein AV, Janin J (1989) The price of lost freedom—entropy of bimolecular complex-formation. Protein Eng 3:1–3
207. Huang S-Y, Zou X (2010) Inclusion of solvation and entropy in the knowledge-based scoring function for protein-ligand interactions. J Chem Inf Model 50:262–273

208. Kongsted J, Ryde U (2009) An improved method to predict the entropy term with the MM/PBSA approach. J Comput Aided Mol Des 23:63–71
209. Bradshaw RT, Patel BH, Tate EW, Leatherbarrow RJ, Gould IR (2011) Comparing experimental and computational alanine scanning techniques for probing a prototypical protein-protein interaction. Protein Eng Des Sel 24:197–207
210. Kollman PA, Massova I, Reyes C, Kuhn B, Huo SH, Chong L, Lee M, Lee T, Duan Y, Wang W, Donini O, Cieplak P, Srinivasan J, Case DA, Cheatham TE (2000) Calculating structures and free energies of complex molecules: combining molecular mechanics and continuum models. Acc Chem Res 33:889–897
211. Baron R, Huenenberger PH, McCammon JA (2009) Absolute single-molecule entropies from quasi-harmonic analysis of microsecond molecular dynamics: correction terms and convergence properties. J Chem Theory Comput 5:3150–3160
212. Srinivasan J, Cheatham TE, Cieplak P, Kollman PA, Case DA (1998) Continuum solvent studies of the stability of DNA, DNA helices. J Am Chem Soc 120:9401–9409
213. Gohlke H, Case DA (2004) Converging free energy estimates: MM-PB(GB)SA studies on the protein-protein complex Ras-Raf. J Comput Chem 25:238–250
214. van der Vegt NFA, van Gunsteren WF (2004) Entropic contributions in cosolvent binding to hydrophobic solutes in water. J Phys Chem B 108:1056–1064
215. Tidor B, Karplus M (1994) The contribution of vibrational entropy to molecular association—the dimerization of insulin. J Mol Biol 238:405–414
216. Hermans J, Wang L (1997) Inclusion of loss of translational and rotational freedom in theoretical estimates of free energies of binding. Application to a complex of benzene and mutant T4 lysozyme. J Am Chem Soc 119:2707–2714
217. Amzel LM (1997) Loss of translational entropy in binding, folding, and catalysis. Proteins-Struct Funct Genet 28:144–149
218. Lee J, Seok C (2008) A statistical rescoring scheme for protein-ligand docking: consideration of entropic effect. Proteins-Struct Funct Bioinform 70:1074–1083
219. Ruvinsky AM (2007) Role of binding entropy in the refinement of protein-ligand docking predictions: analysis based on the use of 11 scoring functions. J Comput Chem 28:1364–1372
220. Ruvinsky AM, Kozintsev AV (2005) New and fast statistical-thermodynamic method for computation of protein-ligand binding entropy substantially improves docking accuracy. J Comput Chem 26:1089–1095
221. Xiang ZX, Soto CS, Honig B (2002) Evaluating conformational free energies: the colony energy and its application to the problem of loop prediction. Proc Natl Acad Sci USA 99:7432–7437

Chapter 12
ADMET Prediction Based on Protein Structures

Ákos Tarcsay and György M. Keserű

12.1 Introduction to ADMET

The most desired objectives of the medicinal chemistry programs are achieving high efficiency with desirable safety profile. In order to reach these goals, candidate molecules have to form optimal interactions with the primary target or targets and should avoid unwanted interactions with antitargets. Interactions with off targets results in undesirable toxicological events. Absorption, distribution, metabolism and excretion (ADME) and pharmacokinetics have a significant impact on both efficacy and safety. The pharmacokinetic behaviour and the toxicology together are usually abbreviated as ADMET.

The majority of the drugs are administered orally, therefore first the drug dissolves in the gastro-intestinal tract, is absorbed through the gut wall and then passes the liver to get into the blood circulation. During distribution, the active pharmacological ingredient passes biological barriers and reaches various compartments, tissues and organs in the body. Central nervous system drugs reach the brain by passing the blood-brain barrier. The distribution enables the drug to bind its molecular target, for example enzymes, receptors or ion channels to exert its pharmacodynamic effect. The compounds are then recirculated into the liver and metabolized in order to increase its polarity. Finally, they are excreted by the renal tract via urine, or in some specific cases via faeces by enterohepatic circulation.

Á. Tarcsay
Discovery Chemistry, Gedeon Richter Plc., 19-21 Gyömrői út, H-1103, Budapest, Hungary
e-mail: akos.tarcsay@gmail.com

G.M. Keserű (✉)
Research Centre for Natural Sciences of the Hungarian Academy of Sciences, Magyar tudósok körútja 2, Budapest 1117, Hungary
e-mail: gy.keseru@ttk.mta.hu

© Springer International Publishing Switzerland 2014
G. Náray-Szabó (ed.), *Protein Modelling*, DOI 10.1007/978-3-319-09976-7_12

Table 12.1 Proteins with distinguished importance to ADMET

Protein	Type	Adverse event	ADMET impact	Structure (example PDB IDs)
Serum albumin	Transport protein	Plasma binding, low free drug level Drug–drug interaction	Distribution	1GNI
P-glycoprotein	Efflux transporter	Efflux transport	Absorption, distribution	3G60
Cytochrome P450 1A2	Metabolic enzymes	Metabolism Inhibition of metabolism Food effect Drug–drug interactions	Metabolism	3TBG, 3QM4, 3QM4, 3TDA
Cytochrome P450 2C9				1OG2, 1OG5, 1R9O
Cytochrome P450 2D6				1TQN, 2J0D, 2V0 M, 3NXU, 3TJS, 3UA1
Cytochrome P450 3A4				2HI4
Sulfotransferases (SULT)	Metabolic enzymes	Metabolism	Metabolism	2D06
Pregnan X receptor (PXR)	Nuclear receptor	CYP, P-glycoprotein induction	Adsorption, distribution metabolism	1NRL
Constitutive androstane receptor (CAR)	Nuclear receptor	CYP, P-glycoprotein induction	Adsorption, distribution metabolism.	1ILG, 1ILH, 1M13, 1NRL, 1SKX, 2O9I, 2QNV, 3R8D
hERG	Potassium ion channel	Cardiovascular side-effects	Toxicity	Homology model
5-HT$_{2B}$ receptor	G protein-coupled receptor	Valvular heart disease	Toxicity	4IB4
α1 adrenergic receptor	G protein-coupled receptor	Vasoconstriction of arteries	Toxicity	Homology model (GPCR templates)

Breakthroughs during the last decade in protein engineering and crystallography resulted in atomic level structural information of the majority of proteins with distinguished relevance in ADMET processes such as plasma protein binding, active transport, cytochrome P450 (CYP) mediated metabolism, its inhibition and induction as well as some toxicity related targets such as the hERG (human Ether-a-go-go Related Gene product) potassium channel. Proteins having the highest contribution to ADMET processes are listed in Table 12.1. Although the list is not comprehensive, it contains the antitargets with most attention and research to date.

Considering the protein modelling aspects of ADMET-related proteins, it is important to emphasize that these proteins are evolutionarily optimized to recognize and bind broad variety of compounds with multiple (HSA) and/or flexible (HSA, CYPs, PXR and P-gp) binding sites. The CYP enzyme family illustratively exemplifies the complexity of ligand binding event, since as few as six isoforms (1A2, 2C8, 2C9, 2C19, 2D6 and 3A4) are responsible for the metabolism of ~75 % of the marketed drugs [1]. The characteristic non-Michaelis-Menten kinetics of co-operativity effects further underlines the intricacy of the ligand binding event. The crystal structure of CYP3A4 with ketokonazole contains two copies of the ligand providing atomic level argument for multiple binding and indicating the associated high level of binding site flexibility [2]. Similarly, crystal structure of the active transport protein (P-gp) contains two stereoisomers of a cyclic hexapeptid ligand [3]. The capacity of binding diverse chemical series is achieved by multiple, non-specific contacts. Therefore the binding site of such proteins is generally lined by numerous hydrophobic residues. Calculating less oriented hydrophobic ligand-protein contacts and estimating the associated hydration effects in a flexible protein environment contributes to the challenges of ADMET modeling at molecular level.

Drug discovery is a multidimensional task that requires an outstanding balance between desirable and undesirable properties to fulfil the efficiency and safety criteria of the target product profile. Structure based rational design of chemical modifications using the atomic resolution structure of ADMET-related proteins is therefore highly demanded. Maintaining the optimal interactions with the primary target limits the freedom for chemical modifications to optimize the pharmacokinetic profile. Therefore in silico ADMET predictions become integrated part of the drug discovery programs such as site of metabolism prediction, modelling the induction of the metabolic and transport proteins (CAR, PXR) or estimating HSA binding.

12.2 Computational Methods for Structure-Based ADMET Prediction

The aim of this section is to give a brief introduction to the general theory and practice of protein modelling methodologies. Prior to the discussion of the computational methods, a brief overview is presented here regarding the fundamental aspects of protein structures.

Atomic level information of the protein conformation is fundamentally important for structure-based drug design. High-resolution protein structures are available from X-ray crystallography, NMR methods or computational approaches. If the protein structure is not available but the structure of sequentially homologous proteins are known homology modelling can be applied. The most important public depository of experimental protein structures is the RCSB Protein Data Bank (PDB, http://www.pdb.org/pdb/home/home.do) that contains nearly 100k structures in 2014. Approximately 89 % of the deposited structures were determined by X-ray crystallography, nearly 10 % of the cases were modelled based on NMR measurements and ~1 % of the structures were based on electron microscopy studies.

Simulations based on protein models derived from experimental observations carries the errors of the original data. The errors in the data sets limit the accuracy of the subsequent modelling irrespectively to the techniques applied [4]. Thercfore quality assessment of protein models has essential impact on the further results. In the case of the X-ray experiments the electron density is deduced with inverse Fourier transformation. Accordingly, the electron density map contains the key information. The resolution of the structure is a widely used global descriptor defining the particular size limits of atomic objects that can be resolved in a particular experiment; the acceptable resolution is typically lower than 3.5 Å. The quality of the structural model can be assessed by the R-factor that quantifies the differences between measured data and data predicted from the model. Consequently lower R-value indicates a more consistent model. Cross validation during the refinement to avoid over fitting to the data (generally ~5 % of the data is selected for this purpose) is measured as R_{free} that should be lower than 0.45. Higher difference between R and R_{free} than 0.05 is indicative for over fitting. These data (resolution, R and R_{free}) are generally available in PDB database. More sophisticated methods are available to assess the global quality, such as the coordinate error [4]. Local metrics carry information on the specific regions of the protein model. These metrics indicates the quality of fit to the electron density for individual atoms or a small set of atoms. First, it is suggested to assess the electron density map in the vicinity of the active site and the ligand if present. Three further methods are commonly used for local analysis such as the occupancy-weighted B-factor (owB-factor: atomic B-factor divided the corresponding occupancy), the real-space R-factor (RSR), and real-space correlation coefficient (RSCC) [5]. In general, B-factors measure relative vibrational motions, thus low B-factor indicates well-ordered motifs. Lower owB-factor, lower RSR and RSCC close to one characterize a good model.

12.2.1 Fundamental Methods for Energy Calculation (Quantum Mechanics, Molecular Mechanics, and Hybrid Methods)

Quantum mechanics (QM) explicitly represents both the nuclei and the electrons during the calculations, therefore solving the core, time dependent Schrödinger equation entirely describes the properties—such as the energy—of the investigated molecular object. However, the exact analytical solution is problematic for larger, polyelectronic systems. Therefore approximations are required in order to make the QM methods applicable for larger systems. One of the most important steps is the Born-Oppenheimer approximation: since the electrons and the nuclei particles have 3 orders of magnitude difference in their masses, their movements can be decoupled. Accordingly, the total wave function can be separated into the product of the electronic and the nuclei wave functions. This approximation has huge impact on the applicability of the QM methods, and as a result, the energy can be calculated for continuous configurations of the nuclei. This means that a potential energy surface (PES) can be described that depends on the coordinates of the nuclei, instead of having distinct vibrational levels. Several other approximations exist, that are not described here, such as the one-electron approximation that finally results to the Slater determinant type electronic wave function and the Hartee-Fock (HF) model. The electron correlation energy can be corrected using computationally more intensive post-HF methods such as the Møller-Plesset perturbation method (MP), configuration interaction (CI), coupled cluster method (CC), to name only some of them. The other widely used approach for QM calculations is based on the Hohenberg-Kohn theorem: the ground state properties of the polyelectronic system are defined by its electron density (DFT approach). Focusing on electron density is very straightforward, since it has only three spatial variables. Recently, QM methods become integrated parts of the drug design, but they are only suitable for a limited number of heavy atoms (some hundreds).

The Born-Oppenheimer approximation and the PES inspired the birth of a much simpler molecular mechanics approach to study as large systems as proteins (see also Chaps. 5, 7 and 10). The principle of the molecular mechanics (MM) model is that the energy of the system relies only on the conformation of the nuclei, and the role of electrons is modelled with simple functions, like the Hook's law in case of bond stretching. The electrons are not explicitly incorporated into the MM model. The nuclei are connected with bonded functions (strings) and further interactions are represented by non-bonded functions. It is important to note here that MM methods are generally accurate in the minima of the PES but less convenient for conformational transition state calculations. Since the role of electrons are modelled with a set of simple mathematical functions studying reactions involving polarization, electron transfer, opened shell atoms or radical formation is out of the scope of MM calculations.

The core bonded terms include stretching, bending, torsion and the non-bonded terms include van der Waals and Coulomb electrostatic potentials. The total energy of the system is calculated as a sum of all these terms (Fig. 12.1 and Eq. 12.1).

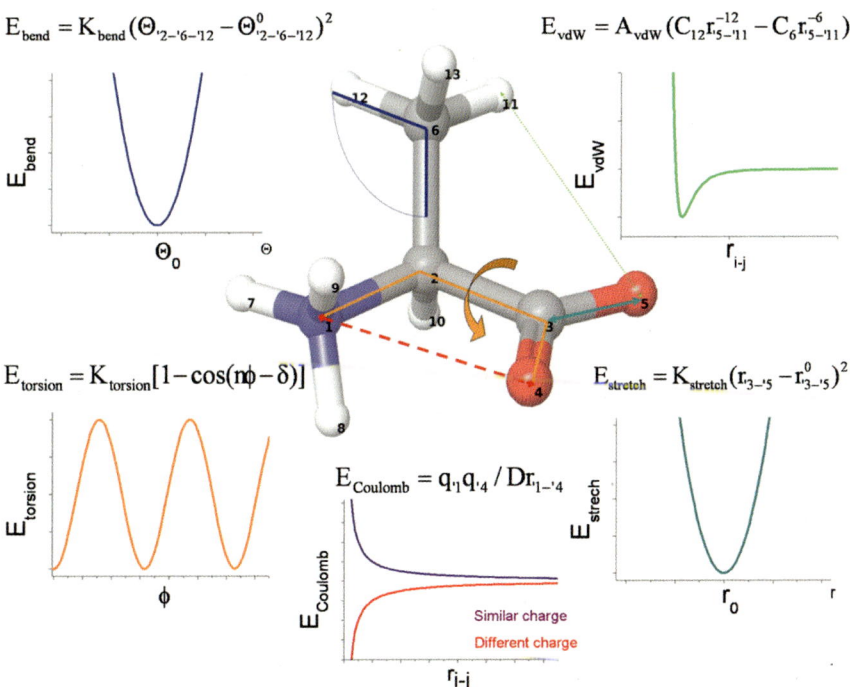

$$E_{bend} = K_{bend}(\Theta_{'2-'6-'12} - \Theta^0_{'2-'6-'12})^2$$

$$E_{vdW} = A_{vdW}(C_{12}r^{-12}_{'5-'11} - C_6 r^{-6}_{'5-'11})$$

$$E_{torsion} = K_{torsion}[1-\cos(n\phi - \delta)]$$

$$E_{stretch} = K_{stretch}(r_{3-'5} - r^0_{3-'5})^2$$

$$E_{Coulomb} = q_{'1}q_{'4} / Dr_{'1-'4}$$

Similar charge
Different charge

Fig. 12.1 Force field terms exemplified on alanine amino acid. Numbers with apostrophes represents the corresponding atom labels of alanine. Further details are described in the text

$$E_{MM} = \sum{}_{bonds} E_{stretch} + \sum{}_{angles} E_{bend} + \sum{}_{dihedrals} E_{torsion} + \sum{}_{ij} E_{vdW} E_{Coulomb}$$

$$(12.1)$$

The symbols r, Θ and δ designate bond length, angels, and torsions, respectively; r_0 and Θ_0 are corresponding equilibrium values; n and φ are torsional multiplicity and phase, respectively. K_{strech}, K_{bend}, and $K_{torsion}$ are bonded constants; $A_{nonbonded}$, C_6 and C_{12} are adjustable Lennard-Jones parameters; D is the dielectric parameter, while q is the net charge on the corresponding atom.

Due to the speed of molecular mechanics calculations, it is the key approach for simulating protein structures and bound ligands. The hydration effect can be incorporated into the model by implicit solvation or using explicit water molecules. In the case of implicit solvation, the work to transfer from vacuum to the medium is calculated using constant or distance dependent dielectric constants.

Hybrid QM/MM methods (see also Chap. 4) embrace the benefits of both approaches to simulate fundamental biological processes with essential contribution of the exact modelling of electrons embedded in the protein structure or solvent environment; the case typical for e.g. enzyme catalysed reactions. During these calculations a subregion of interest in electronic detail is modelled with QM approach, while the rest of the system is calculated with the fast MM method.

Two major QM/MM methods are used for energy expression: subtractive and additive schemes. In case of the subtractive scheme the energy is calculated for the whole system with MM method and the selected subregion is calculated on both MM and QM level. The total energy is then calculated by adding the energy of the entire system at MM level to the subregion QM energy and subtracting the subregion MM energy (ONIOM model) [6]. In case of the additive model, three energies are summarized: (i) the outer region calculated by MM method, (ii) the subregion calculated with QM approach and (iii) a special QM/MM coupling energy term. The hybrid QM/MM interaction energy calculation is the most sensitive part of this approach, since its unique role in describing the boundaries between the systems. Two methods are used to solve the link between the two approaches: (i) link atom method uses artificial dummy atoms to cap the covalent bond at the edge of the quantum system, or (ii) frozen localized molecular orbitals can be used to connect the quantum and classical regions [7].

In the case of the CYP family, as an example, where the catalytic centre consists of a heme iron and the porphyrin structure that coordinates a molecular oxygen to form the highly-reactive compound I state both electronic and steric factors have essential role. Therefore, the atomic level understanding of the catalysed reaction requires the application of hybrid QM/MM approach (Fig. 12.2).

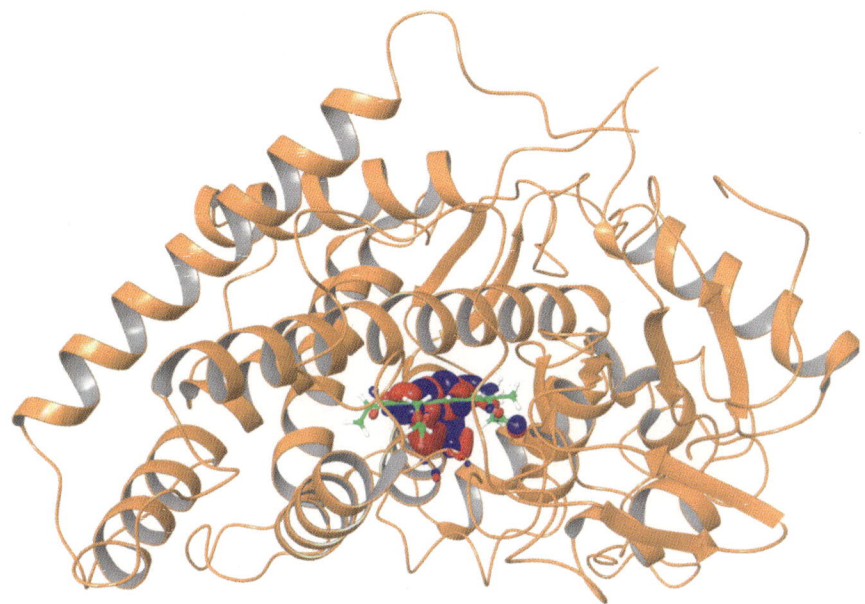

Fig. 12.2 Scheme for QM/MM method application. The protein structure, considered with MM method is represented as ribbons, while the QM subregion incorporating the catalytic centre is represented with *balls* and *sticks*. HOMO orbitals are also shown with *blue/red* surfaces

12.2.2 Molecular Dynamics

An essential synergism exists between the conformational dynamics and the function of proteins (see also Chap. 10). Molecular dynamics simulations links the structure and dynamics to explore the conformational energy landscape accessible to protein molecules to facilitate the atomic level understanding of the subtle atomic details of how proteins work [8]. Molecular dynamics simulation calculates consecutive configurations of the system by integrating Newton's law of motion. The recorded trajectory specifies the positions and energies of the particles in the system as a function of the simulation time.

Numerical solution is based on the finite difference approach in which the integration is partitioned into small steps, separated in time by a specific Δt period [9]. The popular leapfrog algorithm, as an example, uses the positions at time t and velocities at time $t - (\Delta t/2)$ for the update of both positions and velocities via the calculated forces acting on the atoms at time t (see Eqs. 12.2 and 12.3).

$$x(t + \Delta t) = x(t) + v(t + \frac{\Delta t}{2})\Delta t \qquad (12.2)$$

$$v(t + \frac{\Delta t}{2}) = v(t - \frac{\Delta t}{2}) + a(t)\Delta t \qquad (12.3)$$

Molecular dynamics simulations have fundamental role in ADMET-related protein modelling such as to explore protein conformations of the flexible proteins, or to calculate $\Delta\Delta G_{bind}$ using free energy perturbation (FEP) method.

12.2.3 Homology Modelling

In spite of the rapid growth of the PDB, the number of known proteins still enormously exceeds the number of corresponding structures [10]. On the other hand, the structure of the protein is exclusively determined by its primary sequence [11]. During the evolution sequential variations do not translate into significant, domain level structural changes directly. As a result, the known structures with high sequence similarity adopt approximately identical structures. Homology modelling is based on these fundamental observations (see also Chap. 9).

Homology modelling consists of four subsequent steps: (i) fold assignment, (ii) sequence alignment, (iii) model building and (iv) refinement [12]. The known sequence without available 3D structure is called 'target' and the sequentially homologous protein with available structure is called 'template'. The goal of fold assignment is identifying the possible templates based on sequence similarities to that of the target sequence. This step can be aided by in silico secondary structure predictions (several tools are available at www.expasy.org) that are based on extracted correlations of the available structural information. The selection of correct template or templates has crucial impact on the quality of the predicted model. The choice of template structure has another aspect, since many proteins involved in ADMET modelling adopt different conformations. In the case of the CYP enzyme family, for example, the active site cavity offers huge variability among the crystal structures of the same isoform depending on the presence and type of the co-crystallized ligand [13]. Similarly, P-gp templates were crystallized in inward-facing or outward-oriented structures co-crystallized with ATP [3]. Homology modelling is therefore a useful tool, if only the closed apo structure is available, but highly homologous templates exists with more relevant active site conformations.

During the sequence alignment step the manual fine-tuning of the target-template sequences is typically done. The aim is to find the balance between sequence overlap, possible alignment of sequence length differences (location of gaps and insertions) along with the alignment of proposed secondary structural elements of the target with that of the corresponding elements in the template.

The usual procedure for model building includes the construction of the backbone and subsequently the side chains of the homology model. The conformations of the conserved regions in the template are restored, while the insertion and gap regions are being rebuilt. The aim of the refinement step is to increase the validity of the generated coarse model. It can include further loop re-modelling, energy minimization or molecular dynamics simulations. During the refinement step the quality of the model is constantly monitored by stereochemical parameters such as the Ramachandran-plot, statistics-based potentials and physics-based energy functions. If further studies aim at calculating ligand-protein interactions special attention is to be paid for the active site. This can be carried out either during the homology modelling steps (ligand steered homology modelling) or subsequently, by simultaneous optimization of the ligand conformation and the surrounding protein environment (induced fit docking or MD refinement).

12.2.4 Docking

Docking is the prediction of the binding conformation (pose) of the ligand within the active site of the protein along with the estimation of the binding free energy (see also Chap. 11). In order to predict the binding mode, both the internal conformational and the six translational, and rotational degrees of freedom of the molecules are to be sampled. Prediction of the ligand-protein complex can have two different applications. First, either studying individual ligand-protein interactions along with the prediction of accurate binding free energy, or preparing seed structures for molecular dynamics or QM/MM calculations require highly accurate models. In these cases the conformation of both binding partners are considered flexible and thus the docking study is computationally intensive. The second application is virtual screening. Due to the high flexibility of the binding site and the related computational costs of sampling, in most of the high-throughput docking methods the protein is considered to be rigid. Thus docking in virtual screening is realized as a special conformational search of the ligand within the boundaries of the rigid protein binding site. This is a very time-efficient method, a single CPU can calculate approximately 3,000 ligand-protein complexes a day. The most important aim of virtual screening is to sort active molecules over inactives. Accordingly, docking a large (10 K–10 M) compound collection would result in a small set (10–1,000 entries) of top-ranked compounds with significantly higher propensity of actives compared to random selection, called enrichment. Docking typically consists of three steps: (i) posing or sampling, (ii) ranking of possible poses and (iii) binding energy estimation (scoring). First, all the possible ligand locations, orientations and conformations are enumerated. This process can involve systematic sampling or stochastic methods such as Monte Carlo, genetic algorithm or tabu search [14]. In case of the systematic sampling compounds are fragmented into core and side chain regions. After core placement, the remaining parts of the molecule are being rebuilt within the active site, to exploit the benefits of systematic sampling but avoid the costs of generating high number of irrelevant conformations. The possible poses are then ranked according to their complementarity to the binding site using fast-scoring to eliminate the inappropriate conformations. A set of desirable poses are generally minimized using MM methods. The most favourable pose is scored using accurate scoring functions to predict the binding free energy. Scoring can be classified into (i) force-field based, (ii) empirical or (iii) knowledge-based methods [14].

In general, fast docking algorithms results in plausible binding modes. As a recent example, Glide docking reproduced protein-ligand complexes with <1.5 Å RMSD relative to the X-ray structure in 80 % of the cases [15]. The other widespread application of docking is virtual screening. High experimentally validated enrichments prove its utility in the drug design [16]. However, it is important to note that direct binding free energy estimation by fast-docking algorithms have generally limited performance.

12.3 Modelling of ADMET-Related Proteins

Modelling of ADMET proteins will be discussed in the following section grouped into adsorption and distribution, metabolism and toxicity related targets. Selection was based on the relevance to pharmaceutical research and the availability of structural information. After giving a short summary of the significance, structure and function of the targets, we are going to focus on ligand-protein interactions, in terms of the applicability and feasibility of the models in drug design.

12.3.1 Adsorption and Distribution

12.3.1.1 Human Serum Albumin

Human serum albumin (HSA) is the most abundant plasma protein found in 0.6 milimolar concentration in the blood and accordingly serves as an important regulator of osmolarity [17]. It is responsible for binding and transport of endogenous and exogenous substances including non-esterified fatty acids, porphyrins, drugs and other hydrophobic ligands. HSA has both desirable and undesirable effect on drug discovery. Moderate HSA binding can decrease the clearance or poor distribution of hydrophobic drugs. On the other hand, high affinity to HSA might contribute to low free plasma concentration of the drug resulting in suboptimal tissue distributions and the lack of in vivo efficacy.

More than 70 X-ray structures of the apo or ligand bound HSA complexes are available to date (Fig. 12.3). The 585 residue long protein folds into three domains (I-III) consisting of two (A and B) subdomains built form α-helices. It has heart-shaped overall tertiary structure incorporating seven fatty acid binding sites. The tertiary structure is stabilized by 17 disulfide bridges contributing to the observed high thermostability [18]. Both the protein dynamics and the role of the individual disulfide bonds were studied using molecular dynamics simulations recently [19]. The study shed light on the role of selected disulfide bridges such as Cys168−Cys177 and Cys278−Cys289 maintaining the secondary and tertiary dynamics of HSA.

HSA has two major sites for binding drugs. Site 1 is located at domain IIA, and also known as warfarin binding site, while site 2 is at domain IIIA, often identified

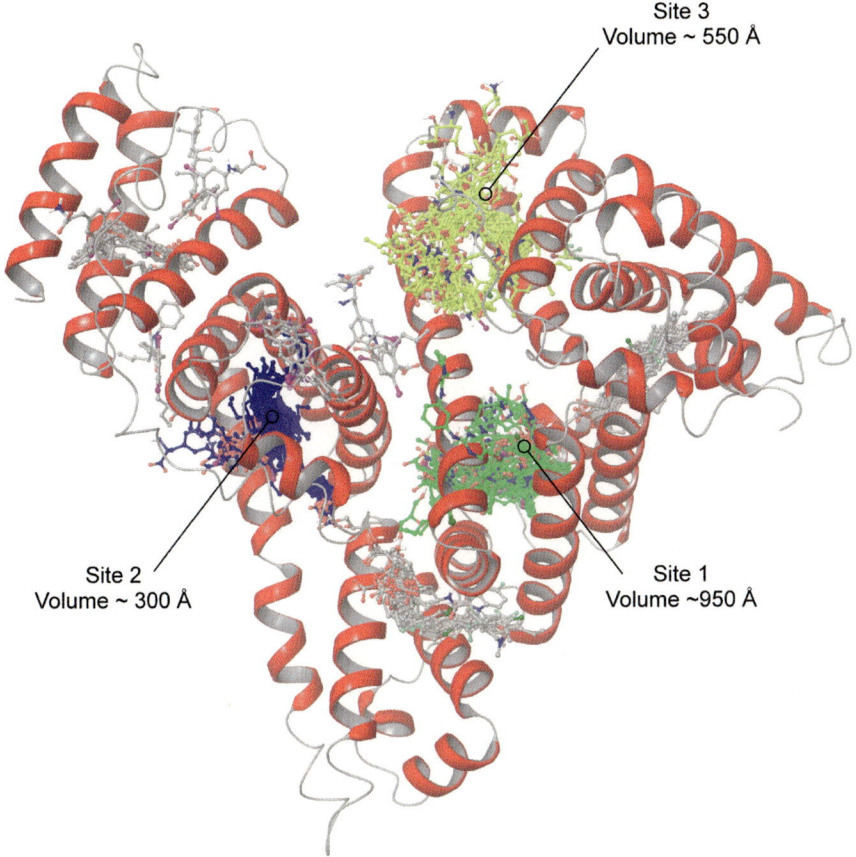

Fig. 12.3 Structure of the HSA. Available holo complexes were aligned to 1BJ5 crystal structure. Ligands within the site 1, 2, and 3 are coloured *green*, *blue* and *yellow*, respectively. The approximate binding site volumes were calculated for site 1, 2 and 3 using 1H9Z, 2BXF and 1NU5 crystal structures, respectively

as the indole-diazepam site. A minor drug binding pocket, site 3, is also described at subdomain IB that transports hemin, bilirubin and fusidic acid. Site 1 prefers large, heterocyclic and negatively charged compounds (warfarin, azapropazone and dansylamide), while site 2 binds smaller carboxylic acids (diazepam, ibuprofen, and arylpropionic acids) [20]. HSA ligand-protein interactions are elusive, since the binding sites are large, open and diverse with high degree of flexibility. It is interesting to note that without a priori information on the binding site of a given ligand, all the possible binding pockets are to be sampled during docking calculations. Moreover, it can host multiple ligand copies simultaneously such as the case of idarubicin [PDB ID:4LB2]. In spite of the underlying complexity, there are documented successful optimizations (e.g. Bcl-2 and COX inhibitors), where HSA binding was decreased by structure based design [21, 22].

Due to the intrinsic difficulties related to modelling HSA ligand-protein interactions, the majority of the methodologies for drug discovery purposes employ combined ligand-based statistical models with structure-based docking. This straightforward strategy results in an acceptable estimation of HSA affinity and provides a plausible binding mode to aid the further design [18].

Aureli and co-workers used docking to categorize 37 compounds as site 1 or site 2 binders [23]. Based on the proposed classification, a site 2 PLS model was established to estimate the binding free energy of the compounds, including 10 novel interleukin (IL)-8 inhibitors. Descriptors derived from the predicted binding conformations of antibiotic drugs were used by Li and co-workers to develop a QSAR model for plasma protein interactions [24]. An automated workflow was published by Zsila and co-workers for the classification of ligands as HSA substrates or non-substrates and to predict the possible site along with the binding mode [25]. Support vector machine calculation was used for affinity prediction representing the molecules with 45 ligand-based descriptors. In the subsequent step chemical similarity is calculated between the query compound and the ligand binding on the predicted binding site to select the most appropriate protein conformation for docking. The X-ray structure with the most similar ligand is used to estimate the binding mode. This innovative workflow aims at minimizing the errors of docking calculations arising from site mismatch and protein flexibility. A very recent study by Hall and co-workers uses a similar workflow with special focus on docking [20]. As an initial step a multivariate linear regression model was built for HSA ligands. This model includes only four ligand-based descriptors referring to the acidic nature, predicted solubility and lipophilicity of the compounds. Secondly, a Bayesian classification method was developed to predict the HSA binding site of the compounds using both ligand based physicochemical descriptors and fingerprints representing the structural patterns of the compounds with a bit string. In order to adequately describe the ligand-HSA interactions, Hall and co-workers combined ensemble docking and induced fit docking protocols. Binding site conformation based clustering was used to select five and three representative experimental structures for site 1 and site 2, respectively. The selected X-ray structures were used for validation. Induced fit docking employs a three step consecutive optimization of the ligand-protein complex. First, the ligand is docked to the rigid binding site with downscaled van der Waals radii to mimic the effect of binding site perturbation. During the subsequent refinement, the protein side chains are optimized in the dihedral space and small backbone movements are allowed around the rigid ligand position. The last stage includes docking to the optimized binding environment with flexible ligand docking and fixed protein conformation. This methodology resulted in 1.8 Å average RMSD for site 1 and 2 binders in a cross docking scenario. The obtained binding modes have acceptable accuracy to alleviate the undesirable HSA binding of the chemical series by structure based design.

Due to the abundant experimental information, HSA is an attractive target for modelling protein-ligand interactions. Although, the associated intrinsic complexity requires the application of orthogonal techniques, such as ligand-based modelling and the incorporation of protein flexibility during docking to meet the quality criteria required for drug design purposes. Recent workflows combining different approaches represent a viable strategy for the prediction of ligand-HSA interactions.

12.3.1.2 P-Glycoprotein

P-glycoprotein (P-gp) was originally discovered due to its preventive effect against cytotoxic drugs by hampering their membrane penetration, a phenomenon known as multidrug resistance (MDR). Further characterization revealed that P-gp is a membrane embedded active transporter that harness energy from adenosine triphosphate (ATP) hydrolysis, as a member of ATP binding cassette (ABC) super-family, to extrude chemicals out of the cell. A hallmark feature of this transporter is its capability to bind an array of structurally diverse molecules ranging from 100 to 4,000 Daltons (Da). P-gp impacts the pharmacokinetics, especially bio-availability and distribution of xenobiotic drugs by mediating their transport in the liver, intestines, and across the blood brain barrier [26, 27]. On top of its profound importance as an antitarget, it serves as a primary target for projects delivering P-gp inhibitors to enhance the effect of the anti-cancer agents, due to the docu-mented overexpression of P-gp in tumour cells.

P-glycoprotein, a product of the mdr1 gene in humans, is composed of 1,280 residues organized into two homologous halves in a single polypeptide chain. The 170-kDa protein contains six helical transmembrane (TM) domains and a globu-lar nucleotide binding domain (NBD) located at the intracellular side at each half, giving a total of 12 helices and two NBDs per protein (Fig. 12.4). The ATP bind-ing pocket has conserved sequences among ABC family including the Walker-A motif (P-loop), Q-loop, Leu–Ser–Gly–Gly–Gln signature sequence (C-loop), Walker-B motif, D-loop and H-loop that coordinates Mg^{2+} ion and ATP. Its hypo-thetical transport cycle is initiated by ligand binding in the TM domain followed by ATP coordination at the NBDs. ATP binding and/or cleavage drives significant conformational rearrangements, the two NBDs forms a dimer (Fig. 12.4). Coupled conformational changes of the TM domains contribute to the release of the ligand at the opposite, outward facing side of the membrane. The release of Pi/ADP con-tributes to resetting the transporter to the initial inward-facing conformation.

Three X-ray structures including the ligand-free mouse P-gp, and mouse P-gp bound to two stereoisomers of a novel cyclic hexapeptide inhibitor, all in the absence of ATP, presenting open-apo conformation were published in 2009 [3]. The crystal structures revealed a large separation between the NBDs spanning approximately 30 Å and a substantial, nearly 6,000 $Å^3$ volume of internal cavity within the lipid bilayer. Due to possible misinterpretations in these structures, a refinement was published in 2014 including ab initio remodelling several parts of the structure. Significant differences were introduced in the drug binding pocket of the mouse structure including a >6 Å shift of the TM4 backbone atoms from the original positions [28]. Nanobody stabilized inward facing conformation of mouse

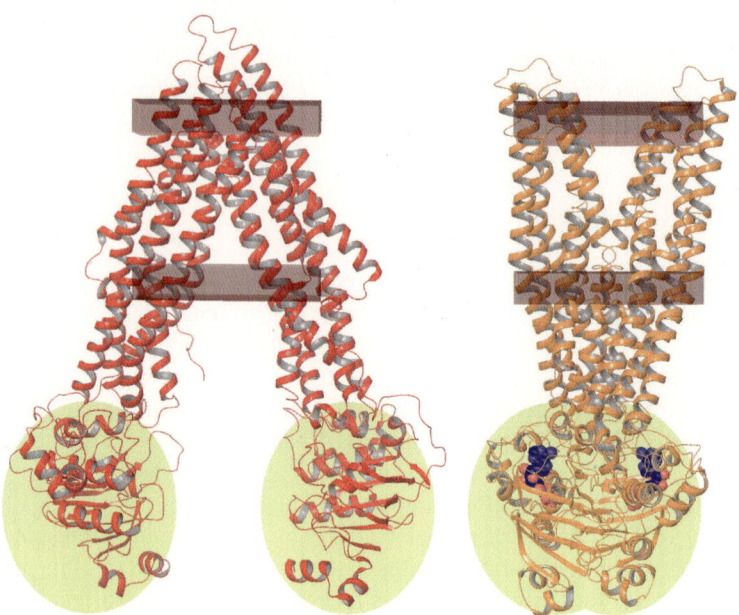

Fig. 12.4 Structures of ABC transporters. Inward facing conformation of mouse P-gp (4KSB), and outward facing SAV1866 (2HYD) are colour red (*left*) and orange (*right*), respectively. Bound ADP is represented with *balls*; carbon and oxygen atoms are colour *blue* and *red* respectively. Position of the membrane is indicated with *brown rectangles*, *top* is the extracellular and *bottom* is the intracellular side. Nucleotide binding domains are highlighted with *yellow* background

P-gp has been published recently [29]. The extent of conformational flexibility of P-gp is highlighted by the wider separation of the NBDs with distance of ~36 Å.

In order to explore the extent of conformational flexibility of P-gp molecular dynamics simulations were carried out for the apo, inhibitor and substrate bound protein conformations [30]. These calculations are extremely resource intensive due to the large number of atoms consisting the explicit solvated, membrane embedded protein system that has approximately 140 K heavy atoms. Simulations at 100 ns time scale highlighted a more extensive separation between the NBDs compared to the crystal structure and depended on the investigated system namely, the ligand-free, substrate and inhibitor bound complexes. Targeted molecular dynamics simulation was carried out between the outward and inward oriented states of bacterial MsbA protein of the ABC family, to gain atomic detailed information of the transition pathway [31]. During the simulation the disruption of the nucleotide binding sites at the NBD dimer interface were observed as the very first event that triggers the subsequent conformational changes, verifying the assumption that the conformational change is driven by ATP hydrolysis.

Prior to the publication of the human P-gp structure homology modelling was applied based on mouse and bacterial templates [32–35]. The applicability of docking based virtual screening using comparative models built on mouse template was assessed in two studies [34, 35]. Both yielded similar, negative results in term of enrichment. The active compounds could not be significantly ranked over the inactives. Several reasons might contribute to the poor prediction power observed: (i) the coordinate errors of the template that were corrected later on, (ii) the large and hydrophobic binding site of P-gp, (iii) multiple binding sites along with the possibility of cooperative binding, and (iv) the observed flexibility of the protein that was validated both experimentally and computationally and underlines the need of incorporating protein flexibility during docking calculations. Sampling difficulties originates from the large binding pocket were studied for propafenone type inhibitors of P-gp [36]. The exhaustive sampling protocol involved the enumeration of 100 binding poses for each ligand within four different protein models. Clustering of the poses revealed that propafenone binding site is located between the TM helices 5, 6, 7 and 8. Analysis of the interacting residues highlighted the essential role of Tyr307 in coordinating the ligands by H-bond. In order to investigate the effect of protein flexibility induced fit docking of four ligands with extensive site directed mutagenesis data were carried out [34]. Considering the induced-fit binding conformations of rhodamine, verapamil, colchicine and vinblastine, the residues located in the binding pocket showed good correlation with the available mutation data.

In conclusion, the binding conformation of known ligands can be predicted at the level of interacting residues using enhanced sampling or incorporation of protein flexibility. Simulations are valuable tools to understand and interpret the experimental data from biochemical and biophysical studies. The experimental results and simulations can cross-validate and improve each other as it was exemplified for the refinement of the mouse P-gp structure. However, numerous difficulties arise for predicting P-gp-ligand interactions that prevent the large-scale application of simulations. To date structure based virtual screening is not suitable to distinguish P-gp substrates form non-substrates routinely.

12.3.2 Metabolism

12.3.2.1 Cytochrome P450 Family

Cytochrome P450 (CYP) enzymes have essential physiological importance and ubiquitous occurrence in almost all living organisms. The catalysed biotransformations have major contribution to phase I drug metabolism in man. The enzymatic activity of CYPs influences wide variety of pathophysiological processes including detoxification of xenobiotic compounds, bioactivation of nontoxic compounds into toxic reactive intermediates and ultimate procarcinogens. Metabolic liability can limit drug exposure and might contribute to lack of in vivo efficacy, meanwhile the produced metabolites arise further safety concerns [1]. The promiscuity of these enzymes and the fact that only some isoforms are responsible for the metabolism of the majority of marketed drugs can contribute to undesirable

Drug–drug interactions (DDIs) upon co-administration [37]. DDI associated modulation of enzymatic activity, like inhibition or induction, results in serious alternations of drug plasma concentrations from the expected value.

The globular tertiary structure of CYP enzymes principally composed of helices labelled A to L commencing from the N terminus (Fig. 12.2). The B–C loop contributes to substrate specificity, the F and G helices form the roof of the active site, and the I and L helices embrace the heme prosthetic group, also known as protophorphyrin IX at the bottom of the pocket. The heme moiety and the residues forming its bassinet are highly conserved among the CYP family, along with the characteristic kink on the I helix above the active site. CYP enzymes are suggested to be membrane anchored by the N-terminal region, particularly by the F-G loop and parts of the F and G helices. According to the binding site volumes the mammalian CYPs can be partitioned into three categories: CYP2C5, 2C8, 2C9, 3A4 > 2B4, 2D6 > 2A6, in decreasing order [13]. The binding pocket of the 1A2 crystal structure, appeared after the publication of the classification system, fits into the smallest category occupying 630 Å^3 volume. In contrast, the binding site of CYP3A4 extends to 1,500 Å^3, indicating the large variability of CYP binding sites.

Oxidative transformations catalysed by CYP enzymes proceed within a precisely coordinated cycle, in which the catalytic heme iron undergoes changes regarding its spin state, coordination and oxidation number. The schematic catalytic cycle is shown in Fig. 12.5. In the initial resting state hexa-coordinated Fe(III) is in low spin doublet state with the axial position occupied by a water molecule. The displacement of the water molecule by the ligand results in a penta-coordinated Fe(III) with a sextet state and the change of the redox potential. Due to the alternation of the redox potential, Fe(III) can accept an electron from the nicotinamide adenine dinucleotide phosphate (NADPH)-P450 reductase to form the high spin Fe(II). The next event is the coordination of the molecular oxygen, resulting in a singlet oxy-ferrous complex. After capturing the second electron from the reductase, the ferric-peroxo anion is formed. In the subsequent steps the anion is protonated by a proton shuffle mediated by the protein environment. The following step is the heterolytic cleavage of the O–O bond. Subsequently, a water molecule is liberated and the highly reactive ferryl-prophyrin-π radical cation is formed. The so called 'Compound I' intermedier has three unpaired electrons: two is located on the ferryl group and one is shared with the sulphur of the distally coordinated cystein residue. This species oxidize the substrate that is followed by the replacement of the product with a water molecule to restore the resting state.

Publication of the first human P450 structure in 2003, the CYP2C9 isoform, proved to be a milestone giving new impetus to protein modelling understanding specific mechanistic details and predicting the site of metabolism. Since then all of the pharmaceutically relevant human isoforms were crystallized including 2C9, 2C8, 3A4, 2A6, 2D6, 1A2, 2A13, 2E1 and 2B6 (PDB). Considering simulation approaches relying on molecular mechanics force fields, it is important to note the CYP enzyme possess the unique prosthetic heme group requiring special parameters for the various states of the catalytic cycle even in apo simulations [38]. These parameterizations are limited for a given force field, thus each force field demands

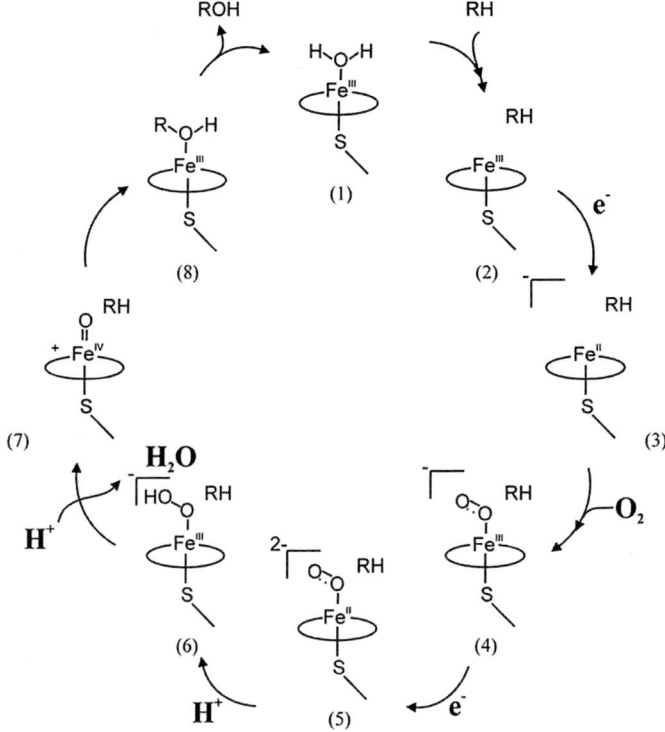

Fig. 12.5 Catalytic cycle of the CYP

an own set of corresponding values. The parameters are typically derived from QM calculations and available for some of the most applied force fields including CHARMM and AMBER [39–41]. As an example, Shahrokh and co-workers recently developed a set of AMBER parameters for a range of heme states found in the catalytic cycle [42].

Molecular dynamics simulations were carried out to understand the molecular basis for CYP substrate promiscuity. Simulation of the CYP3A4 isoform by Park and co-workers revealed that the high-amplitude flexibility of the loop at the substrate entrance in the apo state might have responsibility for the observed broad specificity [43]. Lampe et al. calculated 200 ns long trajectories of the thermophilic P450 CYP119 enzyme. The results suggested that the correlated motions of the active site are relatively independent from the outer protein framework [44]. Further simulations on CYP3A4, 2C9 and 2A6 revealed that protein flexibility is directly related to substrate specificity [41]. CYP2A6 possessing narrow substrate specificity was found to be the less flexible. In contrast, CYP3A4 had more widespread movements in line with its enormous promiscuity. Recent molecular dynamics simulations of P450cam aimed at analysing the hydration of the active site [45]. According to the trajectory recorded over the 300 ns long dynamics 6.4 water molecules were observed in the camphor-binding site of the apo form, compared to zero water molecules in the substrate-bound binding site. Interestingly, up to 12 waters can occupy the same cavity in apo-form, revealing a highly dynamic process for hydration with water molecules exchanging rapidly with the bulk solvent. Camphor binding modifies the free-energy landscape of P450cam channels toward favouring the diffusion of water molecules out of the protein active site. However, the process of ligand binding has crucial impact on the kinetics of the enzyme reactions. Observation of the entry or exit of the ligand or the product during unbiased simulations is generally not possible considering the timescale of a typical MD simulation. To overcome this limitation, biased (steered) simulations are applied. An example of these simulations was published by Fishelovitch and co-workers for CYP3A4. In order to identify the preferred substrate/product pathways and their gating mechanism, pulling the products temazepam and 6-β-hydroxy-testosterone out of the enzyme was carried out [46]. In summary, molecular dynamics simulations nicely complement the static experimental picture observed in X-ray crystallography by giving novel and valuable viewpoints to rationalize the function and activity of the protein family at atomic level.

Both the reactivity and the orientation, driven by steric and energetic factors, influence the site of metabolism of a given ligand [47]. Due to the abundance of human CYP structures, docking methods can be applied as cost-effective approach to estimate the site of metabolism. A straightforward concept is to dock the ligand to the active site and evaluate the atomic positions in terms of the distance calculated form the catalytically active heme iron. In principle, the atom with the lowest distance in the top ranked binding mode is the site of metabolism. Unwalla and colleagues assessed the performance of Glide [48] docking to predict the site of metabolism of 16 CYP2D6 substrates using a homology model structure [49]. According to their results, the observed metabolically liable position was within 4.5 Å from the heme iron in 85 % of the cases considering the top five-ranked docking poses. The observation that the substrate has low flexibility during the catalytic cycle and therefore the position of the substrate and the product overlaps inspired the reverse-docking approach, published by Tarcsay and co-workers [50]. According to this methodology, the possible products were enumerated by using a rule-based expert system and the metabolites were docked to the binding site. Interestingly, the reverse-docking approach corresponds to the Marcus theory, since the lowest energy product is suggested to have the lowest barrier height. The proposed methodology was evaluated on the human CYP2C9 isoform, resulting in 84 % success rate for the docking step that combined with the 82 % success of the metabolite generation yield 69 % in overall, considering the top three ranked sites. Estimation of the ligand binding mode and the free energy of binding are sensitive to the water molecules present in the binding site. As it was previously discussed, in the case of the CYP family, solvation of the active site is of fundamental importance. Recently, Santos and co-workers investigated the role of water molecules in docking simulations for CYP2D6 [51]. The MD simulations identified water hydration sites with an occupation probability at least 30 times larger compared to the bulk water. However, these water molecules had an average residence time below 10 ps, thus possessing quite high mobility and having low probability of forming hydrogen bonds with the protein or the ligand. According to the results of the published MD simulations and subsequent ligand docking the influence of water molecules appears to be highly dependent on the protein conformation and the substrate. Incorporation of hydration site waters into docking had an effect on the reliability of the site of metabolism prediction. The effect size was found to be dependent on the chemical similarity to the ligand by which the water positions were selected. Considering dissimilar compounds, net effect was not observed.

Estimation of the binding affinity is among the most desired objectives of the recent computational chemistry approaches. Docking and rapid scoring functions has serious limitations in this regard; therefore computationally more intensive methods are needed to achieve more accurate estimations. The most straightforward approach is to calculate relative free energies for different thermodynamic states by the definition of a suitable thermodynamic cycle. Free energy perturbation and thermodynamic integration are the most well-known methods connecting the thermodynamic states. A recent study by de Beer and co-workers utilized the computationally effective one-step perturbation (OSP) method to explore the C3 hydroxylation of α-ionone by two CYP BM3 mutants, A82W and L437N [52]. During the MD simulations an innovative reference state was constructed by optimizing the force field parameters describing the substrate to have sufficient transitions between the two enantiomers. This optimized reference state was successfully applied to estimate the relative free energy of binding of the substrate enantiomers and predict the observed trends for A82W mutant. In the case of the L437N mutant the predicted free energies of the four possible hydroxylation products were found to be in line with experimentally observed trends.

Modelling the ligand co-operativity observed for CYP isoforms represents a challenging task with ultimate pharmacological importance. A sequential docking protocol was recently described to model the multiple ligand poses in complexes crystallized with clusters of 2-6 cooperative ligands [53]. The optimized protocol was able to reproduce the two-thirds of the structures for the cytochrome P450 subset within 2 Å RMSD calculated for ligand heavy atoms. Molecular dynamics simulations and free energy calculations were carried out to calculate the binding free energy of ketokonazole based on the observed X-ray structure of CYP3A4 with two copies of the ligand [54]. The apo, the single ligand, and the double ligand complexes were simulated. Binding of the first ligand increased the affinity for binding the subsequent ligand by 5 kJ/mol. The effect of shape complementarity represented by van der Waals interactions explained and quantified the experimental results in terms of the values predicted from simple two-ligand binding kinetic model and successfully reproduced the measured titration curve.

MD simulations and hybrid QM/MM approach were utilized for CYP2D6 by Oláh and co-workers to understand the determinants of selectivity for dextromethorphan oxidation [55]. The calculated reaction barrier of two possible competing routes including aromatic carbon oxidation and O-demethylation were compared. QM/MM calculations were able to demonstrate the crucial role of protein environment in determining reactivity of dextromethorphan and explained the lack of aromatic hydroxylation observed experimentally. This study revealed that in contrast to docking or estimating the product by using Compound I reactivity models, QM/MM calculations could rationalize the very strong preference for O-demethylation over aromatic carbon oxidation. The main drive was appreciated to be the strongly disfavoured nature of aromatic hydroxylation in the protein milieu due to interactions at the active site preventing the aromatic moiety from attaining the preferred transition state. This combined MD and QM/MM approach was introduced as a general protocol for accurate modelling the CYP mediated metabolic reactions of drug molecules [56] by simulating CYP2C9 biotransformation of three drugs, S-ibuprofen, diclofenac and S-warfarin. The introduced protocol has three main components. First, application of QM/MM method for calculating the barrier is necessary. The QM treatment is indispensable due to the role of electrons, and the protein framework has crucial influence on the regioselectivity. The binding site determines the conformation of the ligand around Compound I, thus modulate the transition state (TS) geometry and the corresponding barriers. In case of the aliphatic hydroxylation of S-ibuprofen the QM barriers for hydrogen abstraction from C2 and C3 computed in vacuum resulted in an opposite trend compared to the experimentally validated result of QM/MM calculations realized in the protein environment. In fact, the formation of low-energy TS geometry calculated in the case of the simple QM system consisting from the ligand and Compound I is prevented by the protein environment. TS conformation calculated in QM/MM calculations resulted in higher energy for C2 path and therefore changing the order of barrier heights leading to C3 site preference. Second, dispersion energy should be accounted to get barrier heights that are in reasonable agreement with experiment, for example by use of empirical corrections. The third point raised by the authors is the essential importance of sampling during the calculation of barriers in enzymes, since broad range of values can often be calculated due to the slight fluctuations in the orientation of the substrate and the surrounding residues. To overcome this possible source of errors, MD simulation of the system is suggested to sample the conformational space and to generate different starting points for calculating several reaction profiles at QM/MM level. The estimated activation energies by the proposed methodology agreed nicely with the experimental results for S-ibuprofen and S-warfarin, however for diclofenac the formation of 5-hydroxdiclofenac was preferred in silico compared to the 4′-hydroxydiclofenac produced in vitro. The reason for this discrepancy might be the low free energy of binding. Therefore the favourable energy of salt bridge formation between the carboxylate moiety of the ligand and the Arg108 residue in the putative correct orientation was not incorporated.

12.3.3 CYP Induction

Pregnan X receptor (PXR) and constitutive androstane receptor (CAR) are two nuclear receptors implicated in the transcriptive regulation of CYP3A and CYP2B families along with the ABC transporter genes. Since CYP3A4 is one of the most important drug-metabolizing enzymes, its upregulation can lead to undesirable drug-drug interactions (DDIs) [57, 58].

12.3.3.1 Pregnan X Receptor (PXR)

PXR consists of 434 residues that folds into DNA- and ligand-binding domains (DBD; LBD) and heterodimerize with RXR (9-cis retinoic acid receptor) to bind responsive DNA elements controlling CYP3A and CYP2B transcription. The first structure of the PXR LBD was elucidated in 2001 at 2.52 Å resolution. To date, 12 human crystal structures are available in the PDB, representing the structural details of several ligand-bound complexes including the macrolide antibiotic rifampicin (1SKX), the phytochemical hyperforin (1NRL) and the cholesterol-lowering agent SR12813 (1ILH). The general fold is in alignment with that of the other nuclear receptors (Fig. 12.6). However, the binding pocket is larger (~1,350 Å³), with higher degrees of flexibility (Fig. 12.6). This phenomenon can be rationalized in terms of the structural elements by the beta-sheet insertion between helices 1 and 3 [59]. The substantial hydrophobic momentum is a hallmark feature of the binding site. Out of the possible 28 residues constructing the cavity 20 possess hydrophobic/aromatic character, 4 are polar (Ser208, Ser247, Cys284, and Gln285) and 4 are charged (Glu321, His327, His407, and Arg410). Analysis of the available PXR holo complexes revealed the role of Gln285, Ser247 and His407 residues with high potential to intermolecular H-bonds with the ligand [57]. The multiple possible arrangements of the key residues (such as His407) located in the binding site might contribute to the wide substrate specificity of PXR and demands special attention during the simulation of protein-ligand interactions (Fig. 12.6).

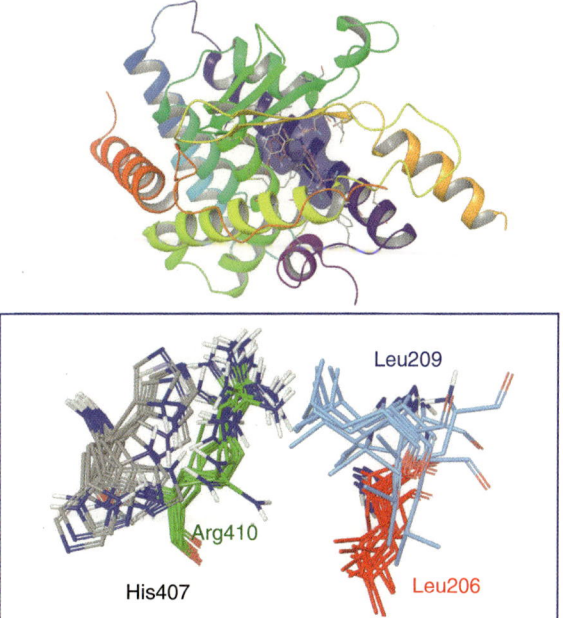

Fig. 12.6 Structure of PXR (2O9I). Conformational flexibility of four selected residues is highlighted

In spite of the modelling difficulties arising from the unique features of the PXR binding site, a successful structure based design against PXR liability was published by Merck researches [60]. Docking the initial PXR activator ligands to the 1NRL crystal structure revealed two energetically equivalent binding modes. In both cases the Trp299 and Phe429 residues had apolar interactions with the aryl moiety of the investigated compounds. The straightforward idea of increasing the polarity of the aryl moiety by the introduction of aryl-nitrogen atoms or a sulphonamide moiety to attenuate the PXR interactions had been proved by reduced PXR activation and accordingly diminished CYP3A4 induction. In contrast, Ekins benchmarked structure-based docking methodology against ligand-based models, and achieved moderate performance for docking [61]. Evaluation of docking scores obtained for 30 activator and 89 non-activator ligands using six human PXR structures resulted in overall accuracies in range of 35–55 %. The hyperforine holo-complex (1M13) was found to yield the highest accuracy (55 %) associated with a modest Matthews coefficient (0.09). In a recent study, the binding mode of nineteen Chinese herbal compounds were predicted using docking approach extended with simulated annealing MD and energy minimizations of several initial binding orientations [62]. Although, the binding mode of the herbal compounds showed different orientations, distinguished interactions were observed with Ser247, Gln285, His407, and Arg401 residues. It was concluded that the binding modes and corresponding binding energies of herbal compounds were not directly related to PXR potency measured.

12.3.3.2 Constitutive Androstane Receptor (CAR)

The 352 residue long CAR protein has in vitro constitutive activity in apo state. However, under in vivo circumstances the ligand-free protein is inactive (silent) and located in the cytoplasm. Binding of activator ligands induces the translocation of CAR into the nucleus where it activates the transcription of ADMET relates genes including CYP members and ABC transporters, therefore influencing the pharmacokinetics of drugs. CAR and PXR share a highly conserved DBD and a moderately conserved LBD, thus the available four crystal structures (resolutions in range 2.6–3 Å) of CAR adopt similar fold to that of the PXR structures [63, 64]. The accessible volume of the CAR pocket is 525 $Å^3$, which is less than half of the PXR pocket. Like other nuclear receptors, the CAR ligand binding pocket is predominantly hydrophobic with only two small hydrophilic patches for possible hydrogen bond interactions. The contribution of apolar contacts to ligand recognition is well exemplified by the holo-complex of TCPOBOP (1XLS), where no direct hydrogen bonds were observed. Meanwhile, the ligand forms an extensive network of hydrophobic interactions. The two pyridine rings are sandwiched between Y234 and F227 residues and Y336 and F244 residues, respectively.

Nuclear receptors require different co-regulator proteins to be active. Ligand binding induces conformational changes to switch between active and inactive states in terms of the conformation of the co-regulator binding domain. The conformation and relative orientation of helix 12 (H12) has crucial role in co-activator protein binding. In order to understand the structural determinants of agonist and inverse agonist induced protein conformational changes, complexes of seven agonists and eight inverse agonists were simulated in the presence of the co-activator protein [65]. According to the analysis of the 10 ns long MD trajectories, the movement of H12 towards H10 seems to favour the binding of the partner protein. In the presence of agonists the H12 conformation was stabilized in the active conformation. In the case of the inverse agonist complexes, the H12 had higher flexibility and shifted away from the active state. The short and rigid helix between H10 and H12, called HX, may retain in the active position and might contribute to the high constitutive activity of CAR. In accordance with the previous finding, the HX region was observed to be destabilized by the inverse agonists compared to apo or agonists bound complexes. Longer, 50 ns MD calculations were carried out to obtain dynamics information on two agonist bound and two corresponding ligand-depleted CAR systems [66]. These simulations resulted in univocal results regarding the contribution of HX stability to the active state. In the case of the ligand-free dynamics simulation, the HX alpha helical region was unfolded while during the simulation of the agonist bound complex it remained stable. In addition, Tyr326 and H12 interactions were ameliorated significantly for the agonist bound complexes versus the apo states. The considerable increase of the van der Waals contacts were mainly based on the interactions with the bound ligands, since both agonists had extensive interactions with the specific tyrosine side chain, thereby limiting its conformational flexibility. As a result of the stabilization effect of the agonist (also in line with the previous MD study) the H12 was found to be more flexible in the apo simulations compared to the agonist bound complex. In summary, these MD studies were able to elucidate the structural changes of human CAR LBD induced by the inverse agonists and agonists compared to the apo structure.

Identification of CAR agonists by 3D pharmacophore screening and subsequent docking was published by Külbeck and co-workers [67]. The initial compound database consisted of ~85,000 entries that were reduced to ~9,700 by the pharmacophore filter. These remaining compounds were ranked according to their scores obtained by docking them to the CAR protein structure (1XVP). Top ranked poses were visually inspected and finally 30 compounds were purchased. This approach was able to identify 17 agonists, corresponding to an excellent hit rate of 56.6 %. Integration of ligand-based and structure-based virtual screening of 753 FDA-approved drugs has been published recently [68]. Similarly to the study by Külbeck et al. ligands were first ranked according to their alignment with pharmacophore models. The authors used four independent pharmacophore models, and the best score for each query ligand was used for ranking. Top 106 compounds were docked to the active site of human CAR, and 19 virtual hits were selected for in vitro testing. Five compounds were identified as moderate activators and two known activators (nicardipine and octicizer) were classified as strong activators, thus significant hit rate (36.8 %) was achieved. Both studies emphasized the benefit of combining orthogonal, ligand and structure based modelling approaches.

12.3.4 Sulfotransferase

The human cytosolic sulfotransferases (SULTs) comprise a 13-member family of enzymes catalysing the transfer of the sulfuryl-group ($-SO_3$) from the donor 3′-phosphoadenosie 5′-phosphosulfate (PAPS) to the hydroxyl or primary amine moieties of vast amount of endogenous and xenobiotic acceptors. Due to its involvement in conjugation type Phase II metabolism, understanding the molecular basis of the enzymatic reaction is highly valuable from ADMET aspects [69]. Recent biophysical studies have revealed that selectivity in binding and catalysis relies in large measure on the plasticity of a conserved active-site cap that mediates substrate interactions by determining their affinities and kinetic behaviour. The SULT active site cap is a dynamic ~30 residue stretch of amino acids interacting with the PAPS donor and the acceptor substrates (Fig. 12.7). Available crystal structures of nucleotide-bound enzymes show a largely ordered framework regardless of whether acceptor ligand in bound. In sharp contrast, the SULT structures without the donor nucleotide has highly disordered cap region (open state).

These observations highlight the importance of the cap region in substrate recognition. Molecular dynamics simulations of the nucleotide-bound and ligand-free structures of SULT1A1 and SULT2A1 gave fundamental information on the function of the cap region [70]. According to the experimental structures and the analysis of atomic root mean square fluctuations (RMSF) during MD simulations, the

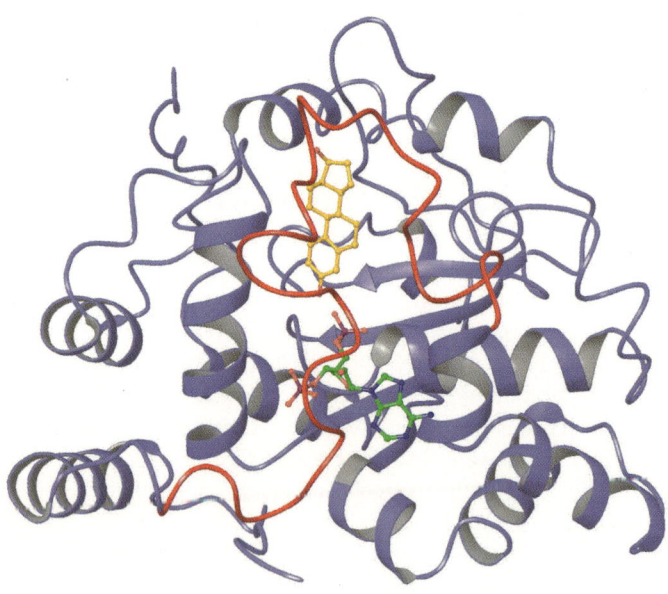

Fig. 12.7 Structure of SULT (2D06). Backbone atoms of the flexible cap region are coloured *red*. Carbon atoms of adenosine-3′-5′-diphosphate (PAP) cofactor and estradiol ligand are coloured *green* and *orange*, respectively

nucleotide free system is in open and disordered state that isomerizes to closed state upon nucleotide binding. Small ligands can bind to the nucleotide free closed enzyme conformation, while large ligands require the opening of the cap prior to binding. This hypothesis was tested with estradiol (small) and fulvestrant (large) ligands. According to the findings of the MD simulations, the nucleotide induced conformational changes affect the large ligand binding, but does not impact small ligand binding, since it can form optimal complex also with the closed state. Binding experiments with or without the PAPS nucleotide were in line with the proposal. In the case of the SULT1A1 enzyme, binding of estradiol was equivalent in cases with PAPS and without, while the apparent on-rate constant for fulvestrant binding decreases 26-fold at a saturating PAPS concentration, which is virtually identical to the 28-fold decrease in K_d obtained from the equilibrium binding measurements [70]. Similar trends were observed for SULT2A1 with corresponding dehydroepiandrosterone (small) and raloxifene (large) ligands [71]. Thorough investigation of the cap region revealed that it has nucleotide and acceptor ligand halves with distinguished protein dynamics. Nucleotide binding induces a conformational change to close the corresponding cap region, but the acceptor ligand binding site can oscillate between closed and open states. According to the proposed gating mechanism, PAPS nucleotide binding induces a preference for closed cap region and thus small ligands are prone to be metabolized by spatial accessibility selection from the pool of possible ligands. In order to test the outlined pore model, as a proof of concept, specific mutations were induced into SULT2A1 uncoupling the flexibility of nucleotide and ligand binding parts of the cap region [70]. The resulted mutant was not able to discriminate between small and large ligands any more, since the flexibility of the acceptor halve of the cap region was increased significantly resulting in the reduction of the closed, small substrate preferring conformation.

The usefulness of structural information was tested in prospective virtual screening scenarios using high-throughput docking method on SULT1A1 and SULT2A1 isoforms [72]. Molecular dynamics simulations were carried out to model the effects of protein flexibility and to convert the closed cap region to open state by simulating nucleotide-free SULT structures. In total, four MD simulations were carried out including nucleotide-bound and nucleotide-free systems for both the SULT1A1 and SULT2A1 isoforms. Frames of the 10 ns trajectories were clustered by using 2 Å root mean square deviation cut-off and cluster centroids were used for docking. The experimental and calculated binding free energy was found to be highly linearly correlated, with correlation coefficients (r^2) of 0.89 and 0.86 for 1A1 and 2A1, respectively. Substrates were distinguished from inhibitors by using a 4 Å distance criterion between acceptor ligand nitrogen or oxygen atom and the catalytic His residue. Those high ranked compounds that could fulfil the criterion were classified as substrates, while in the opposite case they were classified as inhibitors. A set of approved small molecular drugs (1455 entries) were assessed by docking. 76 compounds were predicted to be SULT1A1 substrates. Out of them 53 were known substrates for the enzyme while out of the 23 remaining compounds 21 could be purchased and tested. All of them were identified as true positives with $K_d \leq 100$ μM acceptance criterion. The docking predicted 22 compounds to be SULT2A1 substrates, including 8 novel compounds. Four compounds could be purchased, tested and proved to be substrates. In summary, out of the 22 top-ranked compounds 18 could be investigated by literature or in vitro method, and all were confirmed. Consequently, both SULT1A1 and SULT2A1 docking based substrate prediction resulted in 100 % hit rate. Considering the potential inhibitors, 136 and 35 ligands were identified as virtual hits for SULT1A1 and SULT2A1, respectively. In the case of SULT1A1 17 compounds (12.5 %) were confirmed and additional 34 (25 %) are likely to be positive, while for SULT2A1 19 compounds (53 %) are confirmed and further 8 (23 %) are likely positives. The remaining compounds were not tested. This study represents a remarkable accuracy that is not generally achieved in routine docking-based virtual screening campaigns. In contrast to this outstanding success, no significant enrichment could be observed in the case of the SULT1A1 retrospective enrichment study [58]. In this case, docking was performed into six representatives of the 4,500 structures collected from a 2 ns MD simulation of the ligand and nucleotide bound crystal structure. 60 known, diverse substrates and two different decoy sets were utilized with 49,496 and 13,088 molecules, respectively. According to the analysis, the docking protocol was not able to rank the ligands over the putative decoys. Several reasons might lie beyond this difference: (i) during the prospective study positional constraint was used for the ligands, that might impact the results seriously, (ii) the decoys might contain actives at $K_d \leq 100$ μM acceptance criterion, (iii) in case of the prospective study, not all of the compounds were tested experimentally, therefore false negatives might left unrecognized. However, the prospective study calls the attention to the potential of using structure-based modelling tools exploiting novel ADMET-related antitargets that are not yet integrated parts of the structure-based modelling arsenal.

12.3.5 Adverse Drug Reactions

12.3.5.1 hERG

Ion channels are membrane embedded proteins responsible for the regulation of ion flux through the membrane. The controlled choreography of opening and closing events of the ion channels in cardiac myocytes results in electrical excitations and relaxations as the physiological function of the heart. One of the key components of the cardiac cycle is the native function of the hERG potassium channel. Abnormalities caused by the inhibition of the hERG channel may potentially lead to prolongation of the action potential (specific QT elongation) and considered to be pro-arrhythmic [73]. Accordingly QT elongation is one of the most crucial adverse drug reactions responsible of torsades de point (TdP), a cardial event that can cause sudden death. Several drugs have been restricted or withdrawn from the market due to the hERG inhibition such as astemizole, cisapride, grepafloxacin, sertindole and terfenadine. Therefore, attenuation of hERG inhibition by in silico design has fundamental importance.

hERG is a tetrameric protein with putative C4 symmetry, built up from subunits containing six transmembrane helices in each. The subunits consist of 1,160 amino acids. Helices S1–S4 constitutes the voltage-sensing domain and S5–S6 form the pore domain. The potassium ion selective central pore domain represents well-preserved potassium channel features including selectivity filter to control the movements of the ions. The highly conserved signature S-V-G-F-G sequence is positioned at the C-terminal end of the selectivity filter [74]. Inhibitors bind in the cavity below the selectivity filter composed of the four S6 helices. Extensive Ala-scanning mutagenesis of the hERG channel revealed the impact of Tyr652, Phe656 and Ile647 in S6 helix, Ser620, Ser624, Ser631 and Val625 in the P-loop on inhibitor coordination [75]. Among these residues Tyr652 and Phe656 have the most crucial contribution to binding of high-affinity blockers (Fig. 12.8). The hERG channel, similarly to other voltage-gated channels, possesses three types of conformational states: closed, open and open inactivated. Activation results in

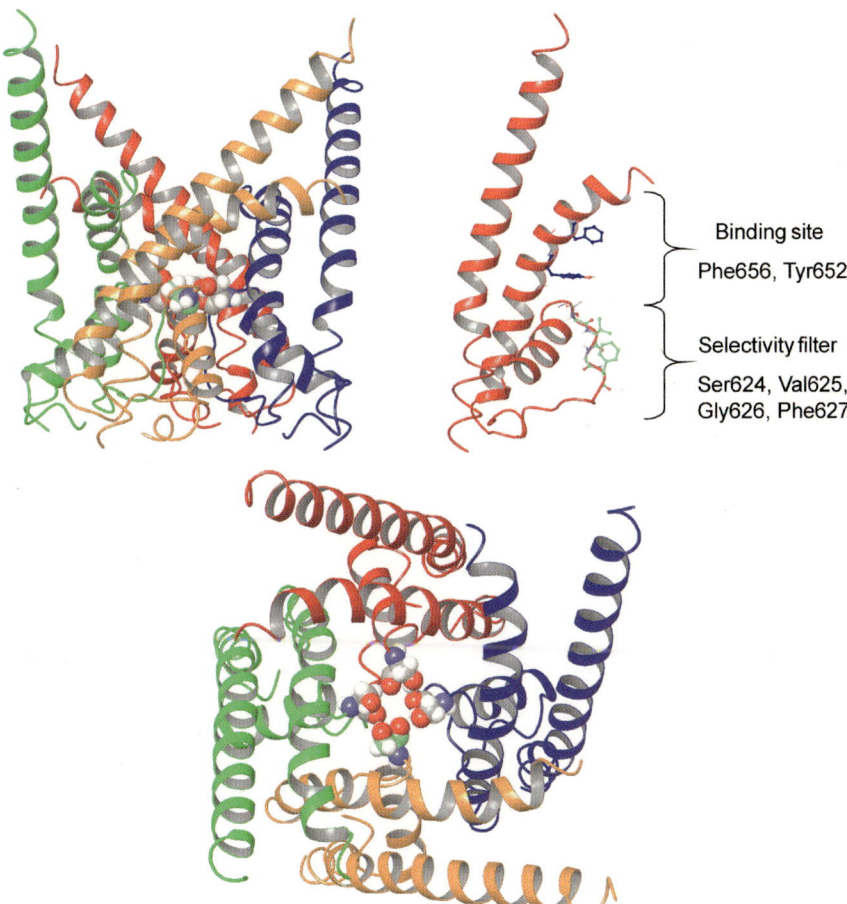

Fig. 12.8 Proposed structure of the hERG ion channel S5-S6 tetramer region. Side view (*top*) and top view (*bottom*), atoms of Ser624 are represented with *balls*. Highly important residues discussed in the text are highlighted on a single protein chain at *top right* position

the opening of the intracellular gate from closed state, while inactivation includes conformational change in the outer pore domain. Consequently, hERG channel represents high complexity from protein modelling aspect, since multiple conformational states can contribute to drug binding.

To date only the structure of an extracellular loop and the cytoplasmic N-terminal domain, responsible for the regulation of channel opening, is known experimentally. Structure based simulations rely on homology models based on fairly low sequence identity (approximately <30 %) to the open or closed bacterial potassium channels KcsA (closed form), MthK (open) or KvAP (open), KirBac1.1 and mammalian channel Kv1.2 [57]. Farid and coworkers published the homology model of a homotetramer hERG structure including S5 and S6 subunits [76]. Backbone torsion angles of residues near the known hinge positions (Gly648 in S6 and Gly572 in S5 helices) of each monomer were manually modified in unison to maintain C4 symmetry and in order to model the opening or closing of the pore. Subsequently, seven high affinity hERG ligands were docked to the open model using the induced fit methodology. Analysis of the interaction pattern of the residues was in line with the mutational data regarding multiple simultaneous aromatic stacking and hydrophobic interactions with Tyr652 and Phe656 residues. Polar groups and basic centres of the ligands were found to interact with Ser624 and nearby polar backbone atoms, also in alignment with the mutational data. Rigid protein docking of five entries from the sertindole compound family resulted in remarkably good correlation ($r^2 = 0.95$). However, correlation obtained on this very limited ligand set does not ensure similarly high accuracy for distinct chemotypes. Thorough analysis of seven homology models based on various alignments of helix S5 was published by Stary and co-workers [77]. The model quality has been assessed from three independent aspects. First, a set of applications relying on knowledge-based statistical potentials were calculated to evaluate conventional geometry, packing and normality descriptors. Second, the stability of the models was examined by using 20 ns MD simulations of the POPC lipid bilayer embedded hERG protein model. Root mean square deviations from the starting coordinates were monitored as a principal indicator of stability. Third, known ligands were docked into the binding site to analyse the protein-ligand interactions. According to the first evaluation two models had severe problems, while one model passed all of the criteria. Although ranking the models was not equivocal. Model 1 and model 6 were found to be the most stable during MD. Interestingly, none of the three investigated ligands (cisapride, MK-499 and terfenadine) could be docked into model 1. Docking into model 6 yielded results that are in good agreement with alanine scan mutational experiments. In summary, none of the single assessments could unambiguously identify a preferred model, but the combination of all three revealed that there is only one model (model 6) that fulfils all quality criteria. This study shed light on the intrinsic difficulties of hERG channel modelling, and arrived to the conclusion that further refinements might enhance the validity of the models, since only the S5 sequence alignments were investigated. The role of incorporating experimental knowledge to aid docking based ranking has been highlighted with the comparison of the prediction power of the native scoring function and number of interactions with experimentally confirmed interacting residues. The interaction count resulted in significantly better correlation with the observed affinity [78] than that of the scoring function.

MD based refinement of the open and closed state homology models were used to predict the binding free energy of 12 inhibitors by De-Cuny and co-workers [79]. PLS analysis was performed on the energy terms to build a model for pIC_{50} values. The obtained model had high correlation coefficient ($r^2 = 0.81$), and the interaction pattern of the top ranked binding modes were found to match the reported hERG mutagenesis data. Enrichment study had also been performed to rank 147 hERG inhibitors out of 498 compounds, but in this case ligand based QSAR models yielded higher hit rate than that of the ranking obtained with docking and scoring. The limited performance of scoring in case of the hERG ion channel underlines the importance of exploiting all the available experimental data to yield acceptable ranking.

The assessment of protein flexibility has been studied on multiple levels including side-chain and domain level conformational changes. In order to incorporate protein flexibility, local conformation space of the cavity of the open state channel was extensively explored leading to 215 models [80]. The predictive power of these models was benchmarked by using a set of sertindole analogs. Both single structure and ensemble docking evaluations were carried out. Solvation effects were studied by using MM-PBSA refinement. Three ensemble calculation schemes were applied: best score, arithmetic mean and Boltzmann-weighted average. The comparison revealed that the best single model could not be outperformed by the multiple structure docking methodology. Furthermore, MM-PBSA refinement could not improve the results obtained. The best single structure model was challenged to show general applicability by docking 14 structurally diverse inhibitors. The correlation between the experimental activity and predicted docking score yielded acceptable correlation ($r^2 = 0.6$). The major outlier was astemizole, which was found to be difficult to place in the crowded binding site due to its relatively large size. Recently, different modelling techniques including homology modelling, de novo design with incorporated experimental constraints and all-atom membrane MD simulations were performed to simulate all the three states (open, closed and open-inactivated) of the hERG channel [81]. Several state-selective blockers were docked to the models of the open and inactivated states using the induced fit docking protocol, and good correlation was observed with experimental affinities for both high- and low affinity blockers. Capturing different states of the hERG conformational space may offer atomic level information to better understand the molecular mechanism of the state-dependent hERG channel inhibition.

12.4 Concluding Remarks

Simulations of protein structures with or without bound ligand have fundamental importance to interpret the results of biochemical and biophysical investigations of protein functions. Since modelling should rely on experimental data, theory and practice must cross-fertilize each other. Approximate models can be very useful tools to design novel experimental conditions to test various hypotheses. Later on the results obtained can be channelled back to fine-tune the calculated models. On top of the benefits of the synergism between in silico and in vitro techniques to understand the intricacy of many biological phenomena, protein modelling has direct impact on drug discovery. Computational approaches contributed to the discovery of several drugs already reached the market [82, 83]. Due to the expanding knowledge on protein structures, ADMET related targets become attractive elements of recent computer aided drug design (CADD) workflows.

Experimental data including protein structural information, affinities of ligands and mutagenesis data have cardinal influence on the accuracy of the in silico approaches. The available knowledge certainly defines the scope of the protein modelling tasks. Without structural information, the first step is the prediction of atomic coordinates of the protein residues, while with accurate structural information on the ligand-bound complex one can estimate the binding free energy of the ligand and its analogues with acceptable accuracy.

ADMET related proteins have unique properties regarding their flexibility, binding site character, and size thus rationally translating the structural information into compounds with improved ADMET properties is very demanding. The collection of the available crystallographic data and the properties of the investigated targets along with the expected scope of the ligand-protein simulations are presented in Table 12.2. With regard to the methodologies, it is evident from the collected cases that MD simulations are generally applied as a reliable tool with distinguished importance in all of the collected cases.

First, we would like to highlight the CYP family since it has the most impressing coverage of high-quality X-ray structures including several isoforms with apo and ligand-bound complexes. Simulations of the CYP family accomplish almost all of the modelling approaches and technologies. Due to the nature of the active site heme QM-MM hybrid methods can be applied to model the reaction pathway. MD simulations and free energy calculations are available to estimate relative binding energies, in some cases within an error of 1 kcal/mol. This target can be considered to be the most feasible for simulations due to the single, closed active site. Accordingly, several successful studies and applications are available [38, 47]. In the case of HSA with similarly broad coverage of high-quality experimental structures the modelling complexity is multiplied due to the numerous and open sites capable to interact with the ligands. Therefore only the binding mode of the ligand might be predicted in order to design modifications to attenuate HSA liability. SULT, the last entry with high-quality crystallographic data is a very recent target for structure-based ADMET optimization. According to the available knowledge docking or enrichment studies on this target might be generally tractable.

PXR and CAR nuclear receptors have acceptable amount of structural information. In these cases binding mode prediction is generally acceptable, or in some outstanding cases even enrichment might be observed. P-gp and hERG are membrane embedded proteins with significant domain level flexibility along with large and open binding sites. As a consequence, modelling of protein-ligands

Table 12.2 Structural information and protein modelling techniques used for ADMET-related proteins

Target	HSA	P-gp	CYP	PXR	CAR	SULT	hERG
Number of structures at given resolution:							
<1.5 Å	2	–	2	–	–	–	–
1.5–2 Å	6	–	20	–	–	2	–
2–2.5 Å	34	–	46	–	–	4	–
2.5–3 Å	46	–	49	8	3	–	–
3 < Å	12	10	8	–	–	–	–
Structural information	High quality	Low quality	High quality	Medium quality	Medium quality	High quality	Homology model
Protein flexibility	SC, BB	SC, BB, D	SC, BB	SC	SC	SC, BB	SC, BB, D
Binding site	Multiple, opened	Single, large, opened	Single, closed	Single, large, closed	Single, closed	Single, closed	Single, large, opened
Modelling methods	MD, docking	Docking, (MD)	QM-MM, MD, FEP, docking	Docking, (MD)	Docking, (MD)	Docking, MD	Homology modelling, MD, docking
Expected outcome	Binding mode	Interacting residues, (binding mode)	Binding mode, enrichment, binding energy, details of mechanism	Interacting residues, (binding mode)	Binding mode, enrichment	Binding mode, enrichment, details of mechanism	Interacting residues, (enrichment)

SC side chain, BB backbone, D domain. Smaller importance is indicated with parenthesis

interactions have the highest level of complexity. The uncertainty arising from the lack of human crystal structures enhances the difficulties further. The abundance of mutational data in both cases provides experimental viewpoints to evaluate the binding poses, and the constructed models serve as useful pivots to design further mutations. However, it is important to note, that some of the built protein models were capable to rank congeneric series of compounds or a limited ligand sets.

Parallel developments on the field of structural biology and computational approaches brought significant achievements to the atomic level understanding of ADMET proteins and processes. Moreover, this collaboration is now mature enough to be able to deliver innovative solutions at different levels of resolution to foster the design of novel drugs.

References

1. Guengerich MK (2008) Cytochrome p450 and chemical toxicology. Chem Res Toxicol 21:70
2. Ekroos M, Sjögren T (2006) Structural basis for ligand promiscuity in cytochrome P450 3A4. Proc Natl Acad Sci USA 103:13682
3. Aller SG, Yu J, Ward A, Weng Y, Chittaboina S, Zhuo R, Harrell PM, Trinh YT, Zhang Q, Urbatsch IL, Chang G (2009) Structure of P-glycoprotein reveals a molecular basis for poly-specific drug binding. Science 323:1718
4. Warren GL, Do TD, Kelley BP, Nicholls A, Warren SD (2012) Essential considerations for using protein-ligand structures in drug discovery. Drug Discov Today 17:1270
5. Davis AM, St-Gallay SA, Kleywegt GJ (2008) Limitations and lessons in the use of X-ray structural information in drug design. Drug Discov Today 13:831
6. Senn HM, Thiel W (2009) QM/MM methods for biomolecular systems. Angew Chem Int Ed Engl 48:1198
7. Ferenczy GG (2013) Calculation of wave-functions with frozen orbitals in mixed quantum mechanics/molecular mechanics methods. Part I. Application of the Huzinaga equation. J Comput Chem 34:854
8. Karplus M, Kuriyan J (2005) Molecular dynamics and protein function. Proc Natl Acad Sci USA 102:6679
9. Adcock SA, McCammon JA (2006) Molecular dynamics: survey of methods for simulating the activity of proteins. Chem Rev 106:1589
10. Schmidt T, Bergner A, Schwede T (2013) Modelling three-dimensional protein structures for applications in drug design. Drug Discov Today. doi:10.1016/j.drudis.2013.10.027
11. Epstain CJ, Goldberger RF, Anfinsen CB (1963) The genetic control of tertiary protein structure: studies with model systems. Cold Spring Harb Symp Quant Biol 28:439
12. Hillisch A, Pineda LF, Hilgenfeld R (2004) Utility of homology models in the drug discovery process. Drug Discov Today 9:659
13. Otyepka M, Skopalík J, Anzenbacherová E, Anzenbacher P (2007) What common structural features and variations of mammalian P450s are known to date? Biochim Bi-ophys Acta 1770:376
14. Kitchen DB, Decornez H, Furr JR, Bajorath J (2004) Docking and scoring in virtual screening for drug discovery: methods and applications. Nat Rev Drug Discov 3:935
15. Repasky MP, Murphy RB, Banks JL, Greenwood JR, Tubert-Brohman I, Bhat S, Friesner RA (2012) Docking performance of the glide program as evaluated on the Astex and DUD datasets: a complete set of glide SP results and selected results for a new scoring function integrating WaterMap and glide. J Comput Aided Mol Des 26:787

16. de Graaf C, Kooistra AJ, Vischer HF, Katritch V, Kuijer M, Shiroishi M, Iwata S, Shimamura T, Stevens RC, de Esch IJ, Leurs R (2011) Crystal structure-based virtual screening for fragment-like ligands of the human histamine H(1) receptor. J Med Chem 54:8195
17. Carter DC, Ho JX (1994) Structure of serum albumin. Adv Protein Chem 45:153
18. Vallianatou T, Lambrinidis G, Tsantili-Kakoulidou A (2013) In silico prediction of human serum albumin binding for drug leads. Expert Opin Drug Discov 8:583
19. Castellanos MM, Colina CM (2013) Molecular dynamics simulations of human serum albumin and role of disulfide bonds. J Phys Chem B 117:11895
20. Hall ML, Jorgensen WL, Whitehead L (2013) Automated ligand- and structure-based protocol for in silico prediction of human serum albumin binding. J Chem Inf Model 53:907
21. Wendt MD, Shen W, Kunzer A, McClellan WJ, Bruncko M, Oost TK, Ding H, Joseph MK, Zhang H, Nimmer PM, Ng SC, Shoemaker AR, Petros AM, Oleksijew A, Marsh K, Bauch J, Oltersdorf T, Belli BA, Martineau D, Fesik SW, Rosenberg SH, Elmore SW (2006) Discovery and structure-activity relationship of antagonists of B-cell lymphoma 2 family proteins with chemopotentiation activity in vitro and in vivo. J Med Chem 49:1165
22. Mao H, Hajduk PJ, Craig R, Bell R, Borre T, Fesik SW (2001) Rational design of diflunisal analogues with reduced affinity for human serum albumin. J Am Chem Soc 123:10429
23. Aureli L, Cruciani G, Cesta MC, Anacardio R, De Simone L, Moriconi A (2005) Predicting human serum albumin affinity of interleukin-8 (CXCL8) inhibitors by 3D-QSPR approach. J Med Chem 48:2469
24. Li H, Chen Z, Xu X, Sui X, Guo T, Liu W, Zhang J (2011) Predicting human plasma protein binding of drugs using plasma protein interaction QSAR analysis (PPI-QSAR). Biopharm Drug Dispos 32:333
25. Zsila F, Bikadi Z, Malik D, Hari P, Pechan I, Berces A, Hazai E (2011) Evaluation of drug-human serum albumin binding interactions with support vector machine aided online automated docking. Bioinformatics 27:1806
26. Eckford PD, Sharom FJ (2009) ABC efflux pump-based resistance to chemotherapy drugs. Chem Rev 109:2989
27. Lee CA, Cook JA, Reyner EL, Smith DA (2010) P-glycoprotein related drug interactions: clinical importance and a consideration of disease states. Expert Opin Drug Metab Toxicol 6:603
28. Li J, Jaimes KF, Aller SG (2014) Refined structures of mouse P-glycoprotein. Protein Sci 23:34
29. Ward AB, Szewczyk P, Grimard V, Lee CW, Martinez L, Doshi R, Caya A, Villaluz M, Pardon E, Cregger C, Swartz DJ, Falson PG, Urbatsch IL, Govaerts C, Steyaert J, Chang G (2013) Structures of P-glycoprotein reveal its conformational flexibility and an epitope on the nucleotide-binding domain. Proc Natl Acad Sci USA 110:13386
30. Ma J, Biggin PC (2013) Substrate versus inhibitor dynamics of P-glycoprotein. Proteins 81:1653
31. Weng JW, Fan KN, Wang WN (2010) The conformational transition pathway of ATP binding cassette transporter MsbA revealed by atomistic simulations. J Biol Chem 285:3053
32. Pajeva IK, Globisch C, Wiese M (2009) Combined pharmacophore modeling, docking, and 3D QSAR studies of ABCB1 and ABCC1 transporter inhibitors. Chem Med Chem 4:1883–1896
33. Becker JP, Depret G, Van Bambeke F, Tulkens PM, Prévost M (2009) Molecular models of human P-glycoprotein in two different catalytic states. BMC Struct Biol 9:3
34. Tarcsay Á, Keserû GM (2011) Homology modeling and binding site assessment of the human P-glycoprotein. Future Med Chem 3:297
35. Chen L, Li Y, Yu H, Zhang L, Hou T (2011) Computational models for predicting substrates or inhibitors of P-glycoprotein. Drug Discov Today 17:343
36. Klepsch F, Chiba P, Ecker GF (2011) Exhaustive sampling of docking poses reveals binding hypotheses for propafenone type inhibitors of P-glycoprotein. PLoS Comput Biol 7:e1002036
37. Bode C (2010) The nasty surprise of a complex drug-drug interaction. Drug Discov Today 15:391

38. Kirchmair J, Williamson MJ, Tyzack JD, Tan L, Bond PJ, Bender A, Glen RC (2012) Computational prediction of metabolism: sites, products, SAR, P450 enzyme dynamics, and mechanisms. J Chem Inf Model 52:617
39. Brooks BR, Brooks CL, Mackerell AD, Nilsson L, Petrella RJ, Roux B, Won Y, Archontis G, Bartels C, Boresch S, Caflisch A, Caves L, Cui Q, Dinner AR, Feig M, Fischer S, Gao J, Hodoscek M, IM W, Kuczera K, Lazaridis T, Ma J, Ovchinnikov V, Paci E, Pastor RW, Post CB, Pu JZ, Schaefer M, Tidor B, Venable RM, Woodcock HL, Wu X, Yang W, York DM, Karplus M (2009) CHARMM: the biomolecular simulation program. J Comput Chem 30:1545
40. Autenrieth F, Tajkhorshid E, Baudry J, Luthey-Schulten Z (2004) Classical force field parameters for the heme prosthetic group of cytochrome c. J Comput Chem 25:1613
41. Skopalík J, Anzenbacher P, Otyepka M (2008) Flexibility of human cytochromes P450: molecular dynamics reveals differences between CYPs 3A4, 2C9, and 2A6, which correlate with their substrate preferences. J Phys Chem B 112:8165
42. Shahrokh K, Orendt A, Yost GS, Cheatham TE (2011) Quantum mechanically derived AMBER-compatible heme parameters for various states of the cytochrome P450 catalytic cycle. J Comput Chem 33:119
43. Park H, Lee S, Suh J (2005) Structural and dynamical basis of broad substrate specificity, catalytic mechanism, and inhibition of cytochrome P450 3A4. J Am Chem Soc 127:13634
44. Lampe JN, Brandman R, Sivaramakrishnan S, de Montellano PR (2010) Two-dimensional NMR and all-atom molecular dynamics of cytochrome P450 CYP119 reveal hidden conformational substates. J Biol Chem 285:9594
45. Miao Y, Baudry J (2011) Active-site hydration and water diffusion in cytochrome P450cam: a highly dynamic process. Biophys J 101:1493
46. Fishelovitch D, Shaik S, Wolfson HJ, Nussinov R (2009) Theoretical characterization of substrate access/exit channels in the human cytochrome P450 3A4 enzyme: involvement of phenylalanine residues in the gating mechanism. J Phys Chem B 113:13018
47. Tarcsay Á, Keserû GM (2011) In silico site of metabolism prediction of cytochrome P450-mediated biotransformations. Expert Opin Drug Metab Toxicol 7:299
48. Friesner RA, Banks JL, Murphy RB, Halgren TA, Klicic JJ, Mainz DT, Repasky MP, Knoll EH, Shelley M, Perry JK, Shaw DE, Francis P, Shenkin PS (2004) Glide: a new approach for rapid, accurate docking and scoring. 1. Method and assessment of docking accuracy. J Med Chem 47:1739
49. Unwalla RJ, Cross JB, Salaniwal S, Shilling AD, Leung L, Kao J, Humblet C (2010) Using a homology model of cytochrome P450 2D6 to predict substrate site of metabolism. J Comput Aided Mol Des 24:237
50. Tarcsay Á, Kiss R, Keserû GM (2010) Site of metabolism prediction on cytochrome P450 2C9: a knowledge-based docking approach. J Comput Aided Mol Des 24:399
51. Santos R, Hritz J, Oostenbrink C (2010) Role of water in molecular docking simulations of cytochrome P450 2D6. J Chem Inf Model 50:146
52. de Beer SB, Venkataraman H, Geerke DP, Oostenbrink C, Vermeulen NP (2012) Free energy calculations give insight into the stereoselective hydroxylation of α-ionones by engineered cytochrome P450 BM3 mutants. J Chem Inf Model 52:2139
53. Vass M, Tarcsay Á, Keserû GM (2012) Multiple ligand docking by Glide: implications for virtual second-site screening. J Comput Aided Mol Des 26:821
54. Bren U, Oostenbrink C (2012) Cytochrome P450 3A4 inhibition by ketoconazole: tackling the problem of ligand cooperativity using molecular dynamics simulations and free-energy calculations. J Chem Inf Model 52:1573
55. Oláh J, Mulholland AJ, Harvey JN (2011) Understanding the determinants of selectivity in drug metabolism through modeling of dextromethorphan oxidation by cytochrome P450. Proc Natl Acad Sci USA 108:6050
56. Lonsdale R, Houghton KT, Żurek J, Bathelt CM, Foloppe N, de Groot MJ, Harvey JN, Mulholland AJ (2013) Quantum mechanics/molecular mechanics modeling of regioselectivity of drug metabolism in cytochrome P450 2C9. J Am Chem Soc 135:8001
57. Stoll F, Göller AH, Hillisch A (2011) Utility of protein structures in overcoming ADMET-related issues of drug-like compounds. Drug Discov Today 16:530

58. Moroy G, Martiny VY, Vayer P, Villoutreix BO, Miteva MA (2012) Toward in silico structure-based ADMET prediction in drug discovery. Drug Discov Today 17:44
59. Orans J, Teotico DG, Redinbo MR (2005) The nuclear xenobiotic receptor pregnane X receptor: recent insights and new challenges. Mol Endocrinol 19:2891
60. Gao YD, Olson SH, Balkovec JM, Zhu Y, Royo I, Yabut J, Evers R, Tan EY, Tang W, Hartley DP, Mosley RT (2007) Attenuating pregnane X receptor (PXR) activation: a molecular modelling approach. Xenobiotica 37:124
61. Ekins S, Kortagere S, Iyer M, Reschly EJ, Lill MA, Redinbo MR, Krasowski MD (2009) Challenges predicting ligand-receptor interactions of promiscuous proteins: the nuclear receptor PXR. PLoS Comput Biol 5:e1000594
62. Liu YH, Mo SL, Bi HC, Hu BF, Li CG, Wang YT, Huang L, Huang M, Duan W, Liu JP, Wei MQ, Zhou SF (2011) Regulation of human pregnane X receptor and its target gene cytochrome P450 3A4 by Chinese herbal compounds and a molecular docking study. Xenobiotica 41:259
63. Suino K, Peng L, Reynolds R, Li Y, Cha JY, Repa JJ, Kliewer SA, Xu HE (2004) The nuclear xenobiotic receptor CAR: structural determinants of constitutive activation and heterodimerization. Mol Cell 16:893
64. Xu RX, Lambert MH, Wisely BB, Warren EN, Weinert EE, Waitt GM, Williams JD, Collins JL, Moore LB, Willson TM, Moore JT (2004) A structural basis for constitutive activity in the human CAR/RXRalpha heterodimer. Mol Cell 16:919
65. Jyrkkärinne J, Küblbeck J, Pulkkinen J, Honkakoski P, Laatikainen R, Poso A, Laitinen T (2012) Molecular dynamics simulations for human CAR inverse agonists. J Chem Inf Model 52:457
66. Windshügel V, Poso A (2011) Constitutive activity and ligand-dependent activation of the nuclear receptor CAR-insights from molecular dynamics simulations. J Mol Recognit 24:875
67. Küblbeck J, Jyrkkärinne J, Poso A, Turpeinen M, Sippl W, Honkakoski P, Windshügel B (2008) Discovery of substituted sulfonamides and thiazolidin-4-one derivatives as agonists of human constitutive androstane receptor. Biochem Pharmacol 76:1288
68. Lynch C, Pan Y, Li L, Ferguson SS, Xia M, Swaan PW, Wang H (2013) Identification of novel activators of constitutive androstane receptor from FDA-approved drugs by integrated computational and biological approaches. Pharm Res 30:489
69. Leyh TS, Cook I, Wang T (2013) Structure, dynamics and selectivity in the sulfotransferase family. Drug Metab Rev 45:423
70. Cook I, Wang T, Almo SC, Kim J, Falany CN, Leyh TS (2013) Testing the sulfotransferase molecular pore hypothesis. J Biol Chem 288:8619
71. Cook I, Wang T, Falany CN, Leyh TS (2012) A nucleotide-gated molecular pore selects sulfotransferase substrates. Biochemistry 51:5674
72. Cook I, Wang T, Falany CN, Leyh TS (2013) High accuracy in silico sulfotransferase models. J Biol Chem 288:34494
73. Durdagi S, Subbotina J, Lees-Miller J, Guo J, Duff HJ, Noskov SY (2010) Insights into the molecular mechanism of hERG1 channel activation and blockade by drugs. Curr Med Chem 17:3514
74. Wang S, Li Y, Xu L, Li D, Hou T (2013) Recent developments in computational prediction of HERG blockage. Curr Top Med Chem 13:1317
75. Du L, Li M, You Q (2009) The interactions between hERG potassium channel and blockers. Curr Top Med Chem 9:330
76. Farid R, Day T, Friesner RA, Pearlstein RA (2006) New insights about HERG blockade obtained from protein modeling, potential energy mapping, and docking studies. Bioorg Med Chem 14:3160
77. Stary A, Wacker SJ, Boukharta L, Zachariae U, Karimi-Nejad Y, Aqvist J, Vriend G, de Groot BL (2010) Toward a consensus model of the HERG potassium channel. Chem Med Chem 5:455
78. Dempsey CE, Wright D, Colenso CK, Sessions RB, Hancox JC (2014) Assessing hERG pore models as templates for drug docking using published experimental constraints: the inactivated state in the context of drug block. J Chem Inf Model 54:601

79. Du-Cuny L, Chen L, Zhang S (2011) A critical assessment of combined ligand- and structure-based approaches to HERG channel blocker modeling. J Chem Inf Model 51:2948
80. Di Martino GP, Masetti M, Ceccarini L, Cavalli A, Recanatini M (2013) An automated docking protocol for hERG channel blockers. J Chem Inf Model 53:159
81. Durdagi S, Deshpande S, Duff HJ, Noskov SY (2012) Modeling of open, closed, and open-inactivated states of the hERG1 channel: structural mechanisms of the state-dependent drug binding. J Chem Inf Model 52:2760
82. Clark DE (2006) What has computer-aided molecular design ever done for drug discovery? Expert Opin Drug Discov 1:103
83. Congreve M, Murray CW, Blundell TL (2005) Structural biology and drug discovery. Drug Discov Today 10:895

Index

© Springer International Publishing Switzerland 2014
G. Náray-Szabó (ed.), *Protein Modelling*, DOI 10.1007/978-3-319-09976-7

Printed by Printforce, the Netherlands